Contemporary Ergonomics and Human Factors 2015

Contemporary Ergonomics and Human Factors 2015

Editors

Sarah Sharples, Steven Shorrock & Patrick Waterson

CRC Press
Taylor & Francis Group
Boca Raton   London   New York   Leiden

CRC Press is an imprint of the
Taylor & Francis Group, an **informa** business

A BALKEMA BOOK

Chartered Institute
of Ergonomics
& Human Factors

Typeset by MPS Ltd, a Macmillan Company, Chennai, India
Printed and bound in Great Britain by CPI Group (UK) Ltd, Croydon, CR0 4YY

ISBN: 978-1-138-02803-6 (Pbk)
ISBN: 978-1-315-68573-1 (eBook PDF)

Contents

Preface

This book contains the proceedings of the International Conference on Ergonomics and Human Factors, held in April 2015 at Staverton Park, Daventry, United Kingdom. The conference is a major international event for ergonomists and human factors professionals and attracts contributions and delegates from around the world. It holds particular significance this year as we celebrate the award of Chartered status to the Institute.

Papers are subject to a peer review process before consideration by the Programme Committee. The Programme Committee selects papers for publication in proceedings as full or short papers and identifies groups of papers to form dedicated sessions at the conference. In addition to the open call for papers, many authors respond to calls for symposia. These papers are still subject to peer review but are included within dedicated specialist sessions. This conference included symposia on transport, healthcare and manufacturing, as well as a doctoral consortium.

This is the second year that Sarah Sharples and Steve Shorrock have acted as editors of *Contemporary Ergonomics* and we have been joined this year by the Chair of the conference organising committee, Pat Waterson. Again we are pleased to have received papers from a diverse range of application domains, including transport, healthcare, manufacturing, defence, office work, and various design applications. There were also various workshops on issues pertaining to the discipline and profession of ergonomics and human factors. A range of methodological approaches is reflected, spanning both quantitative and qualitative techniques, as well as field and laboratory based research.

Transport is again a major theme at the conference, with road, rail, aviation and maritime represented. Automotive ergonomics papers reflect a number of perspectives. From the driving perspective, papers address safety issues such as young drivers, peer influence and risky driving, and usability issues such as gestures for in-car touch screens, and real time transport information. From the design and manufacturing perspective, there is a paper on test rig lab trials concerning preferred postures and another on virtual simulations and physical validation for new car development. From the perspective of control and communication, another paper addresses visual sampling in a road traffic management control room task. The pedestrian perspective is reflected in a paper that explores where are our street safety for female users, and how everyday harassment effects mobility. Rail is a strong area of research in ergonomics and human factors, and papers include the application of ergonomics and human factors to rail safety issues such as compliance with speed restrictions, non-technical skills, wrong side door release at stations, and slips, trips, and falls among older rail passengers. Ergonomics methods are also applied to the design of the next generation intercity express train and remote condition monitoring systems for predictive maintenance. In the maritime domain, communication on the bridge of a ship is considered. Aviation-related papers include an analysis of pilots' fixation distribution for performing air-to-air and air-to-surface tasks, and an investigation of air traffic control students' attitude to risk.

Papers on *human factors in manufacturing* reflect several emerging themes, including limitations of using only computer aided design (CAD) and digital human modelling (DHM) in design relating to high value manufacturing, user requirements and expectations for cloud manufacturing, and, as mentioned earlier, virtual simulations and physical validation for new car development. Job design is addressed in terms of sustainability, working with robots, search strategies in human visual inspection, novel human skills capture methodologies, and redesign of order-picking. There was also a workshop on manual handling.

Healthcare human factors and ergonomics papers consider a range of issues for design of work, systems and environments for patients and clinicians. Usability and safety are the focus of papers on improving accessibility to hospital clinical guidelines, the 'smart powerchair', and human factors that influence the performance of telecare and safety checklist design for general practice. Safety is also the focus of a paper on the measurement of patient safety culture. Quality is considered with respect to quality improvement, and wellbeing is addressed with respect to stress in UK trainee mental health professionals. A keynote lecture addresses ergonomics, accountability and complexity and a workshop at the conference considered the question of how to challenge myths and misunderstandings about human factors in healthcare.

The role of ergonomics and human factors in *health and safety* is addressed in a range of papers. Several of these address general topics of interest for a wide audience, such as human factors in total safety management, learning and resilience, design thinking in health and safety, and hazard perception and reporting. A paper for a keynote lecture addresses issues of ergonomics, accountability and complexity. For the onshore and offshore high hazard industries, papers considered the combined effects of occupational stressors on health and wellbeing in the offshore oil industry, confirmation bias in a routine drilling operation, and fatigue and shiftwork issues in the Buncefield explosion. Another paper considers the performance of plant personnel in severe accident scenarios. There is also a paper on preliminary findings of focus groups on the aging construction workforce. A workshop discussion on the theme of health and safety posed the question, "is safety culture still a thing?"

Stress emerges as a theme in several papers. There is an update on stress and wellbeing at work, and two papers address cultural issues, including culture as a buffer against occupational stress and a paper on ethnicity and work-related stress in migrant workers in southern Italy. Specific occupational contexts are reflected in papers on call centre employees, and police, as well as trainee mental health professionals, as mentioned earlier.

Several papers covered the practice and teaching of *design*. One paper addresses specifications in the context of innovation, looking at how an existing ergonomics method can be used for a new purpose. Some of these addressed learning issues, such as virtual learning environment design, and case study approaches. Specific design applications included co-design for wearables adoption, wheelchair cushion design, and visual accessibility misconceptions held by graphic designers and their clients. There were workshops on shared collaborative work spaces and the challenges of new hand held technologies, and working within cross-disciplinary teams. Papers also explore a diverse range of *mobile*-related themes, such as usability heuristics for specific domains, task workflow design for citizen science users, attitudes of young users to mobile phones, and sharing by webcomic artists.

Job design is addressed in many papers. Methodological papers include a 'good job score' to explore positive and negative outcomes, assessing multiple factors of well-being using single-item measures, and a standards framework to support job synthesis associated with HCI. Office ergonomics papers explore thermal comfort, as well as sources of risk to health associated with new technologies in the office environment. New technologies and mobile working emerged within this theme, with papers on mobile working with respect to physical discomfort, shared collaborative work spaces and the challenges of new hand held technologies, and the psychology of mobile working. European perceptions of workplace inclusion and the application of ergonomics are also explored.

Defence-related papers address the human component in defence acquisition process, policy and guidance information, standards for risk assessment and human factors at design stage, and initial evidence of a multi-dimensional control task model.

Other papers include domain knowledge and interaction with visual analytics, and understanding the effect of cultural and societal factors on conflict.

This year, the conference included several workshops and discussions on the discipline and practice of ergonomics and human factors. Three workshops considered issues relevant specifically to the Institute and its members – a chartership panel session, including a question and answer session by the CIEHF Chief Executive and President, a workshop on continual professional development, and a simulated disciplinary hearing. Other workshops included discussions on a prototype 'human performance standard of excellence' and on the nature of ergonomics/human factors: art, craft or science?

As in recent years, our conference includes an ever popular 'Just-a-minute' poster session; the short papers from this session are published in this volume. In addition to the papers contained within the proceedings, the conference will include a Doctoral Consortium, where PhD students from across the UK and Europe present and discuss their research activities.

Collectively, the papers cover a range of key ergonomics criteria, including safety, health, wellbeing, comfort, productivity, efficiency, and sustainability, demonstrating the continued wide applicability of ergonomics and human factors and the range of domains for which we must ensure there are appropriate methods and tools.

Editors

Sarah Sharples, Steven Shorrock & Patrick Waterson, January 2015

Invited and Plenary Lectures

Donald Broadbent Lecture: All Systems Great and Small? Chris Baber
Institute Lecture: Big Data Analytics: Why should HF/E Care? Colin Drury
Keynote Lecture: Ergonomics, Accountability and Complexity.
   Sidney W. A. Dekker
Keynote Lecture: Bryn Baxendale
Plenary Lecture: Jane Reid

Workshops

Several workshops were held, reflecting the need to maximise interactivity and engagement at conferences. This year, there were workshops and debates on the following:

- Chartership panel session and Q&A (Steve Barraclough [Chief Executive] and Roger Haslam [President])
- A disciplinary hearing (or two) (David Rumens, IEHF, UK)
- Ergonomics/human factors – art, craft or science? (Sarah Sharples and Peter Buckle, University of Nottingham, UK)
- Mobile Generations – Shared collaborative work spaces and the challenges of new hand held technologies (Jim Taylour, Orangebox Ltd, UK)
- How do we challenge myths and misunderstandings about human factors in healthcare? (Paul Bowie, NHS Education for Scotland, UK)
- Working within cross-disciplinary teams: How can we help bridge the gaps? (Becky Mallaband, Loughborough University, UK)
- CPD Workshop (Adrian Wheatley)
- Manual handling – when is it OK or not OK? (Wendy Morris, Jaguar Land Rover, UK)
- Is safety culture still a thing? (Steven Shorrock, EUROCONTROL, France)
- Towards A Human Performance Standard of Excellence (Barry Kirwan, EUROCONTROL)
- Safety-I, Safety-II, and Human Factors (Mark Young, RAIB, UK and Steven Shorrock, EUROCONTROL, France)

Programme Committee

Professor Sarah Sharples (Chair), University of Nottingham
Dr Steven T. Shorrock, Eurocontrol and University of New South Wales

Dr Patrick Waterson, Loughborough University
Dr Sue Hignett, Loughborough University
Dr David Golightly, University of Nottingham
Dr Mike Tainsh, Krome
Dr Will Baker, Cranfield University
Dr Sarah Fletcher, Cranfield University

Organising Committee

Dr Patrick Waterson (Chair)
Dr Nora Balfe
Dr Chris Vincent
Dr Ruth Sims
Professor Sarah Sharples
Dr Steve Shorrock

IEHF Organising Team

James Walton, Tina Worthy

DONALD BROADBENT
LECTURE

ALL SYSTEMS GREAT AND SMALL

Chris Baber

*School of Electronic, Electrical and Systems Engineering,
University of Birmingham*

In this talk, I want to consider how some concepts inspired by cybernetics could be useful in helping to focus our understanding of 'systems'. In doing so, the size of a system can become less important than its complexity, and approaches which work for small systems can be scaled to the study of larger systems. The reason why this argument is central to ergonomics is that, for the discipline to develop further, it is important to refine the scientific basis and application of the systems concept.

A science of systems

This talk follows the recent tradition of Broadbent (and plenary) lectures addressing 'systems'. The reason why the question remains pressing is because the nature of the systems that Ergonomics confronts seems to be changing. Ergonomics has always been concerned with the human-machine system on both a micro- and a macro-scale, but has often treated these as distinct species of problem requiring different thinking and methods. In this talk I am interested in exploring what they have in common.

A further theme stems from what Broadbent (1979) referred to as 'civilian problems' (in contrast to the sort of military problem that the founders of the Society were addressing during the 1940s), and which he felt were characterised by *"... multiple objectives, frequently conflicting with each other, and often having no administrative link with each other"*. This is a nice encapsulation of the sort of 'systems' that face ergonomics today, particularly in domains such as rail or healthcare. The question is whether such domains are unique, or whether a theory of systems could be equally applicable to micro- and macro-ergonomics. In particular, it raises the question of how systems meet multiple, conflicting objectives which, in cybernetics terms, raises the question of 'control'.

Systems thinking

Ergonomics has always been concerned with systems. In part this reflects the shared roots that the discipline had with cybernetics (indeed, many Universities had departments which were jointly staffed by ergonomists and cyberneticists back in the 1950s). Over the years, the relationship between ergonomics and cybernetics has diminished, with each appearing to go its separate way. I wonder if it might be timely to reconsider and re-critique concepts from cybernetics rather than consigning these to history?

In his overview of systems thinking in ergonomics, Wilson (2014) offered the following definition: "*A system is a set of inter-related or coupled activities or entities …, with a joint purpose, links between the entities which may be of state, form, function and causations, and which changes and modifies its states and the interactions within it given circumstances and events, and which is conceptualised as existing within a boundary;*". This definition helps us to appreciate key features of the systems approach, such as: System focus (describing the system of interest); Context (recognising the context in which the system operates); Interactions (understanding how the parts of the system work together); Holism (appreciating how the parts of the system combine into a unified whole); Emergence (understanding how behaviours arise from the interactions); Embedding (ensuring that ergonomists engage with the system effectively).

A science of systems should allow us to move from describing the structure of systems to predicting their behaviour. This involves not only a shift in focus in the way we think about systems but also the questions we ask of systems. Put simply, the shift is from asking what does it do to asking how is it doing it? From the list of features above, I suggest that while ergonomics applies a system focus to appreciate context, interactions and embedding, it is less adroit at dealing with holism and emergence (beyond recognising that these are good things to consider). From Wilson's (2014) definition, my focus is on the 'joint purpose' of the system and the ways in which it 'changes and modifies its states and the interactions'. This is not meant to be an either-or argument so much as an encouragement to elaborate on our thinking about systems.

Complexity versus size

When we study systems, we are continually seeking to reduce the size of that system to something manageable. Looking through recent papers in the journal *Ergonomics*, diagram of 'systems' had between 3 to 20 nodes, with a median of 6, and between 3 to 20 links, with a median of 11. Given that these systems ranged from manufacturing lines to hospitals to battleship operations rooms, there is no inherent reason why they should have similar sizes. However, as analysts I think that we are likely to seek a 'human-scale' representation of the system which we can study. The reason why the diagrams are small is not a reflection of the workplace under consideration so much as a reflection of the analytical lens that we apply in order to describe the 'system' as a static entity. You could argue, for instance, that a system of several hundred nodes would be difficult to read and so the smaller scale is better suited to presentation in a paper.

On the other hand, these small diagrams provide us with the opportunity for qualitative analysis – we make small diagrams so that we can talk about them because we do not have techniques for analysing systems quantitatively. If one considers the Internet as a 'system' of nodes, then reducing it to a handful of nodes that we can talk about is not likely to lead to insights into the structure of the network (although it could provide interesting material for a specific case study). In order to study a system as large as the internet, analysis focuses of specific aspects of overall performance or structure, e.g., highlighting the 'scale-free' nature of the internet's structure. My point is that the size of a system, while it might lead one to assume that bigger systems are more difficult to study, is less important to current Ergonomics practice than one might think. This is because we have a tendency to ignore size (when it gets too big) and to focus on a 'human-scale' version of the network which can be talked about.

Control, variability and attractor states

A (simple) cybernetic model of system activity involves the use of a feedback loop comparing the current output of the system against some desired state. Variability could be seen as noise which needs to be corrected out of performance by a comparator. The comparator which makes use of the feedback, therefore, requires a model of the desired state (and, most likely, an associated model of the world in which the system operates). This reflects Ashby's (1956) law of requisite variety, which claims that for a system to be stable, its control mechanism must have a set of states greater than the number of states that the system is capable of obtaining. Not only did cybernetics apply such models to 'technical' systems but it also used this to describe 'control and communication' (Weiner, 1961) in animals (including humans). Indeed, Broadbent (1958) suggested that *"... servo systems seek goals by negative feed-back ... [Human] nervous systems are networks of the type shown ... and of no other type."*

Variability and consistency

Consider a micro-ergonomic system consisting of a person, a hand-tool (say a saw), an object (say, a piece of wood), and the environment in which activity occurs. One could view this 'system' from a human-centred perspective as being wholly driven and controlled by the human. Claiming that the system is *not* controlled by the human might feel perverse and wrong-headed. However, any movement that people can make involves a hugely complex collection of degrees of freedom of the various limbs involved in that movement. As Turvey (1990) pointed out, the concept that movement arises from a connected system of nodes of movement is much more efficient, simpler and elegant than any model which appeals to central executive control. This points to the notion of coordinative structures proposed by Bernstein (1967). Rather than these collections resulting in a centrally stored program they are the steady state of the 'system' during that type of movement. In this case, the system self-organises into an appropriate, movement-specific structure. Such a structure can ensure consistency and repeatability in movement.

Nonlinear dynamics (Guastello and Gregson, 2011) allows one to dispense with the central controller assumed in cybernetics and to consider instead the attractor states towards which a system gravitates. These attractor states represent local minima in terms of system variability. As an example of this, consider the murmuration of starlings; a large flock of birds wheels and swoops in formation, seemingly following complicated choreography. Starlings do not seem to simply follow their nearest neighbour so much as follow a set of neighbours (6 or 7), which seems to be a more efficient way of avoiding predation (King and Sumpter, 2012). Thus, rather than relying on a central comparator (involving, say, predator location against some concept of threat-level), the flock continually adapts its activity to the changing context, resulting in sudden shifts which are difficult to predict locally (but which seem to follow system-level 'rules').

Affordance as an attractor state

If one accepts that it is possible to describe activity in terms of attractor states, then it is possible to also question some basic assumptions about how people interact with their environment. Take the concept of 'affordance' which is used to describe the perception-action coupling between a person and an object. Conventionally, the properties of an object 'afford' (or support) a particular action on the part of the user. Thus, a cup affords (the action of) picking up. The point is that we tend to view the concept of affordance anthropocentrically, with the user being the active perceiver of the (inactive) object's properties. This accords with the several definitions of affordance that Gibson (1979) offered. However, an alternative perspective is to the see affordance as the result (rather than the cause) of the enacted relationship between person and object. Following Malafouris' (2013) conception of 'material engagement', one can see how interactivity between a person and the object in pursuit of a goal is not simply a matter of the person imposing meaning on the object but a reciprocal process. The object influences the actions that the person performs, the person adapts her actions to the changing state of the object, and the overall pursuit of a goal arises from these relationships. For Malafouris (2013) *"Agency is the relational and emergent product of material engagement"*, and it makes sense for both the object and the human to swap agency as the process unfolds.

Elasticity, rigidity and resilience

Resilience is was defined by as *"… the intrinsic ability of an organization (system) to maintain or regain a dynamically stable state …."* [Hollnagel et al., 2006]. From the perspective of the preceding discussion, one can see resilience as the ability of a system to return to an attractor state following some form of perturbation. The question of what defines the attractor state, therefore, need not depend upon what goals the system has (or what functional purpose it is meant to have) so much as on the activities that characterize 'steady state' behavior. When the system is disturbed, performance could be expected to return to this steady state through stabilization of the aspects which were perturbed. This means that resilience is not locally stable but changes in terms of the nature of the perturbation and the ease with which the system can move to the more appropriate attractor state (which need not be the original state of the system).

Conclusions

I want us to be less concerned with the size of a system and more concerned with its complexity, to be less concerned with the structure of the system and more concerned with its behaviour. In doing this, I also want ergonomics to move from a view of the world which places humans at the centre (rather like ancient cosmology placed the earth at the centre of the universe) to one which seeks to better understand lawful relations between elements in a system (rather like explaining how gravity keeps planets in an ordered arrangement). As I see it, the core to this understanding is the purpose of the system and way in which the system is controlled to maintain this purpose. To complicate matters, the 'purpose' is not always pre-determined, but arises (emerges) from the activity within the system (sometimes in ways which are counter to the intentions of the systems designers or managers) and that 'control' is not always something which is exercised by specific system elements but which reflects the stability of the system in operation.

References

Ashby, W.R. (1956) *An Introduction to Cybernetics*, London: Chapman and Hall.
Bernstein N.A. (1967) *The Coordination and Regulation of Movements*. Oxford: Pergamon.
Broadbent, D.E. (1979) Is a fatigue test possible?, *Ergonomics, 22*, 1277–1290.
Broadbent, D.E. (1958) *Perception and Communication*, London: Pergamon Press.
Gibson J.J. (1979) *The Ecological Approach to Visual Perception*. Boston, MA: Houghton-Mifflin.
Guastello, S.J. and Gregson, R.A.M. (Eds.). (2011) *Nonlinear dynamical systems analysis for the behavioral sciences using real data*. Boca Raton, FL: C R C Press.
Hollnagel, E., Woods, D.D. and Leveson, N.G. (2006) *Resilience Engineering: concepts and precepts*, Avebury: Ashgate.
King, A.J. and Sumpter, D.J.T. (2012) Murmurations, *Current Biology, 22*, R112.

Malafouris, L. (2013) *How Things Shape the Mind: a theory of material engagement*, Cambridge, MA: MIT Press.

Turvey, M.T. (1990) Coordination, *American Psychologist, 45*, 938–953.

Weiner, N. (1961) *Cybernetics, or control and communication in the animal and the machine*, New York: MIT Press [2nd edition]

Wilson, J.R. (2014) Fundamentals of systems ergonomics/human factors, *Applied Ergonomics, 45*, 5–13.

INSTITUTE LECTURE

BIG DATA ANALYTICS: WHY SHOULD HF/E CARE?

Colin G. Drury

University at Buffalo: SUNY, Buffalo, NY, USA

Advances in sensor technology, connectedness and computational power have together produced huge data sets. The treatment and analysis of these data sets is known as Big Data Analytics (BDA). Computational methods have been developed to derive meaningful, actionable conclusions from these data bases. This paper examines BDA, often characterized by Volume, Velocity and Variety, giving examples of successful BDA use. This examination provides a context for considering examples of using BDA on human data, using BDA in HFE studies, and studies of how people perform BDA. Significant issues for HFE are the reliance of BDA on correlation and the ethics of BDA.

Characteristics of big data analytics

If we are to understand how human factors/ergonomics (HFE) does, can and should interact with Big Data Analytics (BDA) then we need a more exact understanding of what it is and how it is being used. There are a number of books on the subject, and research and survey papers covering aspects of BDA, but a starting definition is useful;

> big data n. data of a very large size, typically to the extent that its manipulation and management present significant logistical challenges; (also) the branch of computing involving such data (www.oed.com).

Foremost, obviously, is the issue of size – the bigness of big data. Hilbert and Lopez (2011) provide useful estimates of how much data is being accumulated and how rapidly this stock is growing. This paper uses information theoretic concepts to estimate the world's information capacity. The annual rates of growth of data storage, data transmission and data processing are 28%, 23% and 58% respectively. A major implication of Big Data Analytics (e.g. Mayer-Schonberger & Cukier, 2013, page 26) is that the data set has moved from a sample size (n) to close to the entire population size (N). If the data set is large enough we do indeed have the whole population rather than a sample. Thus instead of carefully sampling a few (hopefully well-chosen) data points to find representative data, we can potentially analyze the whole set of data. Traditionally we have used inferential statistics to allow us to predict population parameters and effects from a sample but with the whole population measured, then there is logically no requirement for inferential statistics. If an effect, however small, is found then it does indeed exist in the population, with no need for inference.

One example of BDA in practice is Ginsberg et al. (2009) who used Google search data to identify rapid predictors of the onset of influenza by geographical region in the 2007–2008 influenza season. This data set was comprised of all search terms across the USA's nine public health regions, whether influenza-like illness related or not. Validation was against the Centers for Disease Control and Prevention (CDC) data on past levels of influenza in each region. The resulting logistic regression using the top 45 search terms gave a correlation of 0.90 against CDC data. Further validations were performed, then the regression was used to predict future CDC data, resulting in good predictions "1–2 weeks ahead of" CDC reports (Ginsberg et al., 2009, page 1013).

Finally, a characteristic of BDA is that it often combines disparate data sets from different sources. This enables much richer hypotheses to be tested but at a cost of typically messier data. Our usual attempts to have complete sets of data across all conditions for all measures may not be possible here. Instead of a complete tableau, typical of our data for a complex ANOVA following a designed experiment, we may only have a relatively sparse matrix or tableau of data, albeit one rich in both independent and dependent variables. In our traditional work we have tried hard to obtain complete data for our sample, using estimators for missing data if only a tiny fraction of points are missing, or using a General Linear Models ANOVA that can deal with a reasonable amount of missing data. If we have considerable missing data, we can combine data from related categories (levels of a factor) to perform a less detailed analysis. Now, however, we need statistical methods that are not just scaleable to very large samples, but ones that can deal effectively with quite messy data. This usually implies methods based at their heart on correlations.

Human factors/ergonomics implications of BDA

Big Data Analytics may be more lasting than most business fads: large data sets are not going away, and neither is the lure of obtaining "meaningful insights" from already-collected and hence relatively inexpensive data. This technology is also affecting our lives as professionals and as citizens, so that on both levels we need to be aware of potential impacts.

The end of theory?

A number of authors (e.g. Kuhn and Johnson, 2014) have suggested that BDA's reliance on correlations at almost the population level of archival data rather than the more traditional search for causation has great implications for the scientific method itself. If we find any correlations among variables in a data set, then that is equally an inherently-significant finding (whole population), a useful outcome for prediction, and also a valid substitute for a prior theory. The example quoted in the previous section relied on the data to suggest what was happening rather than prior hypotheses that were then tested on the data. Compare this "black-box" approach to the methodology that we in HFE have relied upon for most of our insights, models and predictive abilities.

Traditionally in HFE, we have moved from case studies ($n \sim 1$) where we develop sensible hypotheses, to a sampling methodology ($n << N$), where we test those hypotheses by running an observation study, a questionnaire study or field or laboratory experiment. We even use such hypotheses to guide our examination of archival data, testing specific aspects rather than embarking upon the "fishing expedition" that our statistics educators warned us against. This is often what we mean by "understanding" an issue, derived ultimately from the physical sciences where the aim of science is to deduce a set of first principles or axioms that reduce the complexity of nature to at least a perceived orderliness. In HFE we still have "understandable" models from thermal physiology, through signal detection theory and naturalistic decision making to biomechanics. Big data uses a sample size of at least the same order as the population size ($n \sim N$) and thus represents a relatively new situation bringing with it the ability to perhaps dispense with "understandable" models and rely on the actual relationships among population data for our predictions. Perhaps prediction is enough for the utilization of HFE in a practical context, but it would be a large philosophical step to renounce theory/models/axioms entirely.

Or would it? In HFE we already use correlations extensively, although usually being careful not to interpret them a causal. The inter-correlations between multiple dependent measures in a study can be subject to analyses such as Principal Components Analysis or Factor Analysis to provide fewer, and often intuitively appealing, concepts or Factors for further analysis or understanding. This works well on an empirical level, but that does not stop us from assigning intuitive names to factors to aid "understanding". Going further, HFE theorists and practitioners have used neural network methods as aids to modeling. Such a representation of data takes multiple system inputs and relates them to several outputs using a network of intermediate layers that function like neurons. One advantage is that the methodology encompasses non-linear as well as the more usual linear relationships, e.g. of multiple regression models. It can produce accurate results, see Hou, Lin and Drury (1993), but has no transparency as to what the intermediate layer nodes represent. It is truly a black-box approach, showing that examples exist of HFE practitioners being comfortable with an approach that does not attempt to assign understandable labels to its models.

Much BDA is performed on human data

Despite the data deluge from digital sensors, much of the data used in BDA for insights comes from human sources or is collected on human behaviour, giving HFE a role that is perhaps necessary but rarely acknowledged. Much use of BDA is in the form of bioinformatics data mining, where researchers attempt to find specific sections of DNA that are associated with specific diseases. When the

DNA is human, as is typical, this research can provide links between those who have contacted a disease or inherited a condition and parts of the overall DNA sequence that may influence that condition. The aim is to eventually associate specific treatment protocols with DNA sequences so as to enhance outcomes, a process popularized as individualized medicine.

A novel use of BDA on humans comes from self-worn fitness devices (Hernandez 2014). As these become more widely used, the internet-based software that provides the user interface to their data could become a source for wide scale studies of activity and sleep patterns. Such devices, know as personal biosensors, have been used for some time in HFE research for example the Actiwatch®. These have usually been carefully calibrated. However, recently a number of consumer-grade devices such as Fitbit® have come into widespread use. These devices contain a tri-axial accelerometer and have been tested under limited conditions such as treadmill walking with fit adults (Takacsw et al., 2014) and found to be reliable and valid for step count but not distance when validated against direct observation. Despite shortcomings in validity, such devices could fit the BDA profile of large but messy data sets, providing research access to potentially huge amounts of physical activity data collected under realistic conditions of every-day living. There would of course be biases, as only self-selected people buy and wear such technology, and they may well leave them unused at times (e.g. bathing or even sleeping). But again, perhaps messy data is of less concern that then the overall data set size (Hernandez, 2014).

How do people perform BDA?

There have been many studies of BDA and the closely-related field of Data Mining in the literature of statistics, with at least five authors advocating multi-step approaches to understanding how BDA or Data Mining proceeds. A comprehensive review of these five authors (Ma & Drury, 2003) found that their steps could be combined into an overview to provide an initial task analysis of BDA and eventual cognitive task analysis. These steps were labeled five Stages of Data Mining: (1) Understand Business Domain; (2) Identify Data Mining Goals; (3) Collect and Prepare Data: (4) Building Data Mining Models; and, (5) Evaluate and Deploy Data Mining.

Following this analysis, ten published data mining case studies from the literature were examined to derive a hierarchical task analysis model of the process. The Applied Cognitive Task Analysis framework was used with five experienced data mining researchers, refining the stages above into a more detailed exposition. Finally, two sets of observational studies were performed on eight experienced researchers using individual observations and recording of collaborative meetings (Ma & Drury, 2005) to develop a comprehensive model.

BDA used in HFE studies

One example of the use of a large data base in HFE, again from work with which I am familiar, was the study of behavioural modeling in automobile seat belt wearing

(Drury, Drake and Thomaz, 2002). They had a data base of about 40,000 observations of front-seat occupants wearing or not wearing seat belt, collected at 113 sites across the state of Indiana during 2001. All the data points that recorded gender and age of front seat occupants (15,249 observations) was used to determine whether behavior of one occupant was the same as or different from the other occupant. The study found the modeling existed and that negative modeling was almost twice as prevalent as positive modeling. Also, modeling was stronger where the age and/or gender of driver and passenger matched. In this study, no statistical significance values (using Chi-square tests) were reported as "All comparisons made in this report are highly significant ($p < 0.001$)" (page 1731). Also the authors again pointed out that observational studies cannot prove causation. This example uses a large data set and addresses some of the BDA issues noted earlier, but it would not be classified as Big Data because it was based on a large sample not "almost the whole population". Also it selected only data points where complete information was available, so as to form a complete matrix of observations rather than trying to use the original, much more sparse, matrix of all observations.

BDA was used in one HFE study by Zhu et al. (2013). Although they did not use the words "Big Data" the authors used a very large data base of 182 million Wikipedia edits from the inception of the site until 2008. This was essentially the whole population to that date. They also used typical BDA software (Hadoop) and machine learning to automatically code the text data into features that could then be used to test hypotheses about responses of Wikipedia editors to different management styles in a way that is quite familiar to HFE researchers. Their findings were based on between 1 and 2 million observations that met their criteria, depending on the specific hypothesis tested. They were explicit about correlation analyses not implying causality, and about the ethics of conducting this first study and a second direct intervention experiment. The Zhu et al paper tested specific HFE hypotheses (type of management feedback and editing performance) in a way totally familiar to HFE professionals.

Privacy and ethics

The extent to which the accumulation and analysis of Big Data affects our lives can be seen in this quote from King (2011): "*date of birth, gender, and ZIP code alone are enough to personally identify 87% of the U.S. population.*" It can also be seen whenever we receive mail, either email or paper mail, from so many sources directly related to our own past behaviour. As a relatively benign example, charitable giving in the USA generates solicitations not just from the charity originally supported but other similar ones. A quote from a charity evaluation web site is instructive: "*Small donations, such as $25, barely cover the costs the charity incurred in soliciting the gift. To recoup those costs, many charities will simply sell the donor's name to another charity doing similar work*" (http://www.charitynavigator.org/, accessed November 2014). Data bases and whole markets exist to compile, monetize and trade this, and other behavioral information (Mayer-Schonberger & Cukler; 2013, p. 99).

Perhaps the most salient technology for data collection without individual consent is image-based surveillance. This is in widespread use in the streets of cities and in retail establishments as well as in and around sensitive government installations and business properties. A recent edition of Fortune magazine (Proctor, 2013) estimated around 10,000 cameras in Chicago, 4,000 in Manhattan and "half-a-million" in London. Aerial surveillance has been practiced since World War 1, but has become far more prevalent as technology has improved to unmanned aerial vehicles (UAVs) which can now be bought by consumers. A major HFE question is who interprets the surveillance data? This is the old Allocation of Function issue in HFE, brought about in this case by the increasing capabilities of algorithms to detect and recognize/classify people and events. With video interpretation allocated to a human operator, there would never be enough people to view the videos. Our world has protected by this logic because of this fundamental allocation of function problem. Now that much of the interpretation can be performed by rapidly-evolving computer hardware and algorithms, society must look to new ways to balance public safety and individual privacy.

Ethics has been taught in engineering for many years, and is part of professional HFE training. For example, it is part of the core competencies specified by the Board of Certification in Professional Ergonomics (BCPE) which also has a code of ethics. Ethics is taught with reference to key philosophers for ethical theories, with case studies used to discuss ethical dilemmas that can arise in engineering and HFE. A typical dilemma of interest in this paper is the balance between the good of society (e.g. John Stuart Mill in Western philosophy) and individual rights (e.g. Immanuel Kant, John Locke). Either can be the dominant ethic of a particular case. Even the IEA definition of HFE required both system performance (societal good?) and human well-being (individual good?) to be jointly improved or optimized. As Hancock and Drury (2011) point out, while the IEA's mission statement is to improve the quality of life, the real question is who's life?

Conclusions

The first conclusion is that BDA could be useful in the practice of HFE. We already know how to use archival data to answer HFE questions and test HFE hypotheses, so in one way BDA is an expansion of this knowledge. The data bases may be much larger, and certainly less clean, but there are tools available for compiling the data base, analysing it and turning the findings into useful information. We have presented examples of HFE use of large data bases and BDA tools, and demonstrated our ability to find meaningful conclusions. Even so, we are not yet exploiting all aspects of BDA such as using data on almost the whole population, or analysing sparse data matrices. HFE practitioners and more academic researchers need to be alert to what data is, or may become, available that would potentially yield useful insights unobtainable elsewhere.

The second conclusion is that HFE could be useful in BDA. Because BDA is typically a team enterprise and highly mediated by software and hardware tools, HFE has an obvious place in enhancing system BDA performance while improving the work environment of BDA practitioners (who could of course be HFE practitioners, as above). We should be able to improve system performance by not ignoring the user, although BDA practitioners do not deliberately ignore the user, but often lack our specialist tools and techniques for improving the users' work. HFE is all about interactions (Wilson, 2000) and here is a field where human-system and human-human interactions are the essence of the work (Ma and Drury, 2005).

The third conclusion is that BDA may well change how scientists, including HFE practitioners, view the world and their own work philosophy. The ability to collect near-population-sized data, combine a number of data sources and perform correlational analysis on this potentially useful information trove could lead us to a point where we analyse data before formulating hypotheses. This is exactly what is advocated by BDA proponents: "let the data speak". We are exhorted to use correlations, knowing that they do not imply causation, as if they were enough in themselves. With almost the whole population to analyse, we do not even need statistical inference to reach conclusions, merely assessing practical significance of our correlations. Perhaps this approach may have advantages, but we need to debate this carefully in the profession.

Finally, although not the least important, is how to integrate ethics into BDA. We are all citizens of the world, and need to be ever-mindful of the possibility that BDA will be misused, deliberately or inadvertently, in our professional and private lives. Examples abound of data collected on people without their permission, or collected for one purpose with permission and later combined with independently collected data to allow new analyses unimagined at the time permission was granted. Our ethical training needs to be examined and debated to determine whether it can keep up with rapid changes in what is technically possible. When the use of personal data is for obviously detrimental purposes to the population, then the choice may be easy, although finding a way to protest this choice effectively may not be as easy. When the use provides some collective good, we need to actively balance the needs and rights of society and the individual. This problem is always with us in collecting and analysing data on our fellow humans, but BDA may make the finding of solutions more urgent.

References

Drury, C. G., Drake, M. L. and Thomaz, J. E. (2002) Demographic effects of behavior modeling in seat belt use: analysis of 15,000 observations. *Proceedings of the Human Factors and Ergonomics Society Annual Meeting,* Baltimore, MD, October 2002.

Ginsberg, J., Mohebbi, M. H., Patel, R. S., Brammer, L., Smolinski, M. S. and Brilliant, L. (2009) Detecting influenza epidemics using search engine query data. *Nature*, 457, 7232, 1012–1015.

Hancock, P. A. and Drury, C. G. (2011) Does human factors/ergonomics contribute to the quality of life? *Theoretical Issues in Ergonomics Science*, 12 (5), 416–426.

Hernandez, D. (2014) Gadgets Like Fitbit Are Remaking How Doctors Treat You, *Wired Magazine,* 06.14, Accessed November 2014 at http://www.wired.com/2014/03/internet-things-health/

Hilbert, M. and López, P. (2011) The World's Technological Capacity to Store, Communicate, and Compute Information, *Science* 332, 60–65.

Hou, T.-S., Lin, L. and Drury, C.G. (1993) An Empirical Study of Hybrid Inspection Systems and Allocation of Inspection Function. *International Journal of Human Factors in Manufacturing*, 3, 351–367.

King, G. (2011) Ensuring the Data-Rich Future of the Social Sciences, *Science* 331, 719–721.

Kuhn, M. and Johnson, K. (2014) Who's afraid of the big black box?: Statisticians' vital role in big data and predictive modelling *Significance*, 11.3, 35–37.

Ma, J. and Drury, C. G. (2003) The human role in data mining. *Proceedings of the Human Factors and Ergonomics Society Annual Meeting*, Denver, CO, October 2003.

Ma, J. and Drury, C. G. (2005) Analysis of Collaborative Meetings in Developing Data Mining Models, *Human Factors and Ergonomics Society 49th Annual Meeting,* Orlando, FL. 686–690.

Mayer-Schonberger, V. and Cukier, K. (2013) *Big Data: A Revolution That Will Transform How We Live, Work and Think*. Houghton Mifflin Harcourt, NY.

Proctor, K. (2013) The great surveillance boom, *Fortune Magazine*, April 26, 2013, http://fortune.com/2013/04/26/the-great-surveillance-boom/ accessed December 2014.

Takacs, J., Pollock, C. L., Guenther, J. R., Bahar, M., Napier, C. and Hunt, M.A. (2014) Validation of the Fitbit One activity monitor device during treadmill walking, *Journal of Science and Medicine in Sport*, 17, 496–500.

Zhu, H., Kraut, R. E. and Kittur, A. (2013) Effectiveness of Shared Leadership in Wikipedia, *Human Factors* 55.6, 1021–1043.

KEYNOTE

ERGONOMICS, ACCOUNTABILITY AND COMPLEXITY

Sidney W. A. Dekker

Safety Science Innovation Lab, Griffith University, Brisbane, Australia

The last twenty years in ergonomics have seen a gradual move away from the terms than denote individual rationality and agency. From arguments for a system approach that still radiate outward from errors and violations as their core (Reason, 1995), a more wholesale shift is evident in recent work that promotes new vocabularies and eschews individual attributions of success and failure. In my lecture I discuss these changes and the implications these have for our views of personal accountability within an increasingly complex world.

The human: From cause to recipient of trouble

"Accidents," said Oxford psychologist Vernon in the early 20th century, "depend in the main on the carelessness and lack of attention of the workers" (Burnham, 2009, p. 53). The "human factor," at the time, was thought of as the "physical, mental or moral defects" that predisposed certain people to accidents (p. 61). In the proto-ergonomics world, the major source of risk to the well-being and safety of people in workplaces was sought in the individual human. It was the human who did not comply, took risk, did not pay attention. And these behaviors could be linked to physical, mental and moral shortcomings of the individuals involved. The human factor was seen as the problem to control. And we exercised that control by intervening in people's behavior: telling them to pay more attention, to try harder, to follow the rules or directives, to use proper body posture. Coming, as they did, at the dawn of the 20th century, these ideas reified and represented late-Victorian thinking about the role of the human (or, more specifically, the worker) in the world. Deficient individuals could be held accountable for their own failures and made to bear the consequences.

In ergonomics, we like to believe we have come a long way since then. What is often seen as the revolution that birthed our field, after all, flipped this model. The "human factor" was not the cause of trouble, but the recipient of it. Most in the field will have heard about the tinkerers of WWII—engineering psychologists like Alphone Chapanis, Paul Fitts—who were able to eradicte whole families of incidents and problems not by telling people to try harder and take more responsibility, but by intervening in the conditions that made up people's working environment, including its engineered and organizational features. Rather than seeing human errors as the brain bloopers of morally or mentally deficient people, ergonomics' insights and interventions showed how "errors" are systematically connected to features of people's tools and tasks. Change the conditions of work, and you change the behavior that goes on inside of that workplace.

"It should be possible to eliminate a large proportion of so-called '[human]-error' accidents by designing equipment in accordance with human requirements," Fitts and Jones concluded (1947, p. 3). Their work, as much after it, showed that the ideal of the fully rational, regulative human is constrained by design and operational features. The latter should be our target, not the former. The field seemed to have found its central guiding ideology in this: people may be a problem to control, but you don't exercise that control by telling them to try a little harder, by telling people to be more rational. You assume that they are locally rational vis-à-vis the conditions under which they are made to work. So we change the conditions; we change the relationships between people and working environment, between people and technology. If problems keep occuring, it is because we have not understood the collaborative relationship between human and environment or technology well enough.

And then, not long ago, we could read about the probable cause of a recent accident that killed three people on a large airliner. If we take some of the jargon out, it was determined that the crash was caused by human mismanagement of the aircraft's descent path to the runway, the human operators' unintended deactivation of automatic speed control, human inadequate monitoring of speed and their delayed decision about whether to even continue the approach or not, in the face of evidence that the system was out of tolerances. The conclusions are late-Victorian thinking redux. Through two decades of research into automation surprises (Sarter & Woods, 1995; Sarter, Woods, & Billings, 1997), ergonomics assembled enough of an empirical corpus and conceptual apparatus to retrospectively model (if not help redesign) the confusions that come from an occasionally unexpected forest of automation mode combinations that these airplanes can end up in. Early on in this research, we realized that the ergonomic issues of WWII were being reproduced on a massive scale, *in silico*. The mode errors of today are Chapanis' switch selection confusions of 1943. What we call "human error" is still, or again, normal behavior systematically connected to features of the design, operational and training environments in which we expect people to work. And yet, today, an expensive, government-sanctioned investigation simply brings it all back to human deficit. To a few erratic, unreliable humans who don't try hard enough. To humans who didn't see, didn't pay attention, didn't understand, didn't decide.

In an optimistic reading, such findings may be the result of the practical and political constraints imposed on the investigation that produces them. And they probably are—in part. The problem is, the ideas carried in such findings are found too widely, and come from too many different domains, to believe that. In the wake of the Mid Staffordshire scandal in the UK, Health Secretary Jeremy Hunt announced a new offence of wilful mistreatment or neglect of patients, an offence that could mean five years of jail for nurses and doctors (Mason, 2013). Like most stories of drift into organizational failure, cases of neglect came to light in Mid Staffordshire that were the result of years of budget problems, organizational changes, cutbacks, leadership challenges, changing demographics, training issues and more—all pretty much amounting to a "standard" organizational disaster (Reason, 1997). A focus on the supposedly rational, criminal choices made by care providers at the sharp end severely oversimplifies and limits the depth and breadth of our consideration of the complex causal web beneath the disaster.

The role of ergonomics in asserting personal accountability

But what is ergonomics' own role in this? A recent review of the patient safety literature (Waterson, 2009) showed that although some of the studies actively looked at teams (17%) or organizations (14%), 98 of the 360 articles reviewed addressed the individual level of analysis, focusing for example on human error. Similarly, medical adverse event investigations are known to stress individual agency and responsibility (Berlinger, 2005; Cook, Nemeth, & Dekker, 2008; Sharpe, 2004). Even fields of practice that have been open to ergonomics from the beginning, like aviation (Roscoe, 1997), have difficulty escaping this focus. Between 1999 and 2006, 96% of US aviation accidents were attributed in large part to the flight crew. In 81%, people were the *sole* reported cause (Holden, 2009).

Perhaps subtly but pervasively, the possibility of full rationality and the human's moral obligation to attain it, gets reproduced and reconfirmed in the language of ergonomics itself. We typically explain ourselves in terms of the Western regulative moral ideal, where full rationality is possible and desirable. Testimony to this is the asserted usefulness of a concept like "situation awareness" in ergonomics (Parasuraman, Sheridan, & Wickens, 2008). Whatever limited awareness was present inside an operator's mind, it can always be contrasted against a sort of "ground truth" (p. 144) or fully rational ideal that the operator could or should have achieved. This represents a naïve Cartesian-Newtonian scientism: total knowledge of the world is achievable, the world is "out there" as an object entirely separable from us observers.

And there is more. Terms such as "error" and "violation" have been popularized by ergonomics research itself. And they remain useful handmaidens in a variety of organizational, political and psychological processes (Cook & Nemeth, 2010). The snare is that such terms at once deny and invoke full rationality. Nyssen's and Cote's (2010) ergonomics studies in pharmacy, for example, show that the rationality of operators is constrained by cognitive and social influences—a finding consistent with ergonomic orthodoxy. Yet the results are counted and expressed in terms of "non-compliance" and "violations:" the gap between procedure and actual work can be closed by taking into account people's motivational factors (Nyssen & Cote, 2010). Thus we help shape the world we and others can know, in ways that both facilitate and constrain our understanding and our possible actions.

Ergonomics and a complex world

But does that commitment about the rational individual, whether we make it consciously or not, still have value for ergonomics in today's and tomorrow's world?

If there is one agreement that can be across scientific fields, it is that the world is becoming an increasingly complex place. This means a world of ever more inter-connections and interactions. A world with fewer linear cause-effect relationships, where small, single events can get amplified into disproportional consquences. It is a world in which we, also in ergonomics, can see to it that technologies, products, postures or procedures work well in relative isolation. But when these same things get placed in dynamic, complex, competitive environments, with finite resources and multiple goal conflicts, their connections can proliferate, their interactions and interdependencies multiply, their complexities can mushroom.

Take the aircraft accident briefly mentioned above. Western assumptions about lan-guage competency, operational hierarchy and interpersonal communication and collaboration are woven into the design of the aircraft's cockpit and operating procedures. Airliners that are designed and built in a culture that has no polite-ness differentiation in interpersonal address are flown in cultures that have such differentiation, like the two levels of many European languages, or like the six politeness levels of interpersonal address in some Asian languages. Western tech-nology operated by multiple people also typically assumes a transmitter-orientation in interpersonal communication. That means that lack of clarity is mainly the responsibility of the sender, not the receiver. In some cultures, such transmitter initiative may be seen as insensitive, impolite and career-jeopardizing.

Assumptions about cross-checking and challenging, baked into features of displays, procedures and checklists, may fall flat when subjected to a culture that takes a different view of command, power, politeness and seniority. The result is that original designs, when opened up to such influences, get adapted in ways that are hard to foresee by those who built it, by those who put the original parts together (Dekker, 2011). Is this a world that is still suited to early 20th century thinking about humans as the single largest source of risk in otherwise well-designed and safe systems? Is this still a world that has a place for 17th century assumptions about completely rational choices that get made by supposedly unconstrained, fully informed and free-to-choose human operators? Perhaps we, and they, never lived in such a world.

Take another example—from healthcare. Ergonomic interventions abound there, such as the introduction of new monitoring technology. This, as shown by ergonomics research, can create ripple effects across clinical relationships, power hierarchies and gender gaps, whereby different agents in the system manipulate each others' sensitivities to evidence and decision criteria about when to intervene. These are aspects typical of complexity. Introducing new technology into such a world is not about manipulating one variable and keeping the rest constant. Rather, implementing new monitoring equipment in obstetrics, for example, showed not only how complexly sensitized human practitioners' reading of the evidence is (relative to which the equipment's reading is blunt and severely underdetermined), or how there are predictable interactions between the intervention criteria set by the machine and those developed by its human monitors. But the system is more open and complex than that. Nurses' and midwives' decision criteria were adjusted on the basis of what they believed other clinicians' decision criteria were. The presentation of evidence was in fact actively tuned to either exceed or remain below relevant clinician's decision threshold, something that multiplied across the medical hierarchy (evidence was sometimes presented so that a clinical decision maker would call a more senior colleague, which was precisely what nurses or midwives wanted all along).

Understanding such a complex system is about understanding the intricate web of relationships a new piece of technology weaves, its interconnections and interdependencies, and the constantly changing nature of those as people come and go and the technology gets adapted in use. Failure in such worlds occurs not because individual components break, but because normal interactions between components give rise to emergent behaviors that fall outside of what we could model and predict. Such behavior does not fit inside the limitations of our own fundamental ideas about cause and effect and human action that were once provided to our field by thinkers like Rene Descartes and Isaac Newton. The system we seek to help, ultimately fails because we left it with only one way to succeed.

Ergonomics applies the "systems approach" as one useful pathway out of this. A systems approach identifies the many and various factors that impact a complex situation or event (this may include remuneration schemes, pre-existing hierarchical relationships, cultural inclinations, time and financial pressures, the accuracy of available information about a patient or procedure, and much more). A system, such as a hospital, is a dynamic and complex whole, interacting as a functional unit. One system may be nested within another system. For example, a hospital is nested within a larger healthcare system that helps procure its new technologies. The behavior of a system is understood in terms of the linkages and interactions between the elements that compose the entire system. All medicine is practiced within a 'system.' The makeup of that system can help or hinder medical professionals from doing their job. The goal of a systems approach should not be to reduce human behavior to thoughtless following of rules but to design a system in which individual responsibility and competence can be most effective in creating desirable outcomes. That requires a complex understanding and careful design of the system in which individuals perform their work, and a willingness to revise our models and understanding of how work really gets done as new evidence comes in (Dekker & Leveson, 2014).

Yet even the systems approach (though only after being severely misconstrued) has met with pushback and headwind from those concerned about an erosion of "personal accountability" (Levitt, 2014); a concern that is not foreign from within ergonomics' own ranks either (Reason, 1999). A "systems approach" is mistaken as just more standardization and checklists, or the opportunity for individual operators to eschew their responsibility and blame all problems on "the system." It is neither. In fact, complexity forces ergonomics to make clear that its research is not about humans *or* systems, but about humans *in* systems.

Conclusion

The last twenty years in ergonomics have seen a gradual move away from the use of terms that denote individual rationality and agency. From arguments for a system approach that still radiate outward from errors and violations as their core (Reason, 1995), a more wholesale shift is evident in recent work that promotes new vocabularies and eschews individual attributions of success and failure. This includes resilience engineering (Hollnagel, Woods, & Leveson, 2006), which models how systems adapt to their changing environments and absorb challenges to their continued functioning. And, as shown above, there are exciting applications of complexity theory that see system failure as growing out of relationships, not failed components (Dekker, 2011; Leveson, 2011). Questions and challenges abound for an ergonomics that tries to take this seriously. What about accountability in a networked world, for example? What about the impossibility to even design for complexity, as we can really only engineer and design *complicated* systems (Dekker, Cilliers, & Hofmeyr, 2011)? But these questions are invitations rather than obstacles to our field in the coming years. One thing is certain: making the problems we face simpler by denying complexity, and by reverting to people as a problem to control, is not going to help anyone.

References

Berlinger, N. 2005. *After harm: Medical error and the ethics of forgiveness.* Baltimore, MD, Johns Hopkins University Press.

Burnham, J. C. 2009. *Accident prone: A history of technology, psychology and misfits of the machine age.* Chicago, The University of Chicago Press.

Cook, R. I., C. Nemeth and S. W. A. Dekker 2008. What went wrong at the Beatson Oncology Centre? *Resilience Engineering Perspectives: Remaining sensitive to the possibility of failure.* E. Hollnagel, C. P. Nemeth and S. W. A. Dekker. Aldershot, UK, Ashgate Publishing Co.

Cook, R. I. and C. P. Nemeth 2010. "Those found responsible have been sacked: Some observations on the usefulness of error." *Cognition, Technology and Work* **12**: 87–93.

Dekker, S. W. A. 2011. Drift into failure: From hunting broken components to understanding complex systems. Farnham, UK, Ashgate Publishing Co.

Dekker, S. W. A., P. Cilliers and J. Hofmeyr 2011. "The complexity of failure: Implications of complexity theory for safety investigations." *Safety Science* **49**(6): 939–945.

Dekker, S. W. A. and N. G. Leveson 2014. "The systems approach to medicine: Controversy and misconceptions." *BMJ Quality and Safety* **0**(1–3).

Fitts, P. M. and R. E. Jones 1947. Analysis of factors contributing to 460 "pilot error" experiences in operating aircraft controls Dayton, OH, Aero Medical Laboratory, Air Material Command, Wright-Patterson Air Force Base.

Holden, R. J. 2009. "People or systems: To blame is human. The fix is to engineer." *Professional safety* (12): 34–41.

Hollnagel, E., D. D. Woods and N. G. Leveson 2006. *Resilience engineering: Concepts and precepts*. Aldershot, UK, Ashgate Publishing Co.

Leveson, N. G. 2011. "Applying systems thinking to analyze and learn from accidents." *Safety Science* **49**(1): 55–64.

Levitt, P. 2014, 19 March. "When medical errors kill: American hospitals have embraced a systems solution that doesn't solve the problem." *LA Times* Retrieved 23 March 2014, from latimes.com/opinion/commentary/la-oe-levitt-doctors-hospital-errors-20140316,0,4542704.story.

Mason, R. (2013, 16 November). Doctors, nurses and managers to face five years in jail if they neglect patients. *The Guardian*. London, The Guardian.

Nyssen, A. and V. Cote 2010. "Motivational mechanisms at the origin of control task violations: An analytical case study in the pharmaceutical industry." *Ergonomics* **53**(9): 1076–1084.

Parasuraman, R., T. B. Sheridan and C. D. Wickens 2008. "Situation awareness, mental workload and trust in automation: Viable, empirically supported cognitive engineering constructs." *Journal of Cognitive Engineering and Decision Making* **2**(2): 140–160.

Reason, J. T. 1995. "A systems approach to organizational error." *Ergonomics* **38**(8): 1708–1721.

Reason, J. T. 1997. *Managing the risks of organizational accidents*. Aldershot, UK, Ashgate Publishing Co.

Reason, J. T. 1999. Are we casting the net too widely in our search for the factors contributing to errors and accidents? *Nuclear safety: A human factors perspective*. J. Misumi, B. Wilpert and R. Miller. London, Taylor & Francis: 210–223.

Roscoe, S. N. 1997. The adolescence of engineering psychology. *Volume 1, Human factors history monograph series*. S. M. Casey. Santa Monica, CA, Human Factors and Ergonomics Society: 1–9.

Sarter, N. B. and D. D. Woods 1995. "How in the world did we get into that mode? Mode error and awareness in supervisory control." *Human Factors* **37**(1): 5–19.

Sarter, N. B., D. D. Woods and C. E. Billings 1997. Automation Surprises. *Handbook of Human Factors and Ergonomics, 2nd Edition*. G. Salvendy. New York, Wiley: 1926–1943.

Sharpe, V. A. 2004. *Accountability: Patient safety and policy reform*. Washington, D.C., Georgetown University Press.

Waterson, P. 2009. "A critical review of the systems approach within patient safety research." *Ergonomics* **52**(10): 1185–1195.

PLENARY

KEEPING HUMAN FACTORS ON TRACK – THE DESIGN OF THE NEXT GENERATION INTERCITY EXPRESS TRAIN

Daniel P. Jenkins & Carl Harvey

DCA Design International, UK
Hitachi Rail Europe, UK

This case-study discusses the design and development of the UK's new Intercity Express train, to be introduced to the East Coast and Great Western mainlines in 2017. The paper describes the iterative development approach adopted to ensure stakeholder engagement and regulatory compliance throughout the design process. This includes the use of prototype evaluation, from very low to very high fidelity, and the application of human factors tools such as Hierarchical Task Analysis and an innovative approach to glare assessment. The paper highlights how the described multi-method approach considers both the cognitive and physical requirements of the user population, ensuring successful delivery.

Introduction

The approach described in this paper was developed to support the design of the Intercity Express Programme (IEP) Driver's Cab (see Figure 1). As described in Jenkins et al. (in press), the project involves the design and manufacture of 122 new trains for the UK's East Coast and Great Western mainlines, increasing capacity and reducing journey times. The first trains are planned to go into service in 2017. While the project involved the design of all the interior areas of the train, the scope of this paper has been limited to the design of the cab.

The link between cab design and driver performance is clear and well established. A well designed train cab provides reach to all equipment, good visibility of well organised controls and instruments, and a suitable view of the external environment. Historically, guidance on appropriate driving postures and control layouts have been dominated by the anthropometry of the drivers (e.g. Dreyfuss, 1955; Diffrient et al., 1973). On a physical level, controls should be arranged so that they are clearly visible and readable by the target population. Likewise, they should also be arranged to ensure that driving tasks do not require poor postures that may result in driver discomfort or musculoskeletal injuries. Additionally, from a cognitive perspective, cab layouts should ensure that controls are easily identifiable, and that the mode of actuation is highly intuitive, with minimal risk of confusion or inadvertent operation. Controls can be assessed individually based on simple heuristics. However it is also important to assess the cab in its entirety against its anticipated scenarios of use.

Figure 1.   Internal view of the final cab mock up.

Many different people will interact with a passenger train across its lifespan on a physical and a cognitive level. As such, there are many stakeholders that have a vested interest in the final design. These include the representatives for the train drivers, train crew, maintenance staff, cleaners, passengers, cyclists, and persons with reduced mobility. They also include representatives from the train operating companies and the UK Department for Transport. Alongside stakeholder engagement, there is also a need to demonstrate compliance with a number of contractually specified standards and mandatory regulations.

To ensure that the physical and cognitive needs of the stakeholders are considered throughout the design process, it is very useful to create a human factors integration plan to highlight the activities that are required at different stages of the design process. This process is explained at a high-level in the following section.

Approach

Development of product requirements

Contractual and regulatory requirements formed the basis of the human factors requirements. For example, for passenger areas, guidance exists on the location of buttons above floor height, their maximum actuation force and arrangement. The relevant requirements were presented in tabular form in order to capture the description and their origin, along with a column allowing compliance to be recorded at each stage of the development process. Anthropometric datasets of the user population (e.g. Adultdata, 1998) were also used to create additional requirements where appropriate. The requirements list remained a living document throughout the project and was updated as new requirements were identified.

Desk based evaluation of concept

The initial layout of the cab was constructed and evaluated using Computer Aided Design (CAD) models. In the initial stages, 2D projections provide an efficient way of considering the design. Reach envelopes and mannequins based on anthropometric data can be used to optimise the layout for reach and visibility of controls. Likewise, in terms of external visibility, standards exist (GM/RT2161, 1995) that describe largely unambiguous test criteria for assessing forward visibility using sightlines. 3D CAD models are also used to optimise the cab layout as the design matures.

The initial cab control layout was also informed by the guidance provided in GM/RT2161. This standard also classifies each of the commonly used controls within the cab as either primary or secondary in terms of reach and visibility. By arranging the controls based on this guidance and the familiar layout of legacy trains (such as the Class 395 train), a basic design was derived. The initial control layout for the cab was also heavily influenced by the requirement to operate the combined power brake controller (the 'T' shaped handle to the left of the driver; see Figure 1). As the actuation of this control requires the use of the driver's left hand, the majority of controls and screens with touch screen functionality were moved to the right side of the cab, while indicator lamps and displays without controls (CCTV screens) that do not require actuation, were located on the left side of the cab. Functional grouping was also employed to cluster similar controls to aid the task of identification (e.g. grouping all controls to do with the control of diesel engines in one place). Likewise consideration was made of the consistency with current rolling stock and the need for future upgrades.

Physical evaluation and refinement of initial layouts using part prototypes

Once a basic cab arrangement had been formulated, this was then assessed with representative train drivers from each of the train operators (three from each). To support this, a low-fidelity prototype was constructed (see Figure 2). This 1:1 scale mock up provided a low-cost representation of the cab panels along with controls printed on paper, allowing them to be easily repositioned. Each of the drivers performing the assessment was encouraged to rearrange the controls to best represent their ideal control layout. As expected, different drivers favoured different layouts, partly due to individual preferences and experiences and partly due to different working procedures between the two Train Operating Companies.

Figure 2.   Validation of early control layout.

However, through a process of rapid iteration, the group was quickly able to achieve a consensus of opinion that closely matched the original proposition.

The early engagement of stakeholders provided an extremely useful method for validating the design and accommodating changes before significant design work had been undertaken. It is, however, important to draw the distinction between stakeholder-informed design and user-led design. The low fidelity mock up allowed the physical space constraints of the cab to be communicated. In addition, a human factors specialist was involved throughout the process to communicate the importance of control grouping and spacing.

Evaluation and refinement of full-sized spatial mock ups

Once an agreed design had been established, the design was revised in CAD and represented in a full-size spatial ergonomic mock up (see Figure 3). A second phase of assessment with train drivers was conducted to further refine the design. The addition of a fully functional production chair allowed a more accurate assessment to be made. As such, further refinements were made based on reach and visibility and the comfort of the posture required to actuate each of the controls.

Evaluation and refinement of full-sized visually representative mock up

Following the ergonomic mock up, a 1:1 scale visually representative mock up of the cab was constructed (see Figure 1 and Figure 4 for exterior view). This mock up adds an additional level of fidelity by using production versions of the driver's seat and controls, as well as providing representative colours and surface finishes.

Figure 3.   Validation of control layout in ergonomic mock up.

Alongside fitting trials assessing comfort and reach, the full mock up was used to conduct a glare study (see Jenkins et al., 2015). Unlike assessment of other factors, such as forward visibility, there are no standardised approaches for performing assessments of glare. While it is unrealistic to evaluate every possible lighting condition that may potentially occur in the vehicle cab in service, a pragmatic and practical approach was taken to provide a good level of indicative information about the cab design's likely glare performance against internal and external light sources. The assessment of internal light sources involved blacking out the cab windows and assessing the impact of glare from internal lights and illuminated controls. The impact of external lights was assessed by simulating external light sources (e.g. the sun, other trains' headlights) by illuminating the cab mock up windscreen, side and door windows with a single light source manually located in a sequence of discrete positions and orientations and assessing the resulting glare impacts (see Figure 4). As a result of the glare study, modifications were made to the cowling along the top of the cab control console, internal lights were recessed, control panel angles were adjusted, and a patterned film was added to the side windows. The glare study was repeated following these modifications to the cab and found to confirm the effectiveness in terms of reduced instances of instances of direct and indirect glare.

Formal assessments of the train operating tasks were also conducted in the full mock up. A structured approach for assessing a train cab against task requirements was developed. The assessment is divided into two stages; (1) the first assessed the location of each of the cab controls in turn against their frequency of use, functional grouping, and risk of inadvertent operation. (2) The second assessed the

Figure 4.   Arrilite light source on a Hague CamCrane K16DV aligned to one of the positions on the train side window.

cab against routine tasks based on a Hierarchical Task Analysis (HTA) model. For the initial static assessment, a list of all the controls within the cab was compiled in tabular form (87 controls). Columns were added to the table to capture the control type, location, frequency of use, and whether the control was used while driving or stationary (e.g. door controls). Three train drivers (recruited to represent both Train Operating Companies, both genders and a range of statures) were asked to actuate and assess each of the controls in the cab in turn, observed by a human factors expert. For each control, the driver was asked to report any concerns or issues with visibility, reach, risk of inadvertent operation and suitability of posture while actuating. The driver's comments were recorded in the table along with any additional observations from the human factors expert.

In order to assess a cab against common tasks, or sequences of operation, some form of task description is required. Ostensibly, task analysis involves breaking down a task into smaller sub-tasks or operations. Arguably, the most commonly used and well-known task analysis technique is Hierarchical Task Analysis (HTA; Annett et al., 2007). HTA involves breaking down the task under analysis into a nested hierarchy of goals, operations and plans. The end result is an exhaustive description of task activity, which, importantly for the train driving task, can be distilled down to modelling the actuation of individual controls. Despite HTA being one of the most commonly used human factors approaches, as reported by Rose & Bearman (2012) there are few examples of HTAs that cover train driving in the public domain. As such, the option of adopting an existing HTA model for the purpose of this analysis was not available. Rose & Bearman (2012) present a task analysis model of train driving for the purposes of identifying human factors issues in new rail technology. However, the model they discuss is based on goal-directed task analysis, a variant of HTA that places a focus on situation awareness. As a result of this focus on situation awareness, the model is primarily concerned with the cognitive aspects of the task, and does not contain the detail of the physical control manipulations required for this analysis. Accordingly, the first stage of the process was to create a task model that included individual manipulations of controls. Initially, an HTA model was built based upon a Class 395 operating manual and cross-referenced against the model created by Rose & Bearman (2012) and a report by Haworth et al. (2005; Based on the Australian railway) to ensure its completeness. The overall goal of the train driving specified at the top of the hierarchy is broken down into sub-goals (for example, start-up, drive train, manage communications). In turn, these goals were decomposed further until an appropriate operation was reached (e.g. place foot on DSD pedal, depress plunger, check for alarm, and check CCTV). The first draft of the HTA model was validated with two train driver experts on two occasions to ensure its suitability and completeness. The validation process involved stepping through the model task-by-task (in a tree view format), adding additional detail and validating the plans. Once an agreed task model was finalised, the task steps (nodes) were coded to indicate which of the tasks would be explicitly assessed in the cab. Omitted tasks included elements that were not supported by the mock up (for example, data entry on the train management system, or using the key to unlock the door). In addition, sub-routines that had been previously assessed a number of times were also omitted. The resultant model contains a total of 513 nodes (360 base level operations) of which 187 tasks were explicitly tested.

The task model was taken into the cab in list form and the drivers were asked to perform each of the tasks in the order dictated by the HTA (read aloud by the human factors specialist). After each task step, the driver was asked to report any concerns or comments about the current layout. These were recorded in an additional column in the HTA table along with additional observations from the human factors specialist. Detailed assessments included an assessment of ingress and egress, an assessment of emergency evacuation of the driver's seat and the second person's seat, assessments of standard driving tasks, and an assessment of emergency procedures. The adopted approach proved to be an effective mechanism for validating the cab control layout. Specifically, the system design and the associated number and location of controls were challenged and in some cases simplified as a result of the process. The static assessment ensured that each control was considered and evaluated in turn. In addition, the sequenced assessment identified a number of issues that are unlikely to have been detected from a static assessment alone. Moreover, the clear structure of both assessments has allowed them to be readily communicated to the wide range of stakeholders involved in the project, thus supporting prompt and well considered decision making.

Conclusions

The outcome of the project is a stakeholder-informed cab design that is compliant with the identified project requirements. The two more innovative assessment approaches discussed in this paper, (1) the glare assessment and (2) the task-based assessment, were found to be very useful additions to the wider development process – producing new insights and allowing the cab to be further optimised. The glare assessment was found to be an efficient means of optimising the cab design to minimise the impact of glare – yielding practical recommendations for improvement. With regards to the task assessment, the sequenced task-based assessment was found to reveal additional insights which were not detected in the static assessment of the controls. This sequenced assessment is, however, considered to be a complimentary approach, rather than a replacement for the static assessment. There are, of course ,clear benefits to assessing each of the controls in turn, since this ensures that each control is operable regardless of the way the cab is to be used.

References

Annett, J., Duncan, K.D., Stammers, R.B., & Gray, M. (1971). *Task Analysis*. London: HMSO.

Department of Trade and Industry (1998). *Adult Data, Handbook of Adult Anthropometric and Strength Measurements: Data for Design Safety*.

Diffrient, N., Tilley, A.R., & Bardagjy, J. (1974). *Humanscale 1/2/3*. MIT press.

Dreyfuss. H., (1955) *Designing for People*. Simon and Schuster, New York.

GM/RT2161 (1995). Requirements for driving cabs of railway vehicles. Accessed 19.11.14. Available at http://www.rgsonline.co.uk/Railway_Group_Standards/Rolling%20Stock/Railway%20Group%20Standards/GMRT2161%20Iss%201.pdf.

Haworth, N., Salmon, P.M., Mulvihill, C., & Regan, M. (2005). Assessment of driver vigilance systems on Metropolitan Melbourne trains. *Monash University Accident Research Centre Report*.

Hitachi Rail Europe (2011). Class 395 train operating manual. HRM/395/TOM issue 4 December 2011.

Jenkins, D.P., Baker, L.M., & Harvey, C. (2015). A practical approach to glare assessment for train cabs. *Applied Ergonomics*. 47, 170–180.

Jenkins, D.P., Baker, L.M., & Harvey, C. (in press). A practical approach to evaluating train cabs against task requirements. *Journal of Rail and Rapid Transit*. Accepted 20th September 2014.

Rose, J.A. & Bearman, C. (2012). Making effective use of task analysis to identify human factors issues in new rail technology. *Applied Ergonomics* 43(3), pp. 614–624.

DEFENCE

THE AIRCREW'S TASK CUBE: INITIAL EVIDENCE OF A MULTI-DIMENSIONAL CONTROL TASK MODEL

Nikolaos Gkikas

BAE SYSTEMS, Military Air & Information

This paper describes the initial development of a compact model for the tasks performed by aircrew in civil and defence platforms. Using a pool of tasks performed while controlling an airborne system towards a goal, the author suggests a series of dimensions, which can eventually be combined into a three-dimensional model. Such model could derive toolsets which the human factors engineer or HMI designer could employ to rapidly define the specific and ubiquitous properties of the system of interest across the development lifecycle.

Introduction

The human task of taking a vehicle airborne, navigating and landing in a safe manner has traditionally been studied comprehensively at micro-level (e.g Chapanis early work on cockpit displays and human error), or at a more generic macro cognitive and behavioural level, leading to the development of concepts such as Wicken's (1983) mental resource theory and Endley's (1993) situation awareness theory. The latter theories, in particular, have been widely accepted as valid models of internal processing and response from a human operator during a given task in a given technical environment, and have greatly expanded their application outside the remit of aviation human factors and ergonomics.

The foundations of a comprehensive model for the task of successfully controlling an airborne system

Despite great success and acclaim within and outside the discipline of ergonomics, such models essentially deal with a single, albeit fundamental, dimension of the pilot's task. During the application of these models to inform the design of a human machine interface (HMI) or a human factors engineering specification of a given airborne system, it is often difficult to consider and even harder to apply those models ubiquitously. As such models deal with a single facet of the human task, in effect, they can only be iteratively applied to individual HMI loops (see Figure 1) for each element of the design.

In an attempt to counteract this recurring limitation, the author collected task data from series of interviews with aircrew, in support of a generic task analysis.

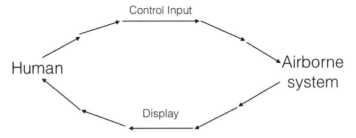

Figure 1.   Fundamental HMI loop.

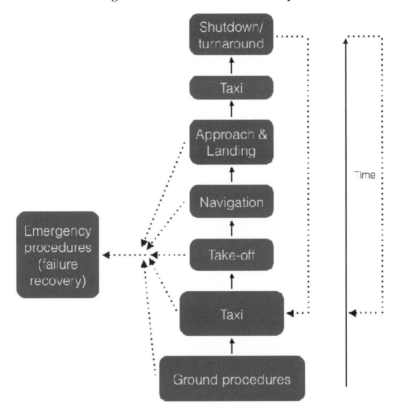

Figure 2.   Common task groups across platforms.

The subsequent pool of tasks included a significant number of common tasks between military and civil platforms. Figure 2 illustrates a common cross-platform framework for grouping those tasks.

That was followed by discussions around the key objectives and determinants of the success of each task: what it is that the aircrew is trying to achieve and to what level of "something". The answers to the first question were never conclusive, with the exception of the two terminal answers: "to complete my mission" and "to arrive

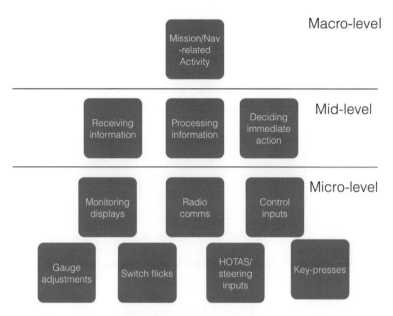

Figure 3. The task pyramid.

at my destination". Every other answer to the question initiated a pyramid of tasks (Figure 3) that led all the way to the aforementioned terminal answers.

With regards to the second question, that criterion 'something' was either a time-interval (which could be from the present to a determined deadline), a level of accuracy (error margin/linear measure), a binary criterion in the form of correct/incorrect actions, or a binary criterion of do/do not at a micro task level – such as single key presses that could be the 'right' or 'false' control input. Even the latter, however, could still be viewed as part of the former correct/incorrect action where there is a single possible 'incorrect' action rather than a plural group of them (see Figure 4).

Revisiting those answers, it appears that the above characteristics/criteria can reasonably be split in temporal demands and level of accuracy. The temporal demand can be prompt (<1 s–1 m), short-term (1 m–1 h), or long-term (>1 h). The level of accuracy required can be below mili-margins (<0.001), within small-medium margins (0.001–0.01) or large margins (<0.01) of the performance/error criterion (Figure 4). Combining the two scales of success criteria for aircrew tasks, results in vector-like illustration such as the one in Figure 4.

Integrating the task elements

By contrasting and logically combining all the aforementioned aircrew-task characteristics towards a compact descriptive model, one could suggest a tridimensional

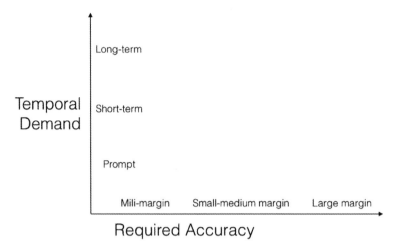

Figure 4.   The task success vector.

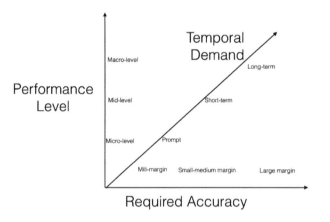

Figure 5.   The integrated multi-dimensional aircrew task model.

model, where the large task-groups of Figure 2 are broken down by the steps in the task pyramid illustrated in Figure 3 to provide the level of control exerted by each task. That control level of the task is linked, albeit not matching, to a temporal demand. Success and failure for each task however, is also determined by the accuracy margins of task performed, as illustrated in Figure 4. The combined model is summarised in Figure 5.

Why bother?

The present article has visited, analysed and proposed a number of concepts as descriptive to a vast and complex array of tasks involved in successfully controlling an already complex system such as an aircraft towards a goal. It is acknowledged that the proposed concepts and the ones upon which they are based will require further iterative cycles of supportive evidence and detail, which currently fall outside the scope of this short article. Nevertheless, even in this early iteration, the multi-dimensional task model provides the human factors engineering practitioner a compact toolset of the key characteristics they should look for when determining technical targets, design against them and/or finally assess them during the development of an airborne system.

References

Endsley, M. R. 1995, Toward a theory of situation awareness in dynamic systems. *Human Factors* 37: 32–64.

Wickens, C. D., Sandy, D. L. and Vidulich, M. 1983, Compatibility and resource competition between modalities of input, central processing, and output. *Human Factors* 25(2): 227–248.

'CULTURE CLASH' – UNDERSTANDING THE EFFECT OF CULTURAL AND SOCIETAL FACTORS ON CONFLICT

Ryan Meeks

Frazer-Nash Consultancy Ltd.

We are all products of our culture and society. They influence what we value, what we believe, what we feel and how we behave. However, they can also cause friction, disputes, and divisions. As the world becomes more connected, people of different cultural and societal backgrounds are living and working together, where these differences are sometimes a source of conflict. In difficult high-stress environments, such as an oil platform or even an airport, cultural and societal factors can influence the escalation of conflict between people. Frazer-Nash has been conducting novel research into how different cultural and societal aspects affect human behaviour in conflict situations, with the aim of providing industry with an informed cultural awareness training template.

Introduction

Frazer-Nash has conducted research into how cultural and societal (C&S) factors affect the volatility of human behaviour. The challenge was to develop a cultural training template for people entering unstable environments, regions and establishments. Different C&S factors were researched to determine their effect on instrumental and hostile human conflict behaviour, and whether they increased the probability of it occurring. The training template includes a series of 'cultural maps', which present various C&S influencers that both pre-dispose people to conflict, and act as 'triggers' for violence and aggression.

Methodology

The research project consisted of the following overarching methodology:

- Conduct literature review on the effect of certain C&S factors on human behaviour, conflict and violence.
- Review the influencers associated with each C&S factor to conflict.
- Identify some of the generic C&S aspects that influence conflict, as well as those traits specific to certain types (such as individualistic, collectivistic, shame-based or guilt-based cultures).

- Produce a series of 'cultural maps' that describe the effect that certain C&S factors have on pre-disposing and 'triggering' people to engage in conflict behaviours.
- Use the 'cultural maps' as a template for the development of more effective cultural awareness training.

The ultimate aim of the work was to develop 'cultural maps' that communicate how different C&S factors influence high risk conflict behaviours, to help industry trainers develop more effective cultural awareness training packages.

Reviewing the impact of different cultural and societal factors

C&S factors manifest themselves in different guises around the world. In different societies, each factor exerts different levels of influence over individuals and groups. This may be linked to the 'tightness' or 'looseness' of the society (Pelto, 1968), the hardships that the communities face (Matsumoto, 2007), whether the society is individualistic or collectivistic (Hofstede, 2001), or shame-based or guilt-based. For example, insults to honour are taken as a personal insult in individualist societies, whereas they are considered an insult to an entire family, tribe or community in collectivistic societies.

Literature reviews were conducted in six C&S areas; social status and hierarchy, honour, ethnicity, historical factors, language, and codes of conduct/social values. The reviews focussed on understanding how the factor influenced both an individual's pre-disposition to engaging in conflict behaviour (planned instrumental aggression), and its effect as a behavioural 'trigger' – where it can spark unplanned hostile aggression. The review also sought to understand how the factors manifest in different cultural environments, such as collectivistic, individualistic, shame and guilt based societies.

Reviewing the nature of human behaviour in relation to conflict

A literature review was also undertaken to understand the nature of human behaviour in conflict, and the effect that culture and society has on this. This primarily involved understanding the main differences between those aspects that affect one's pre-disposition to conflict, and one's vulnerability to be triggered into hostile aggression. This then allowed aspects of the C&S factors to be categorised as having an effect on pre-disposing people to conflict or presenting a 'trigger' for hostile aggression in the 'cultural maps'. Understanding these differences is important for cultural awareness training and in the identification of suitable mitigation strategies.

Weighting the influencing aspects of culture and society

Each C&S factor has a number of influencers. Subjective weightings were provided for each influencer, so that the maps present the aspects with higher weightings as

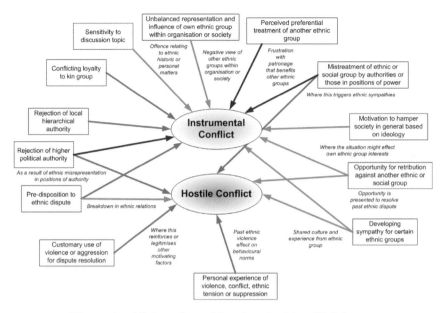

Figure 1.   'Cultural map' for the ethnicity C&S factor.

more important learning points for training. Influencers were given a weighting based on a simple three point scale; red = high, yellow = medium, and green = low. This allowed the weightings to be visually represented to the reader. While it is understood that different cultures vary enormously in the weighting that they give certain influencers, the purpose of this exercise was to provide a generic template which can be used by training developers to adapt to suit specific cultural regions and environments.

Developing the 'cultural maps'

All of the findings from the work, including the definition of each C&S factor, the identification of the specific influencers and the weighting of each of these in relation to both instrumental and hostile conflict behaviours, were incorporated into the development of a 'cultural map' for each of the six areas; social status and hierarchy, honour, ethnicity, historical factors, language, and codes of conduct/ social values.

Results

The 'cultural map' presented at Figure 1 below shows how the findings were presented in an effective visual format, for training developers to incorporate into cultural awareness training. In addition to the 'cultural maps', this work identified

Table 1. Key learning points for each C&S factor.

C&S Factor	Key Learning Point
Social status and hierarchy	Social status and hierarchy play an integral societal role in collectivistic societies. Direct questioning, challenging or confrontation by those with perceived lower status is likely to trigger a volatile response. Social status dictates acceptable behaviours and norms within society, and any deviation from this is likely to become a source of conflict
Honour	Honour codes play a vital role in defining acceptable behaviour in shame-based cultures, and are a mechanism for the maintenance of cultural values and practices. Breach of honour code is likely to result in degradation of social status, interpersonal conflict and even community expulsion
Ethnicity	Ethnicity is likely to be a source of conflict in societies that have suffered historic social tensions. Ethnicity is seen as a source of status and ascribed honour in some cultures, and can be used to suppress other ethnic groups. Ethnic equality must be practiced as a mitigation for ethnic-based conflict
Historical Factors	History is a source of pride, passion and nationalism in some cultures. Ideologies and attitudes associated with historic conflicts can permeate into the mindset of individuals today, pre-disposing them to an inherent distrust, disapproval and wariness of others
Language	Language can be a major source of hostile aggression, especially when it is considered to be dishonouring, disrespectful, insulting and/or rude. Some cultures can be extremely sensitive to certain types of language, and so it must be considered as a potential key influencer to conflict
Codes of conduct/ social values	Some cultures, especially 'tight societies' (Pelto, 1968), implement strict codes of conduct that dictate acceptable behaviours and promote cultural practices. Deviations from these codes of conduct can cause great insult, offence and be perceived as a threat to society, which is often accompanied by violent and aggressive reactions or the dispensing of severe punishment

key learning points related to the effect of each C&S factor on conflict behaviour. Some of these are presented briefly in Table 1.

Conclusion

Culture and society can play a central role in interpersonal conflict, especially for those working and living in diverse, high-stress and multi-cultural environments. It is therefore essential that people understand the effect that culture and society has on their own and other's behaviour so that they can aim to integrate more effectively, build better interpersonal relationships and reap the rewards of cultural diversity. The 'cultural maps' produced from this work provide trainers with a simple and effective visual template in order to further develop more effective cultural awareness training packages.

References

Pelto, P. J. 1968, The differences between "tight" and "loose" societies. *Transaction*, 5 (5): 37–40.

Matsumoto, D. 2007, Culture, context and behaviour. *Journal of Personality*, 75 (6): 1285–1319.

Hofstede, G. 2001, *Culture's Consequences: Comparing Values, Behaviours, Institutions and Organisations Across Nations*. (Sage, London)

ASSESSMENT OF AVAILABLE STANDARDS FOR RISK ASSESMENT AND HUMAN FACTORS AT DESIGN STAGE

Farzad Naghdali, M. Chiara Leva, Sam Cromie & Nora Balfe

Centre for Innovation in Human Systems,
Trinity College, Dublin, Ireland

A thorough consideration and implementation of risk assessment and human factors knowledge into systems at design stage should return the initial investment in terms of enhanced operational performance. The question is whether the standards and guidelines currently available for the industry are able to provide effective guidance to be practically used in the design process by human factors practitioners, project managers, and/or people in a design team. This paper will introduce the relevance of the issue, provide a review of some of the existing standards and guidelines and present early findings from part of a survey designed to collect information about the current perceived gaps and the areas of potential improvement.

Introduction: Importance of human factors engineering (HFE) in risk analysis at design stage

The demand of industries for safer and more efficient operations has changed the human role in many systems from primary actor to supervisor of an automated process. This is particularly true for rapid transport systems, manufacturing production lines and computerised systems; however a certain degree of attention towards human-machine interaction is always required even if it is just for maintenance, commissioning and sporadic supervision (Leva et al., 2012). Additionally, when the complexity of the system increases, the reluctance of the designers to substitute the operator with automated functions will increase as well. This is because the ability of the human to control the system in unforeseen circumstances can help the system to keep functioning normally (Hale et al., 2007) i.e. designers cannot foresee all eventualities and people have good capability – if well supported by design, training and understanding of the system and system state – to diagnose and deal with such problems. Computers do not have this ability and therefore cannot be considered as the only available source of control. Therefore there is a general preference to have the human as a final authority working with the computerized system. Hence, the human operator's task, to operate and control the system, is considered crucial (Nazir et al., 2013). In accident investigations, design inadequacies are often mentioned as a major contributing factor (Hale et al., 2007) and human error is almost always described as a major cause of accidents (OGP, 2010).

International standards are available to guide the application of HFE in the design phase (discussed in the following section), but these are by necessity general and might not be sufficient to support the HFE assessments of the design. This is illustrated by the fact that high reliability organisations in safety critical areas, such as the aviation or nuclear industries, have often developed their own internal standards to provide more specific guidance on HFE assessment and safety by design issues. There are several attempts in the field of design for improved approaches such as Human-Centred Design (Maguire, 2001), intelligent human-machine interface design (Tendjaoui et al., 1991), user needs analysis (Lindgaard et al., 2006), Safety by Design (Kletz, 1996), Cognitive Work Analysis (Vincente, 1999) and Human Factors Integration (Widdowson and Carr, 2002).

There is also the possibility to provide more support in the design phase for human factors by learning from accidents and incidents. Different industries can learn from each other and historical data can contribute to increasing safety but the learning mechanism needs to have a good understanding of the culture, constraints, objectives and the design procedure of the target industry (Drogoul et al., 2007). Although learning from accidents and analysing them is very important for improving knowledge, on the other hand the learning procedure and the loop to give feedback to the engineering strategy is slower than technology developments, thus the learned lessons may lose their value and become inefficient (Rasmussen, 1997). In addition, new technologies introduce new challenges for the safety by design techniques in that there may not be any historical data available for them. Similarly market and financial issues are putting more pressure on the designers, thus the opportunity for safety through design studies is decreasing (Leveson, 2011). In the past the development of a new technology was much slower than at present and it could allow enough time for the hazards to emerge (Leveson, 2011) but hazards have to be foreseen before emerging during operational phase of systems to prevent potential catastrophic consequences (OGP, 2011).

Available standards for human factors engineering: Possible issues and gaps

To support the challenging task of the design team regarding safety and reliability, there are number of standards providing some guidance on the minimum requirements in terms of human centred design (see Table 1). ISO 6385 – Ergonomic Principles in the Design of Work Systems (2004) outlines how technological, economical, organisational, and human factors can affect the work behaviour and well-being of people within a work system. The general principle underlying the standard is that interactions between people and the components of the work system (e.g. tasks, equipment, workspace and environment) should be considered during the design stages. Each design stage is described and appropriate ergonomic principles and methods for each stage are listed. ISO 11064 – Ergonomic Design of Control Centres (2006) provides nine principles for the ergonomic design of control centres and guidance on specific aspects of control room design, including layout, workstation design, controls and displays, and environmental requirements. ISO

Table 1.   Summary of HF issues in system design standards (Naghdali, et al., 2014).

HFE Area of Design	Related existing standards/ best practices	Possible issues/gaps
Design of physical built environments	ISO 6385 (2004) Ergonomic principles in the design of work systems	The standards do not provide any practical guidance on how to actually review the built environment at the design stage involving users (such as 3D reviews)
Design of machinery/ electrical systems	ISO 12100 (2010) Safety of machinery/EEMUA 178 (1994) A design guide for the Electrical Safety of Instrument Control Panels	The standards are seldom applied in the industry and they do not specify to what machinery they should apply
Design of control rooms, HMI for information systems	EEMUA 201 (2010)/ISO 9241-210/ISO 11064 (2006) Ergonomic design of control centres	How to review the mimics of control centres is not specified and the use of task analysis is not clearly suggested
Design of information systems and alarms	EEMUA 191 (1999)/ ISO 11064 (2006)	As above
Workload assessment for design	ISO 11075-3 (2004) Ergonomic principles related to mental workload	Not really applied in the industry
Design of manuals and procedures	ISO 12100 (2010)/ISO 18152 (2010) Ergonomics of human-system interaction – Specification for the process assessment of human-system issues	The standards specify how to assess processes but not how to translate them in to good instructions and procedures
Risk assessment at design stage	ISO 31010 (2009) Risk management – Risk assessment techniques	Little guidance on what standards are available for human reliability analysis

12100 – Safety of Machinery (2010) suggests a five step methodology to perform risk assessment at design stage and the overall strategy requires designers to take into account the safety of machinery for their whole life cycle, considering usability, maintainability and cost efficiency. OGP 454 (2011) is an example of a standard that recommends principles and approaches mentioned above for a specific industrial application (Oil & Gas). Other standards (e.g. ISO31000 [2009] and ISO 31010 [2009]) are more general and intended for application in any domain and across the entire system lifecycle. The generality of these standards, both in terms of domain and lifecycle stage, can be a barrier towards their successful implementation.

EEMUA 191 (1999) is an industrial guideline developed by the Engineering Equipment and Materials Users' Association to support the design of alarm systems taking into account the requirements of the human operator receiving and responding to those alarms. EEMUA 201 (2010) focusses on the design of HMIs and gives guidance on areas such as display hierarchies, the design of the screen format, and the attributes of the environment that may affect the use of the HMI. These guides

Table 2.   Some of the standards mentioned by participants.

OHSAS 18001 – Occupational Health and Safety Management	45.9%
ISO 14001 – Environmental Management	43.9%
ISO 9001 – Quality Management	48.0%
IEC 61508 – Functional safety of electrical/electronic/programmable electronic safety-related systems	10.2%
IEC 61511 – Functional safety – safety instrumented systems for the process industry sector	11.2%
ISO 12100 – Safety of Machinery	11.2%
ISO 6385 – Ergonomics principles in the design of work systems	7.1%
ISO 9241 – Ergonomics of human-system interaction	9.2%
ISO 11064 – Ergonomic design of control centres	5.1%
Recognised industry standards	66.3%
Internal company standards	68.4%
National standards	52.0%

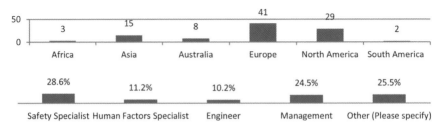

Figure 1.   Demographics.

define only the minimum requirements. This systematic approach is fairly generic and does not provide technical support for the designers. As a result companies must introduce internal standards to tackle the problem. This may not be an option for SMEs where the allocation of budget and time for development of standards might not be feasible (Leva, et al., 2014).

Survey results: Risk assessment and human factors standards

A survey was designed to collect information about the user-friendliness of standards and their content. The survey was designed to collect a wider range of information regarding the current industrial use of risk assessment and human factors methodologies (INNHF, 2014). This study was funded by European Commission under the INNHF Marie Curie Initial Training Networks action. The survey was still on line at the time of writing and the results presented are preliminary and were collected from a specific section of the survey regarding the opinion of participants about available standards. The survey was administered online and participants were primarily recruited from safety related LinkedIn groups. The sample size was 98 participants for the evaluation presented in this article. The participants were from a wide range of industrial branches and the demographics in Figure 1 gives more information about the role of participants within their industry.

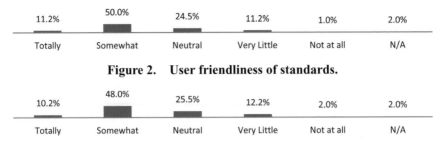

Figure 2.   User friendliness of standards.

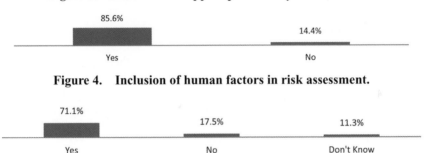

Figure 3.   Information support provided by the standards.

85.6%

14.4%

Yes   No

Figure 4.   Inclusion of human factors in risk assessment.

71.1%

17.5%   11.3%

Yes   No   Don't Know

Figure 5.   Requirement for better HF tools and methodologies.

The overall scope of the survey was not limited only to risk assessment and HFE at design stage but it was aimed at the overall HF issues during risk assessment process regardless of the stage of the work in which the risk assessment is being conducted. In fact some of the HF standards are generic and they can be implemented at different stages throughout systems' lifecycle and they are not bound to a specific stage even though that they can be better applied to the system at design stage. The questions discussed in this paper were assigned pre-defined answers to structure the study in a general form. The questions asked were as below:

1. Which standards (if any) do you apply to manage safety and risk analysis?
2. Did you find the standards user-friendly?
3. Did the standards provide sufficient information to support your work?
4. Do you include Human Factors in your risk assessment?
5. Do you think there is a need for more HF input to risk assessments?

The goal of this study was to evaluate the quality of available standards in terms of the amount of support, user-friendliness and HF inclusion. The demographics show that the sample covered four main groups of industrial practitioners involved in safety assessment (see Figure 1).

The following graphs (Figure 2 to Figure 5) show the results obtained from the responses to questions 2, 3, 4 and 5 of the survey. Unfortunately the preliminary data collected does not give a clear picture of whether the standards are sufficient or not, however it points out that the available standards are broadly matching users' expectations with 10% of participants stating they totally support their work and 50% indicating they are somewhat supportive. With respect to the user-friendliness of standards, 11.2% (Figure 2) of participants found the standards totally user-friendly and 50% of them settled for declaring them "somewhat" user friendly.

During earlier interviews, conducted as part of the overall research project, with industrial engineers involved in design projects or safety issues, the issue of the user-friendliness of standards was raised. Some engineers believe that the human factors standards are not well designed as a communication tool.

Possible areas of improvement in HFE at design stage

Systems are becoming increasingly more complex and there is a need for a new approach for systems' safety to better support the difficult job of achieving safety through design in complex projects. Although a large number of standards and guidelines are available to engineers and design teams (often it is specifically targeted at these groups), the ongoing need for detailed review sessions reveals that these groups are not fully assimilating HFE information.

Integration of HFE principles within broader engineering and design standards may be one way to achieve a more integrated picture. But this cannot be achieved by more input only from human factors specialists. Too often, only human factors specialists are aware of the existence of HFE standards and the principles contained within them. Industrial engineers and practitioners who participated in the study are divided into three groups. The majority (approx. 58%) believe that the available standards in risk assessment and HFE (to their knowledge) provide an adequate level of support for industrial practitioners (see Figure 3) with the remainder either neutral (approx. 25%) or dissatisfied (approx. 14%) with the level of support with the standards. This dissatisfaction may be due to the generality of some of the standards and lack of a mature and concrete overall framework to support industrial engineers in understanding what aspects needs to be taken care at each stage of systems' lifecycle.

A solution may be a set of clear and very well structured guidelines for engineers that provide more support on specific methods and tools for screening projects at early design stage (i.e. OGP 454 [2011]). A new HFE method that can provide operations feedback to future design teams could provide very valuable design input, increasing overall safety and efficiency. In this paper the authors have tried to point out the gaps to be addressed by HFE at design. In future the objective of the study is to collect more data regarding the available methods and practices in industries. A survey will be designed to collect this information only regarding HFE at design stage and limited to the process industry. Afterwards a new framework will be suggested based on the best available practices and the research team will develop a simple communication and information sharing method to establish a connection between the design team and end-user to complete the missing element in the design loop.

References

EEMUA, 2010, *Process plant control desks utilising human-computer interfaces*, Publication 201, EEMUA, London, UK.

Hale, A., Kirwan, B. and Kjellén, U., 2007, Safe by design: where are we now?, *Safety Science* 45, 1–2: 305–327.

INNHF, 2014, www.innhf.eu.

ISO, 11064, 2006, *Ergonomic design of control centres*, The International Standards Organisation, Geneva, Switzerland.

ISO, 11075-3, 2004, *Ergonomic Principles Related to Mental Workload*, The International Standards Organisation, Geneva, Switzerland.

ISO, 12100, 2010, Safety of machinery – General principles for design – Risk assessment and risk reduction, *The International Standards Organisation*, Geneva, Switzerland.

ISO, 18152, 2010, *Ergonomics of human-system interaction – Specification for the process assessment of human-system issues*, The International Standards Organisation, Geneva, Switzerland.

ISO, 31000, 2009, *Risk Management – Principles and guidance*, The International Standards Organisation, Geneva, Switzerland.

ISO, 31010, 2009, *Risk management – Risk assessment techniques*, The International Standards Organisation, Geneva, Switzerland.

ISO, 6385, 2004, *Ergonomic principles in the design of work systems*, The International Standards Organisation, Geneva, Switzerland.

ISO, 9241-210, 2010, *Human-centred design for interactive systems*, The International Standards Organisation, Geneva, Switzerland.

Kletz, T., 1996, Inherently safer design: The growth of an idea, *Process Safety Progress* 15: 5–8.

Leva, M. C., Balfe, N., Kontogiannis, T., Plot, E. and Demichela, M., 2014, Total safety management: What are the main area of concern in the integration of best available methods and tools. *Chemical Engineering Transactions* 36: 559–564.

Leva, M. C., Pirani, R., De Michela, M. and Clancy, P., 2012, Human Factors Issues and the Risk of High Voltage Equipment: Are Standards Sufficient to Ensure Safety by Design?, *Chemical Engineering Transactions*, 26: 273–278.

Leveson, N. G., 2011, *Engineering a safer world*, MIT Press, Massachusetts, USA.

Lindgaard, G., Dillon, R., Trbovich, P., White, R., Fernandes, G., Lundahl, S. and Pinnamaneni A., 2006, User needs analysis and requirements engineering: Theory and practice, *Interacting with Computers* 18 (1): 47–70.

Maguire, M., 2001, Methods to support human-centred design, *International Journal of Human-Computer Studies* 55 (40): 587–634.

Naghdali, F., Leva, M. C., Balfe, N. and Cromie, S., 2014, Human factors engineering at desig nstage: is there a need for more structured giudlines and standards? *Chemical Engineering Transactions*, 36.

Nazir, S., Colombo, S. and Manca, D., 2013, Testing and analyzing different training methods for industrial operators: an experimental approach, *23rd European Symposium on Computer Aided Process Engineering*, 667–672.

OGP, 2010, *Risk assessment data directory, International Association of Oil and Gas Producers*, Report No. 434-5, International Association of Oil and Gas Producers, London, UK.

OGP, 2011, *Human factors engineering in projects, international association of oil and gas producers*, Report No. 454, International Association of Oil and Gas Producers, London, UK.

Rasmussen, J., 1997, Risk management in a dynamic society: a modelling problem, *Safety Science*, Vol. 27, 2–3, 183–213.

Tendjaoui, M., Kolski, C. and Millot, P., 1991, An approach towards the design of intelligent man-machine interfaces used in process control, *International Journal of Industrial Ergonomics* 8 (4), 345–361.

Vincente, K. J., 1999, *Cognitive Work Analysis: Toward safe, Productive, and Healthy Computer-Based Work,* Lawrence Erlbaum Associates Inc, Mahwah, New Jeresy, USA.

Widdowson, A. and Carr, D., 2002, *Human factors integration: implementation in the onshore and offshore industries*, Health and Safety Executive, Camberley, UK.

UPDATING HF INTEGRATION PROCESS, POLICY AND GUIDANCE INFORMATION FOR UK DEFENCE ACQUISITION

Gareth Shaw[1], Tim Hughes[1] & Chris Kelly[2]

[1]*BAE Systems Defence Information, UK*
[2]*Symbiotics, UK*

This paper reports on a programme of work contracted by the United Kingdom (UK) Ministry of Defence (MOD) via the Defence Human Capability Science & Technology Centre (DHCSTC) to update MOD human factors integration (HFI) policy, standards and guidance used in defence acquisition. The paper provides an overview of the programme, describes the outputs and also the potential benefits that may be realised through the new documentation suite. The concept of the Human Factors Integration Management System (HuFIMS) is also introduced.

Introduction

Objective

This short paper reports on work currently being undertaken through the Defence Human Capability Science & Technology Centre (DHCSTC), to update UK Ministry of Defence (MOD) human factors integration (HFI) policy, standards and guidance information, constituting the principle HFI documentation used in UK defence acquisition.

The requirement for change

MOD recognises the need to ensure that the human component of capability is appropriately considered by ensuring that HFI is fully embedded within the defence acquisition process. The key HFI policy and process document (Joint Service Publication (JSP) 912 and the principal defence standard containing guidance material used in contracting for HFI (DEF STAN 00–250) are significantly out of date with respect to HFI best practice, MOD acquisition processes, and new developments in technology. Furthermore, despite HFI having been part of UK MOD acquisition policy for more than 20 years, it continues not to be fully embedded in the defence acquisition process. The current work seeks to refocus policy, process and guidance documentation such that the JSP provides a clear mandate for HFI (i.e. it directs and Informs MOD staff at all levels) and the DEF STAN facilitates a clear contracting

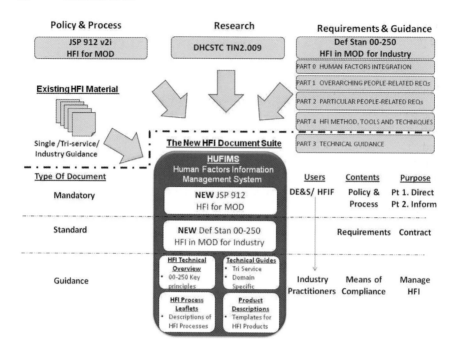

Figure 1.   Transition of Existing HFI Material into New HFI Document Suite.

mechanism for HFI. Appropriate consideration of the human component in defence acquisition is critical in ensuring that delivered systems are safe to operate, meet defined performance criteria, meet user and customer requirements and minimise through life support costs. The first phase of the programme began in December 2013 and it will run through to August 2015.

Summary of work in progress

Figure 1 provides a synopsis of the work being undertaken.

Revision of the HFI process

The existing HFI Process was reviewed in a series of workshops with MOD and industry representatives. The workshops focussed on applying the process against a range different acquisition types from developed capability (e.g. aircraft carriers) to commercial off the shelf (COTS) procured items (e.g. hydration backpacks) to ensure it could be tailored appropriately. The HFI Process has been distilled down to six high-level process steps mapped to the principal acquisition lifecycle. Beneath this a further 52 process sub-steps have been defined. A summary of the revised process will be included in both the new JSP and DEF STAN documents and will be presented in full in the HuFIMS (described later in this paper).

Revision of Joint Service Publication 912

The revised JSP will be delivered in two parts in accordance with a recent MOD initiative for revising JSP documentation. The aim is to improve the accessibility and clarity of MOD HFI documentation. Part 1 (the "Directive") will be succinct, containing a statement of the revised HFI policy and a clear unambiguous directive that HFI must be undertaken in all capability acquisitions. Part 2 ("Inform") will contain a summary of the new HFI Process (referencing the HuFIMS), a description of roles and responsibilities involved and the expected competencies of MOD HFI personnel, plus a number of relevant appendices.

Revision of Defence Standard (DEF STAN) 00–250

The revised DEF STAN 00–250 will be provided in a single part (currently it is in four parts), to improve the accessibility of the documentation and hence increase the likelihood of its uptake in the production of contractual documentation. New human factors (HF) process and system requirements will be generated to replace extant requirements (which have proved to be difficult to contract to in the past). Requirements will be defined in hierarchical sets (through facilitated workshops), enabling users to tailor the number of requirements to be included within contractual documentation. The body of technical content currently comprising Part 3 Section 15 will be reworked as HF technical guides and will reside in the HuFIMS (described later in this paper).

Production of HF technical guides

Figure 2 presents the proposed system for organising HFI technical content.

Single/tri-service information will be structured according to a first principles approach under 10 topic areas, which can be seen in Figure 2. Two exemplar technical guides will be produced under the study namely; anthropometry; and human computer interaction (HCI). Generation of the remaining technical guides is expected to be contracted out by MOD under successive programmes. Each technical guide will contain best practice information, lessons learned and low level requirements traceable to high level DEF STAN 00–250.

Generation of a Human Factors Integration Management System portal

The HuFIMS will be a web-based portal, publically accessible via the MOD's Defence Gateway (requires registration), providing a repository for all available HF information relevant to defence acquisition (whilst also linking to external

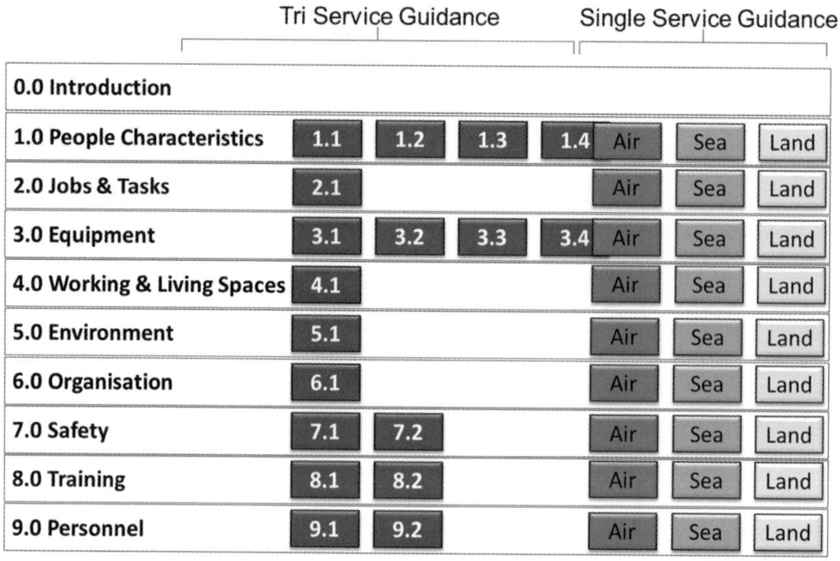

	Tri Service Guidance				Single Service Guidance		
0.0 Introduction							
1.0 People Characteristics	1.1	1.2	1.3	1.4	Air	Sea	Land
2.0 Jobs & Tasks	2.1				Air	Sea	Land
3.0 Equipment	3.1	3.2	3.3	3.4	Air	Sea	Land
4.0 Working & Living Spaces	4.1				Air	Sea	Land
5.0 Environment	5.1				Air	Sea	Land
6.0 Organisation	6.1				Air	Sea	Land
7.0 Safety	7.1	7.2			Air	Sea	Land
8.0 Training	8.1	8.2			Air	Sea	Land
9.0 Personnel	9.1	9.2			Air	Sea	Land

Figure 2.   Proposed Suite of HF Technical Guides.

information). Hosting information in the HuFIMS will facilitate ease of navigation and will enable users to search for information against a number of criteria (e.g. by programme phase, role and responsibilities etc.). HuFIMS structure and content will be agreed through workshops with MOD and industry.

Outcomes

This programme will support effective integration of HFI within defence acquisition through the provision of a mandated process that can be effectively tailored depending on a range of factors. Provision of an integrated HuFIMS containing policy, requirements and guidance information will ensure that both MOD and Industry practitioners have access to, or can locate, all required information. This has the potential to increase the uptake of documentation and also improve integration with other disciplines. The modular arrangement of information in the HuFIMS will also facilitate document maintenance activities (e.g. ensuring currency, removal of obsolete data, etc.).

Acknowledgement

This work has been contracted through the Defence Science and Technology Laboratory (Dstl) and Defence Equipment and Support (DE&S). The work is being performed by a team of experienced HF practitioners from BAE Systems (Defence

Information: Gareth Shaw & Tim Hughes: Advanced Technology Centre: Steve Harmer & Rob Cummings), Frazer Nash (Pam Newman) and Symbiotics (Chris Kelly).

References

UK Ministry of Defence 2008, Human Factors for Designers of Systems, Def Stan 00–250, Issue 1. UK. Ministry of Defence.

UK Ministry of Defence 2013, Joint Service Publication 912: Human Factors Integration for Defence Systems, Version 2.0, 22nd April 2013. UK Ministry of Defence.

DESIGN

A MARRIAGE MADE IN HEAVEN? PRESSURE DISTRIBUTION AND COMFORT IN WHEELCHAIR CUSHIONS

Carol Bartley

University of Salford, Manchester

Permanent wheelchair users can spend approximately 18 hours per day sitting in their wheelchairs, unable to adjust their posture with the weight of the upper body being borne through the ischial tuberosities, coccyx and sacral area. These high load bearing bony prominences are susceptible to pressure ulcers developing if the correct cushion is not prescribed. Comfort for the user is an important consideration in determining compliance. This paper investigates interface pressures and comfort in four commercially accessible wheelchair cushions to ascertain if a relationship exists

Introduction

There are approximately 1.2 million wheelchair users in England (Department of Health, 2004) with varying needs ranging from simple to complex. It has been reported that wheelchair users can spend up to 18 hours a day in their wheelchair thus increasing the risk for pressure ulcer development (Stockton & Parker, 2002). The cost of pressure ulcers in the UK is £1.4–£2.1 billion annually, amounting to 4% of the total UK NHS expenditure (Bennett et al., 2004). There are many different pressure cushions available on the market which can be used for varying risk levels in pressure care from low to high. The majority of these declare that the main benefit is the excellent pressure distribution qualities they possess. However, empirical evidence to support this is scarce with even less reporting on the issue of comfort. Most notably Stockton and Rithalia (2007) studied wheelchair users' perceptions of comfort using interface pressure, temperature and humidity measurements. One of the reasons for the dearth of literature regarding pressure distribution and comfort in cushions could be due to the fact that comfort is a subjective evaluation and lacks a validated assessment tool to address it. Conversely, recording pressure distribution is quite simple using one of the many devices commercially available, however confirming a link between these two concepts is more difficult. In other areas ergonomic research completed in car seat manufacturing supports the need for further investigations concluding that interface pressure is closely aligned to comfort (Fay and Brienza, 2002). Furthermore studies by Kyung and Nussbaum (2007) and Porter et al. (1999) suggest in their findings that there could be a relationship between the two although evidence is currently inconclusive and more large-scale studies are required.

Methodology

The aims of this study were to investigate if a relationship exists between interface pressures and comfort in four of the most commonly prescribed wheelchair cushions in the clinical setting. It also examines if there are differences between ambulant and wheelchair users in terms of pressure distribution and perceived comfort. The cushions used were: pressure reducing foam, air, standard foam and gel. Using qualitative methods purposive sampling was selected for data collection and individual pressure mapping data was captured using the XSensor Pressure Mapping System (XSensor UK). This is a dynamic system involving the use of sensors to quantify the interface pressure between two contacting objects, such as a person and their support surface. It is commonly used by clinicians in healthcare to determine the suitability of a wheelchair cushion and by researchers investigating support surfaces, risk factors for ulceration, and pressure ulcer prevention protocols (Trewartha & Stiller, 2011, Stinson et al., 2013).

Participants

Eight people were recruited for the study and divided into two groups. Two participants withdrew resulting in six participants completing the study. Group A consisted of three ambulant participants, age range 18–50 yrs, two female, one male (m = 26 yrs). Group B consisted of three permanent wheelchair users, age range 55–65 yrs, two male, one female (m = 59 yrs). The weight of each participant was noted only for the purposes of the manufacturer's recommendations for weight limit of each cushion and did not form an integral part of this study. Other studies have been completed which reported the effect on pressure distribution of weight and body mass index (BMI) however comfort was not investigated as part of this study (Kernozek et al., 2002, Stinson, Porter-Armstrong & Eakin, 2003).

Study procedure

Informed consent was obtained and full information given to each participant before the study commenced. Anonymity was ensured throughout the study with the names of each participant known only to the author. Ethical approval was obtained from the University of Nottingham and permission granted from the authors' place of work. Each participant sat on each cushion for a period of two hours over a number of weeks. Participants were pressure mapped on each cushion on four separate occasions for two minutes as the literature suggests (Stinson et al. 2003). Qualitative

Table 1. Mean interface pressures after 2 hours sitting (mm Hg).

Cushion	Mean Average Pressure Group A	Mean Average Pressure Group B	Mean Peak Pressure Group A	Mean Peak Pressure Group B
Pressure Reducing Foam	32.92	39.00	78.69	164.75
Air Cells	33.73	40.82	86.08	122.75
Gel Pad	37.30	41.81	83.44	108.47
Standard Foam	39.54	40.94	91.71	103.45

data was collected after each pressure mapping episode using a comfort rating questionnaire devised by the author along with the Body Part Discomfort Technique (Corlett, 1981).

Results and discussion

The results show the mean highest and lowest interface pressures for the two hour sitting period for the four cushions in both groups (see table 1). The lowest mean pressure readings were recorded for the pressure reducing foam in both groups at 32.92 mm Hg and 39.00 mm Hg respectively. However, the difference between readings for this cushion and the air cells cushion in both groups is minimal with readings at 33.73 mm Hg and 40.82 mm Hg respectively. Overall Peak pressure readings in group B were considerably higher than group A across all four cushions. This could be attributable to the health issues faced by the participants in this group affecting posture such as low trunk tone resulting in increasing downward pressure on the ischial tuberosities and pelvic area. One participant from this group PG6 produced the highest pressure readings across all four cushions in the study. A mitigating factor could be that this participant had a progressive condition characterised by weakening and wasting of musculature. Group A results show the pressure reducing foam produced the lowest peak pressure readings with 78.69 mm Hg, conversely group B produced the highest results for this cushion at 164.75 mm Hg.

Comfort can be a difficult issue to address due to its subjective nature. It is described by Tutton and Sears (2003, pp. 690) as "a state, linked to outcomes such as ease, wellbeing and satisfaction." The ambiguity in addressing comfort in wheelchair cushions may be in part due to the fact that historically seating has been designed in the clinical setting with the sole aim of "equalising and minimising pressure" (Mooney et al., 1979). The rank order of the cushions shown in table 2 illustrates the differences in the two groups. Group A rated the air cells cushion first and gel pad second and group B rating the gel pad first and the air cushion second. The two remaining cushions were rated in the same order for both groups.

Table 2.   Rank order of cushions-comfort.

Rank	Group A	Group B
1	Air cells	Gel Pad
2	Gel Pad	Air cells
3	Pressure Reducing Foam	Pressure Reducing Foam
4	Standard Foam	Standard Foam

The results indicate that comfort ratings did not replicate the pressure readings in both groups with the pressure reducing foam being ranked third in group A and second in group B. Comfort ratings in group B were generally harder for participants to define, although the air cells produced low interface pressures it was described by the participants as "not holding pelvic posture" (PGC5) and "maintenance of cushion would be an issue and transfers difficult" (PGC6). The gel pad received favourable comments such as "well supported, helps keep knees in good alignment" (PGC5), "solid base, gel moulds and warm" (PGC6). Negative aspects of this cushion reported by the participants in group B was that it was too heavy to lift. A comparison of the two groups in this study reveal that interface pressures and comfort can be an ambiguous relationship for permanent wheelchair users, whereas in ambulant people the results suggest that a relationship could exist between the two. The difference could possibly be explained by the ambulant participants not having any pre-existing postural issues and were able to correct any stability concerns by altering their posture subtly to accommodate any feelings of discomfort during the sitting.

Conclusion

This evidence from this study, although limited due to the sample size and potential effect of confounders, highlights the importance that wheelchair users may place on comfort. This is a relatively unexplored field of study where most research has focused on minimising and equalising pressure. The findings from this study concur with previous research which has reported inconclusively about whether a relationship exists between comfort and pressure distribution in wheelchair cushions, where discomfort could be a possible indicator for low compliance. However, in this study a link was found in ambulant people. Wheelchair users in this study rated the gel cushion as most comfortable even though it did not produce the lowest interface pressure readings. Ambulant participants preferred the air cells cushion which did produce low interface pressures although not the lowest. Pressure mapping provides knowledge about interface pressure (Eitzen, 2004) and should be used to assist the assessment process with sound, evidence based clinical judgment. Clinicians, ergonomists and commercial organisations need to be more aware of the importance of comfort to the wheelchair user and collaborate to ensure that this concept is adequately explored.

References

Bennett, G., Delaney, C., Posnett, J. 2004. The Cost of Pressure Ulcers in the UK. *Age and Ageing*, **33** (3): 230–235.

Corlett, E.N., Bishop, R.P. 1976. A Technique for Assessing Postural Discomfort. *Ergonomics*, **19**, 2, 175–182.

Department of Health 2004. Improving Services for Wheelchair Users and Carers-Good Practice Guide. Last accessed 10/10/2014 www.pmg.co.uk

Eitzen, I. 2004. Pressure Mapping in Seating: A Frequency Analysis Approach. *Archives of Physical Medicine and Rehabilitation*. Vol. 85, 1136–1140.

Fay, B.T., Brienza, D.A. 2000. What is Interface Pressure? Paper presented at the 22nd Annual EMBS International Conference, Chicago IL.

Kyung, G., Nussbaum, M, A. 2007. Driver sitting comfort and discomfort part II: Relationships with and predictions from interface pressure. *International Journal of Industrial Ergonomics* **38**, 5, 526–538.

Mooney, V., Einbund, M.J., Rogers, J., Stauffer, E. 1971. Comparison of Pressure Distribution Qualities in Seat Cushions. *Bulletin of Prosthetics Research*.

Porter, M.J., Gyi, D. E., Tait, H. A. 2003. Interface Pressure Data and the Prediction of Driver Discomfort in Road Trials. *Applied Ergonomics*, **34**, 207–214.

Stinson, M., Schofield, R., Gillan, C., Morton, J., Gardener, E., Sprigle, S., Porter-Armstrong, A., 2013. Spinal Cord Injury and Pressure Ulcer Prevention: Using Functional Activity in Pressure Relief. *Nursing Research and Practice* ID 860396.

Stinson, M., Porter-Armstrong, A.P., Eakin, P.A. 2003. Pressure mapping systems: reliability of pressure mapping interpretation. *Clinical Rehabilitation*, **17**, 504–511.

Stinson, M., Porter-Armstrong, A., Eakin P. 2003. Seat Interface Pressure: A Pilot Study of the Relationship to Gender, Body Mass Index and Seated Position. *Arch Phys Med Rehab*, **84**, 405–409.

Stockton, L., Parker, D., 2002. Pressure relief behaviour and the prevention of pressure ulcers in wheelchair users in the community. *Journal of Tissue Viability,* **12**, 3, 84–99.

Stockton, L., Rithalia, S. 2007. Pressure reducing cushions: Perceptions of comfort from the wheelchair users perspective using interface pressure, Temperature and Humidity Measurements. *Journal of Tissue Viability*, **18**, 28–35.

Trewartha, M., Stiller, K. 2011. Comparison of the pressure redistribution qualities of two air-filled wheelchair cushions for people with spinal cord injuries. *Australian Occupational Therapy Journal,* **58**, 287–292.

Tutton, E., Seers, K. 2003. An exploration of the concept of comfort. *Journal of Clinical Nursing*, **12** 689–696.

XSensor UK. 2014. http://www.xsensor.co.uk/ last access November 2014.

VISUAL ACCESSIBILITY MISCONCEPTIONS HELD BY GRAPHIC DESIGNERS AND THEIR CLIENTS

Katie Cornish, Joy Goodman-Deane & P. John Clarkson

Engineering Design Centre, Engineering Department,
University of Cambridge, UK

With an ageing population, Inclusive Design is increasing in importance. However, misconceptions surrounding its practice may be limiting its uptake in industry. The visual accessibility of a product is a key aspect of Inclusive Design, and is particularly important in the graphic design industry due to its heavy reliance on visual communication. To improve the uptake of Inclusive Design in this area, we need to increase our understanding of graphic design practice and the misconceptions surrounding visually accessible design. Interviews were used to identify seven common misconceptions held by graphic designers and their clients, including the belief that the designer will consider visual accessibility, even if it is not in the brief.

Introduction

With an ageing population, Inclusive Design is becoming ever more important in enabling continued inclusion (Waller et al., 2013). For example, around two million people in the UK currently have some form of sight loss, a number that is predicted to rise to four million by 2050 (Access Economics, 2009). However, research reports that in industry, there is a low uptake of Inclusive Design, and its associated tools (Cardoso et al., 2005). This may be due to industry's misconceptions surrounding Inclusive Design, preventing designers from fully recognising its importance and how to put it into practice successfully. Indeed, it has been reported that there is a "widespread misconception within the design world, that Inclusive Design is relevant only to product or environmental design" (Cassim and Dong, 2005). This is not the case as the Disability Discrimination Act prohibits the discrimination of those with disabilities, be it in education, services or employment. In order to prevent discrimination in education, services and employment, several aspects must be considered. This includes graphic and communication design in addition to product and environmental design.

Clarkson et al. (2003) report another false belief that "understanding and responding to user needs is regarded by many designers as an unwelcome curb on creativity, rather than a spur to innovation". This can be considered a false belief as Maguire (2001) reports that personas (a method for understanding and responding to user

72

needs) "are popular with innovative design groups, where they are used to stimulate creativity". Genco et al. (2011) also report that Empathic Experience Design increases creativity in designers. Inclusive Design misconceptions must be identified and addressed in Inclusive Design tool creation and dissemination, to increase their uptake in industry.

Graphic design is a particularly interesting discipline with regard to Inclusive Design, especially in terms of visual accessibility, since it is so heavily reliant on vision. It has to communicate important messages to its audience, from road signage to the ingredients on food packaging. There is also an overlap between graphic design and web design, yet when compared with web design, graphic design has fewer accessibility guidelines. Furthermore, previous research has indicated a need to facilitate the communication of visual accessibility and its importance between graphic designers and their clients (Cornish et al., 2014).

The aim of this paper is to identify misconceptions held by graphic designers and their clients, with regard to visual accessibility. The findings can be used to inform the design and dissemination of Inclusive Design tools, and help increase their uptake in industry. This paper reports on part of a larger study investigating the visual capability loss simulation technique in graphic design.

Definition of terms

Visual accessibility is one aspect of Inclusive Design and is particularly important in graphic design. In this paper, the term 'visual accessibility' refers to the ability of the target audience to see and comprehend the visual aspects of the item. Multiple factors such as legibility of text, clarity of images and contrasts used, and other factors such as glare, impact on the visual accessibility of an item (RNIB, 2006). The level of accessibility affects the number of people who can effectively and efficiently interact with the item in the intended manner. This is vital, as designing in a visually accessible way benefits all members of the population, not just those with visual impairments (Waller et al., 2013). It is particularly important, as designers are often young, and tend to have good vision (Zitkus et al., 2011) making it difficult for them to empathise with some target audiences.

According to Bennett (2006), an important part of graphic design is the communication of visual information, in particular "it is concerned with the efficiency of communication" (Bennett, 2006). It includes many sub-disciplines such as branding, and overlaps with other disciplines such as web design. Graphic designers are responsible for executing the design brief, to meet the needs of the client. They may be in-house or external to the client's company and may have worked on one off projects or longer term with the client. This study focuses solely on graphic design, as there is a lack of research in this area. Furthermore, it is particularly important with regard to an ageing population, as being able to access print based information is paramount for maintaining independence and a quality of life.

Methodology

Aim

To identify misconceptions held by graphic designers and their clients with regard to visual accessibility.

Sample

The sample consisted of 14 participants: seven client-designer pairs (referred to as C1-7 and D1-7). The sample size was limited due to the depth of data required and time constraints. The participants had worked on a print-based graphic design project together in their pairs, and were contacted via email through personal contacts and design directories such as The Dexigner Directory (2014).

A range of subjects were selected to ensure a diverse sample. They differed in terms of the length of the relationship (from a one-off project to multiple projects), the type of the relationship (including an in-house designer and external designers), and the project audience (from a widespread audience to a smaller more specific audience). It is important to recognise the limited sample used in this study, however a more widespread survey is underway.

Data collection and analysis

An individual semi-structured interview, lasting up to one hour, was carried out with each participant, in a location of their choosing. This method was selected to allow the deeper exploration of topics that arose during the interview. The interviews covered the following topics: the participant's awareness of visual accessibility and the associated tools and methods; current practices and the level of consideration given to visual accessibility, including its barriers and drivers; and the design brief and the client-designer communication.

Some of the client and designer questions differed. For example, the clients discussed generating and communicating the brief and receiving the design concepts, whereas the designers discussed receiving the brief and their design process. The client-designer pairs were encouraged to discuss the project they had worked on together to produce results grounded in real world graphic design practice, and the different perspectives that each party has.

Participants gave informed consent and interviews were recorded on a digital audio recorder, and with written notes. The data was transcribed using QSR NVivo Software (QSR International, 2014). The analysis followed the general inductive approach for qualitative data analysis outlined by Thomas (2006). The data were coded and reduced repeatedly to reveal the main themes. A second researcher used the previously developed categories to code one interview, in order to check the clarity of the categories (Thomas, 2006). This lead to the merging of two existing categories, and the refinement of others.

Results and discussion

Seven misconceptions were highlighted through the analysis of the interview transcripts from the graphic designers and their clients. Some of these misconceptions are held only by the designers, or only by the clients, and some are held by both groups. These are discussed below.

Misconception 1 – Shared: visually accessible design is just for the severely visually impaired

One misconception exhibited by both graphic designers and their clients, is related to who will benefit from a 'visually accessible' design. Some participants associated the term 'visual accessibility' with designing solely for people with severe visual impairments, rather than recognising the benefits it has for the larger population with slightly worse vision than that of the designer. When asked about visual accessibility, one designer explained that they "*don't often think about somebody with severe cataracts, or who can just use their peripheral vision, because that's like the 0.00001% of the viewing audience*" (D7). Another designer added "*we don't work with any organisations that are specifically involved with sight impaired people, but I'm sure if we did that would form part of the brief*" (D6).

This can be deemed a misconception as the RNIB (2006) report that there are "two million people in the UK with a sight problem" (RNIB, 2006), and there are even more people with a slight reduction in visual capabilities, such as age-related long sightedness. This is a big problem, as identified by Goodman-Deane et al. (2014) who demonstrate that over 6% of the population were excluded from the simple task of using a digital camera. Furthermore, the environment can temporarily impair people, for example when attempting to read a restaurant menu in dim light. Therefore increased visual accessibility could benefit a much larger proportion of the population than the participants appear to realise.

Misconception 2 – Shared: guidelines alone are enough to create visually accessible graphic designs

A number of participants cited design guidelines as the main method that they use to give consideration to visual accessibility. For example at least one out of every client-designer pair explained that they either provided, or used, design guidelines, although not all guidelines contained information to assist with designing in a visually accessible manner. One client explained that following these guidelines will make their design accessible. For example "*there is an accessibility guideline. With patient information you don't go below 12 point and you also need an accessibility statement*" (C2). A designer suggested that following guidelines ensures legibility, "*12 point font is the smallest we can go down to, so that's the smallest it can go down to and still be legible*" (D1). Furthermore, very few other methods were mentioned by participants, suggesting that they believe guidelines alone could be adequate.

Although participants reported that guidelines are a widely used method of ensuring visual accessibility, many of them recognised the shortcomings with this. One client explained that it is not as simple as just using a certain type font. He said, *"ideally you know it's 12 point in Arial. So you would use that as a guide because some typefaces at 12 point are a little smaller"* (C2), with another adding *"sometimes it's not just the size of the font, it's the weight of the font"* (C4). Another client added *"it's a bit silly really that the NHS colours, one of them doesn't seem to be accessible"* (C2). This suggests that when asked, a number of respondents were aware of the shortcomings of guidelines, but still used them as the predominant method for considering visual accessibility.

Misconception 3 – Shared: providing an alternative means of access to information is enough to ensure an inclusive experience

Some participants suggested that they do not need to pay particular attention to visual accessibility in graphic design because they provide other ways for those with reduced visual capabilities to access the information. Examples of this include large text formats or audio versions of information. For example, one designer explained *"a lot of clients will have their publications in a large text format as well ... we give them the files and they can turn it into a large format which they can print out. So people are catered for"* (D5). He added *"it's a pdf and it's on their [the client's] website so they can blow it up anyway"* (D5).

This is a misconception as it is widely reported that people often experience stigma when relying on devices associated with a loss of function (Parette and Scherer, 2004). This may include relying on an alternative means of access, such as large print versions of information. Furthermore the ADA (2010) states that access for those with disabilities must be equivalent, therefore, providing an alternative means of access should be a last resort for designers.

Misconception 4 – Designer held: considering the user's visual capabilities is more time consuming and difficult than it is worth

A number of designers implied that giving consideration to the visual capabilities of the target audience, would add more time and complexity to the design process, than it is worth. They explained that it *"makes the whole process a lot harder for us"* (D5) with issues to do with time being regularly cited as a reason for not designing inclusively. Another designer added that giving consideration to visual accessibility would add *"cost to our design process. If it meant that we took longer designing something, we would have to pass that cost onto our client, that decision then would have to come from our client because at the end of the day they're the ones paying"* (D3).

Although user trials may take time, there are tools available to assist graphic designers in a quick and cheap manner. For example tools such as the Cambridge Simulation Glasses (Clarkson et al., 2011) have been developed to act as a quick and low cost tool to aid visually accessible design. Other quick and cheap methods available include paper prototyping (Maguire, 2001) and personas (Goodman et al., 2006). If the designers recognise the importance of considering the user's visual capabilities, as well as the quick and cheap tools and methods available, they may realise that this is a misconception.

Misconception 5 – Designer held: you cannot create both stylish and accessible designs

Several designers suggested that there was so much of a trade-off between accessibility and style that both cannot exist together. They explained that certain projects needed to be more stylish or 'edgy' and therefore they could not take visual accessibility into account, otherwise they would lose this aspect. Furthermore it was suggested that the client would be happier with a more stylish design than a more accessible one, as the latter can appear to be 'dull'. One designer stated that "*if you accommodated people's eyesight in every design, everything would be so regimented and so kind of dull*" (D4). Another designer added "*there's quite often a conflict between accessibility and how sort of edgy [a design is]*" (D5), and clients "*quite often want their stuff to be really edgy, and they'll not always follow accessibility guidelines*" (D5).

A recent study and review of usability and aesthetics in HCI concluded that, "poor usability lowered ratings of perceived aesthetics" (Tuch, 2012). Furthermore there are examples of visually accessible and stylish designs such as the Google search engine and the Gov.UK website. Although there are also stylish designs that are not accessible, these examples suggest that in graphic design, style and accessibility are not necessarily mutually exclusive.

Misconception 6 – Client held: the designer will always take visual accessibility into account even if it is not in the brief

The clients suggested that although they deemed visual accessibility to be important, they rarely included it in the brief. For example one client stated, "*I haven't ever written a brief that says it needs to be legible*" (C3) but later added that it is important that their information is clear and legible. The fact that they deem it to be important but don't include it in the brief suggests that they are making the assumption that the designer will take it into account anyway.

However, the designers explained that if visual accessibility was not in the brief, then they would not give it much consideration. One designer stated "*not many clients say to us at the start 'accessibility is important', and I don't go to them and say we need to think of accessibility. That's probably a failing on my part*" (D5). Participants added that time and cost is also factor, explaining why not including it in the brief prevents the designer from taking it into account.

Misconception 7 – Client held: the client's personal judgment is enough to ensure that graphic designs are visually accessible

Another misconception is that the client using their judgment to 'check' the work of the designer for accessibility issues is enough to prevent people from being excluded from the design unnecessarily. Some clients reported that if they don't have any difficulty in seeing a design then they presume others wouldn't, with one explaining, "*if something came back [from the designer] and I thought 'oh that might be difficult to read', then I would reject it. So there are a certain number of years of experience that comes with my assessing things before they actually get into print*" (C6). Another added, "*I think people just rely on their own instinct as to whether or not it is right*" (C7).

Pheasant (1996) identified this as a misconception. He described five fundamental fallacies of design, one of which is that designers think usability is important, but they state "I do it intuitively and rely on my common sense so I do not need tables of data or empirical studies". Although this fallacy is referring to engineering design, it suggests there may be a similar fallacy in graphic design.

Conclusions

The aim of this study was to identify misconceptions held by graphic designers, and their clients, with regard to visual accessibility. Seven misconceptions were identified. This is an important contribution to the field as these previously unidentified misconceptions need to be taken into account when designing and disseminating Inclusive Design tools, to help increase their uptake in industry. One limitation of this study is its sample size. However, a more widespread survey of visual accessibility in graphic design is underway. This aims to determine whether these results are generalizable to a wider audience, and to increase our knowledge of Inclusive Design in the graphic design industry.

The results from this study extend and develop the previous findings from Cornish et al., (2014). Cornish et al. highlighted the importance of facilitating communication between graphic designers and their clients, with regard to visual accessibility. Providing Inclusive Design tools and methods that assist this will help promote the communication and understanding of considering the user and their visual capabilities, and reduce the misconception that "the designer will always take visual accessibility into account even if it is not in the brief".

Acknowledgements

We would like to thank the EPSRC for funding this work.

References

Access Economics. 2009. Retrieved 12th May 2014 from: http://www.rnib.org.uk
ADA (Americans with Disabilities Act). 2010. Retrieved 12th March 2014 from: http://www.ada.gov/

Bennett, A. (Ed.). 2006. *Design studies: theory and research in graphic design.* (Princeton Architectural Press, New York).

Cardoso, C., Clarkson, P.J., and Keates, S. 2005. Can users be excluded from an inclusive design process? *Proceedings of the 11th International Conference on Human Computer Interaction.* Lawrence Erlbaum Associates.

Cassim, J., and Dong, H. 2005. DBA design challenge: engaging design professionals with inclusive design. *ACM SIGACCESS Accessibility and Computing*, (81), 3–8.

Clarkson, P.J., Keates, S., Lebbon, C., and Coleman, R. (eds.). 2003. *Inclusive Design: design for the whole population.* (Springer, London).

Clarkson, P.J., Coleman, R., Hosking, I., and Waller, S. 2011. Retrieved 14th July 2014 from: www.inclusivedesigntoolkit.com.

Cornish, K., Goodman-Deane, J., and Clarkson, P. J. 2014. Designer Requirements for Visual Capability Loss Simulator Tools: Differences between Design Disciplines. *Univ Access in HCI.* (Springer International Publishing, UK), 19–30.

The Dexigner Directory. 2014. Retrieved 12th February 2014 from: http://www.dexigner.com/directory/

Genco, N., Johnson, D., Hölttä-Otto, K., and Seepersad, C. C. 2011. A Study of the Effectiveness of Empathic Experience Design as a Creativity Technique. In ASME 2011 IDETC and CIEC (pp. 131–139).

Goodman, J., Clarkson, P. J., and Langdon, P. 2006. Providing information about older and disabled users to designers. In *HCI, the Web and the Older Population, workshop at HCI.*

Goodman-Deane, J., Waller, S., Cornish, K., and Clarkson, P. J. 2014. A Simple Procedure for Using Vision Impairment Simulators to Assess the Visual Clarity of Product Features. *Universal Access in HCI* (pp. 43–53). Springer.

Maguire, M. 2001, Methods to support human-centered design. *International Journal of Human-Computer Studies*, 55(4), 587–634.

Pheasant. 1996. *Bodyspace: Anthropometry, Ergonomics and the Design of Work.* (Taylor and Francis, London).

Parette, P., and Scherer, M. 2004. Assistive technology use and stigma. *Education and Training in Developmental Disabilities*, 39(3), 217–226.

QSR International: NVivo 8. 2014. Retrieved 7th January 2013 from: http://www.qsrinternational.com/products_previous-products_nvivo8.aspx.

RNIB. 2006, *See it Right.* (RNIB, UK).

Thomas, D.R. 2006. A General Inductive Approach for Analyzing Qualitative Evaluation Data. *American Journal of Evaluation* 27, 237–246.

Tuch, A. N., Roth, S. P., Hornbæk, K., Opwis, K., and Bargas-Avila, J. A. 2012. Is beautiful really usable? *Computers in Human Behaviour*, 28(5), 1596–1607.

Waller, S., Bradley, M., Hosking, I., and Clarkson, P. J. 2013. Making the case for inclusive design. *Applied ergonomics.* (In Press).

Zitkus, E., Langdon, P., and Clarkson, J. 2011. Accessibility evaluation: Assistive tools for design activity in product development. *In SIM Conference Proceedings* (Vol. 1, pp. 659–670).

A USER-CENTRIC METHODOLOGY TO ESTABLISH USABILITY HEURISTICS FOR SPECIFIC DOMAINS

Setia Hermawati & Glyn Lawson

Human Factors Research Group, University of Nottingham

Heuristics evaluation that employs Nielsen's (1994) heuristics is frequently employed for usability evaluation. While these heuristics are generally suitable to evaluate most user interfaces, modification of existing heuristics or the addition of specific heuristics are required for some domains. This paper proposes a general methodology to establish usability heuristics for a specific domain. The methodology encompasses two stages i.e. how to extend heuristics sets for specific domains and how to validate the heuristics sets. Steps in each stage are described.

Introduction

Usability is a key to ensure that the success of a system (Markus and Keil 1994) and is usually improved by performing usability evaluations at different stages of system development (Nielsen 1994a). While the ultimate approach for usability evaluation is user studies, usability inspections are also commonly applied, especially at the early stages of a system's development. Heuristics evaluation, due to its simplicity and low cost (Jeffries et al. 1991; Tang et al. 2006; Hwang and Salvendy 2010), has gained popularity and is frequently employed as part of usability studies. This method was originally proposed by Nielsen and Molich (1990) and requires a number of experts to inspect usability based on "heuristics" which are essentially broad and non-specific rules of thumb.

Nielsen's (1994b) ten heuristics are normally used as the heuristics that guide the evaluation and have been shown to be quite effective. These heuristics consist of: 1) visibility of system status, 2) match between system and the real world, 3) user control and freedom, 4) consistency and standards, 5) error prevention, 6) recognition rather than recall, 7) flexibility and efficiency of use, 8) aesthetic and minimalist design, 9) help users recognise, diagnose, and recover from error, and 10) help and documentation; with short description for each heuristics provided. While these heuristics could be used to evaluate user interfaces for various domains, modification or adjustment are needed to ensure that usability issues that are specific to user interfaces of certain domains are not overlooked (Nielsen 2014).

Ling and Salvendy (2005) reported that modification of heuristics for specific domains could be in three forms: 1) expansion of the heuristics set, 2) alteration of the evaluation procedure, and 3) alteration of the conformance rating scale.

This paper focuses on the first type of modification which generally involves two stages i.e. generation of the heuristics sets and their validation. Interestingly, the methods adopted for the generation and validation of heuristics varied from one study to another (Ling and Salvendy 2005) with a lack of general consensus on how they should be approached.

This paper describes a proposal of a general methodology on how to expand heuristics sets for a specific domain and validate them. Steps required will be explained and anticipated challenges with respect to application of the methodology will be discussed. The methodology is based on the results of a systematic literature review of existing studies which involved a review of a 70 studies from 90 articles (Hermawati and Lawson, 2014). The systematic review indicated that there was a wide diversity of approach to generate and validate heuristics with the majority of them lacks of rigour.

How to expand heuristics sets for a specific domain

Figure 1 shows a flow chart depicting the overall processes that are proposed to expand heuristics sets for a specific domain. As shown in the figure, two main resources that allow for direct expansion of heuristics are documented resources of usability issues and existing guidelines/best practices. Paddison and Englefield (2003) differentiated between the two as evaluation-based and research-based methods. Ideally a balance between the two is achieved. Heuristics that are solely based

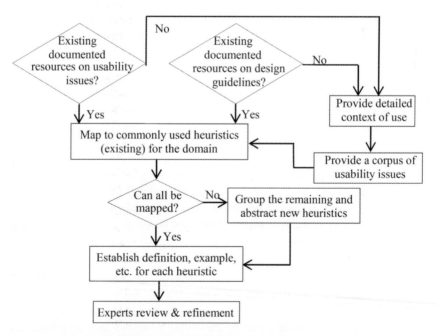

Figure 1.   Processes to expand heuristics sets for a specific domain.

on documented usability issues, while providing a solid ground for heuristics expansion, need to ensure that the usability issues are collated from a wide range of application/system in the domain. On the other hand, heuristics that are solely based on design guidelines, while providing an already synthesised concept, could easily result in increasing the number of heuristics. While there is no limit on the number of heuristics that should be proposed, a lengthy heuristics set should be avoided as they could be cumbersome to use during evaluation (Mckay and Kölling 2012). In some cases, especially for domains which involve new technology, it is likely that documented usability issues and guidelines/best practices are limited. In this circumstance, a combination of understanding the context of use of a system and study(s) to create a corpus of usability issues could be performed. It is important that for both activities, end-users should be involved.

The next step is to map the information above to either existing or commonly used usability heuristics. Most studies employed Nielsen's heuristics (1994) when previous work to establish the heuristics was non-existent (e.g. Rusu et al. 2012). The mapping process should ideally be performed as a team that combines individual work and group work. Core categorisation and reliability check procedure (Jankowicz 2005), an approach introduced to categorise construct in a repertory grid analysis, could be adopted. The procedure requires each individual to produce categorisations and compare the results until a minimum level of disagreement is reached. For the remaining information that could not be mapped, core categorisation and reliability checks could also be employed to group them and abstract new heuristics.

Once the mapping is completed, a definition of each heuristics should then be established. Ideally, the definition should be succinct and easy to understand. Rusu et al. (2011) suggested adopting a formal approach to establish heuristics definition. Each heuristic required definition, example, suggestion, etc. While this could be helpful as part of a training for expert to use the heuristics, during an evaluation, a lengthy heuristics list could be adding an unnecessary burden to an evaluator. Furthermore, the heuristics definition should also not be too specific or general. The last step of the process is to perform a review with experts and use their feedback as a means to refine the heuristics. Any method could be used as part of expert review e.g. interview, questionnaire, online survey. The most important thing is to present the heuristics, along with its detailed definition, to experts and obtain their opinion on suitability, clarity, redundancy, terminology used, etc. for each heuristic. If resources permitted, the review with experts should ideally be accompanied by requesting experts to use the heuristics as part of a mock up evaluation of an application.

How to validate heuristics sets for a specific domain

Figure 2 shows a flow chart depicting the overall processes that are proposed to validate heuristics sets for a specific domain.

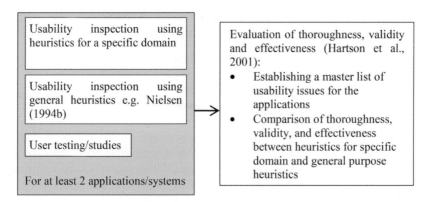

Figure 2.   Processes to expand heuristics sets for a specific domain.

As shown in the figure, the validation method requires an extensive number of studies (at least 6 studies, divided into three sets of studies). A minimum of 2 application/systems are required as part of the validation as this allows performance inspection of the heuristics sets across different systems and reduces the bias that could be resulted from using just one application. Three dimensions, adopted from Hartson et al. (2001), are proposed to be used to evaluate the performance of heuristics sets for a specific domain. Thoroughness measures the proportion of real usability issues found; validity measures the proportion of real usability issues found vs the usability issues identified; and effectiveness is defined as the combined product of thoroughness and validity. A prerequisite for this is the establishment of a master list of usability issues. For this purpose, the result of user testing/studies is used. With respect to the numbers of participants required in each study, ideally at least 5 participants are required in each study (Virzi 1992; Nielsen and Landauer 1993). Another thing that needs to be considered is the need to balance the expertise of participants that are involved in usability inspection studies. This is to ensure that bias due to differences in individual performance is minimised. Furthermore, training and familiarisation for participants that are assigned to use the heuristics for a specific domain would also need to be addressed.

Discussion

The methodology proposed in this paper was different in comparison to previous methodologies (e.g. Rusu et al. 2011, Yeratziotis et al. 2012). The main difference is the double edged user-centric approach which involves both experts' (who will use heuristics sets) and end-users' (who will use a system that is being evaluated) point of view from the start and through-out the process. Rusu et al. (2011) emphasised more on the view of the experts and only proposed user testing during the validation methodology as complementary activities. This is similar to the approach adopted by Yeratziotis et al. (2012). Furthermore, the proposed methodology provides a clear direction on activities required for domains where existing usability studies for relevant technologies are either abundant or lacking.

The proposed methodology emphasises the importance of end-users and domain-expertise while at the same time acknowledging the results of previous research whether they are in the form guidelines, lessons learned, usability issues encountered, etc. As a consequence, the methodology requires a considerable number of activities, which will impact on the cost and time needed. The cost and time issues could likely be offset by the advantage of heuristics sets for specific domain. Provided that heuristics sets are presented such that it is understandable even for the non-domain experts, heuristics evaluation for specific domains could be performed by any usability expert without the risk of missing major usability issues.

The most difficult and time consuming activity of the proposed methodology would likely be the mapping of usability issues, guidelines and context of use of a system to existing heuristics. This was also the part of the methodology in which the outcome could be different from one study to another even if the same information was used. As such, it is important that the refinement of heuristics involves a considerable number of experts who have not only expertise in usability but also domain knowledge.

The proposed validation methodology is envisioned to be applied to establish heuristics sets for cloud computing and engineering simulation domain which will be developed as part of CloudFlow research project (FP7-2013-NMP-ICT-FoF). The application at both domains will allow an assessment of its feasibility and empirically identify the difficulties and challenges associated with its application.

Conclusions

A user-centric methodology to establish heuristics sets for domains where existing usability studies for relevant technologies are either abundant or lacking was proposed. The methodology provided step by step guidance on how to extend heuristics sets and validate them to evaluate usability for a specific domain.

Acknowledgement

This work was supported through the funding from "CloudFlow" research project under grant agreement no. 609100ICT-285176, which is funded by the European Commission's 7th Framework programme (FP7-2013-NMP-ICT-FoF).

References

Hartson, H. R., Andre, T. S., and Williges, R. C. 2001. "Criteria for evaluating usability evaluation methos." *International Journal of Human Computer-Interaction*, 13, pp. 373–410.

Hermawati, S., and Lawson, G. 2014. Establishing usability heuristics for heuristics evaluation in a specific domain: is there a consensus?. Manuscript submitted for publication.

Hwang, W., and Salvendy, G. 2010. "Number of people required for usability evaluation: the 10±2 rule." *Communications of the ACM* 53(5): 130–133.

Jankowicz, D., 2005. *The easy guide to repertory grids*. (John Wiley & Sons, West Sussex, England).

Jeffries, R., Miller, J.R., Wharton, C., and Uyeda, K. 1991. "User interface evaluation in the real world: a comparison of four techniques." *Conference on Human Factors in Computing Systems*. New Orleans, LA.

Ling, C., and Salvendy, G. 2005. "Improving the heuristic evaluation method: A review and reappraisal." *Ergonomia: An International Journal of Ergonomics and Human Factors* 27(3): 179–197.

Markus, M. L., and Keil, K. 1994. "If we build it, they will come: designing information systems that people want to use." *Sloan Management Review* 35(4): 11–25.

McKay, Fraser and Kölling, Michael 2012. *Evaluation of Subject-Specific Heuristics for Initial Learning Environments: A Pilot Study.* Psychology of Programming Interest Group 24th Annual Conference. London, UK.

Nielsen, J., and Molich, R. 1990. "Heuristic Evaluation of User Interfaces." Conference on Human Factors in Computing Systems.

Nielsen, J., and Landauer, T. K. 1993. "A mathematical model of the finding of usability problems." *ACM INTERCHI Conference*.

Nielsen, J. 1994. Heuristic Evaluation, In Nielsen, J. and Mack, R. L. (eds.), *Usability Inspection Methods*, (John Wiley and Sons, New York), 25–62.

Nielsen, J. 1994a. "Usability laboratories." *Behaviour and Information Technology* 13: 3–8.

Nielsen, J. 2014. "How to conduct a heuristics evaluation." Retrieved 8 September 2014, from: http://www.nngroup.com/articles/how-to-conduct-a-heuristic-evaluation/.

Paddison, C., and Englefield, P. 2003. *Applying heuristics to accessibility inspection in a commercial context*. Conference on Universal Usability. Vancouver, Canada.

Rusu, C., Roncagliolo, S., Rusu, V. and Collazos, C. 2011. A methodology to establish usability heuristics. *4th International Conferences on Advances in Computer-Human Interactions*. Guadeloupe, France.

Rusu, C., Roncagliolo, S., Figueroa, A., Rusu, V., and Gorgan, D. 2012. "Evaluating the Usability and the Communicability of Grid Computing Applications." *The Fifth International Conference on Advances in Computer-Human Interactions*.

Tan, W-s., Liu, D., and Bishu, R. 2009. "Web evaluation: heuristic evaluation vs. user testing." *International Journal of Industrial Ergonomics* 39: 621–627.

Virzi, R. A. 1992. "Refining the test phase of usability evaluation: how many subjects is enough?" *Human Factors* 34: 457–468.

Yeratziotis, A., van Greunen, D., and Pottas, D. 2012. A framework for evaluating usable security. *Sixth International Symposium on Human Aspects of Information Security and Assurance*.

THE EFFECT OF DOMAIN KNOWLEDGE ON INTERACTION WITH VISUAL ANALYTICS

Solomon Ishack, Chris Baber & Adam Duncan

*School of Electronic, Electrical and Systems Engineering,
University of Birmingham*

In this paper we consider how domain knowledge affects interaction with Visual Analytics. In a trial involving a particular musical genre (in which we can demonstrate that some of the participants are expert), we show significant differences in the ways in which novices and experts interact with a visualisation. These differences reflect differences in prior knowledge. This suggests that the manner in which people interact with Visual Analytics is not simply a function of what is displayed to them, but a matter of how this display corresponds to their existing knowledge. This relationship is explored through the concept of Distributed Cognition.

Introduction

A repeated claim for Visual Analytics is it can help analysts gain 'insight' into the problem that they are facing (Thomas and Cook, 2005; Pousman et al., 2007). Insight has long been a subject of debate in Cognitive Psychology, i.e., does insight arise as a sudden 'ah-ha' realisation of the solution to a problem or is it the gradual piecing together of disparate information? (Bowden et al., 2005). Klein (2013) suggests that insight arises in response to one or more of five 'triggers': contradictions, coincidences, connections, curiosity, creative desperation. In his examples, people with experience in their work domain 'see' a particular solution that was not apparent to their colleagues.

In the field of Visual Analytics, insight is a looser concept which relates to the capability of the analyst to spot patterns in the data. Indeed, Keim et al. (2010) suggest that Visual Analytics supports "the exploration of information presented by visualisations [in] a complex process of sense-making" (p. 116). This implies that, rather than a solution 'popping out' of the displayed information, the displayed information is combined with analyst knowledge. Saraiya et al. (2006), in a study of bioinformatics researchers, concluded that the majority of 'insights' reported were not revealed by interaction with the visualisation *per se* but by the experience of the analysts. This leads to the question to be addressed in this paper: how does expertise affect people's interaction with Visual Analytics?

Keim et al. (2010) outline an iterative process in which user interaction updates visualisations, which builds and refines underlying models of the data. The visualisation is the window through which the analyst understands the analysis process, and the relationship between visualisation and model leads to the generation of new knowledge about the data. Our work considers how user interactions with the visualisation can lead to an understanding of the data that creates knowledge about a domain. In their discussion of analyst activity, Yi et al. (2008) propose four processes which characterize user interaction with visualisation: Overview to gain an appreciation of the entire dataset; Adjust parameters, modify ranges, or combine data to create different patterns and combinations; Detect patterns and structures in the data; Match Mental Model – seek to interpret the resulting patterns in terms of analyst expectations and experience.

If these processes are influenced by domain knowledge then one cannot assume that a 'good' visualisation will guarantee that the person using it will be able to see patterns in the data. For instance, it might be possible that a 'pattern' could be seen in the data even when no such pattern exists, i.e., to see a visual pattern when there is no statistical pattern. This problem is made worse by the quantity of data which are being displayed which, coupled with the selection of parameters by the analyst, could create correlations which are statistically reliable but practically meaningless (or at least misleading). Thus, allowing the analyst to play around with parameters could lead to visually compelling patterns which are artefacts of the data rather than reflections of reality. This also implies that manner in which someone interacts with a given visualisation *could* be affected by their domain knowledge. For example, assume that insight might arise when disparate pieces of information are presented together to present an interesting and unexpected relationship in the information. This highlights a role for Distributed Cognition in Visual Analytics, i.e., interaction between the person and the visualization becomes a form of thinking (rather than simply the manipulation of a display). Furthermore, if the person already knows some of this information, they are less likely to search for or query a database for this information (even if the information might be useful for comparison). As such, the expert could miss connections simply because information was not displayed in relation to other information, or the novice could miss connections simply because they were unable to appreciate the meaning of different pieces of information.

Method

In this study, participants were presented with an interactive visualisation and asked to make judgements about a set of statements. The domain was US Hip-Hop artists. This domain was selected to provide a sufficiently narrow range of interest to make it easy to recruit 'expert' and 'non-expert' participants. Initially, participants were recruited through personal contacts of the first author, and then screened for expertise. An 'expert' was self-declared to have an interest in this genre. This interest was tested through a short questionnaire which asked about the number of albums they owned, the number of gigs they attended, which music magazine they read, together with some questions, such as who had a hit with a given song. From this we ensured that the 'expert' group was highly enthusiastic and knowledgeable about the topic. Of 15 people approached, we selected 10 who met the criteria of 'expert'.

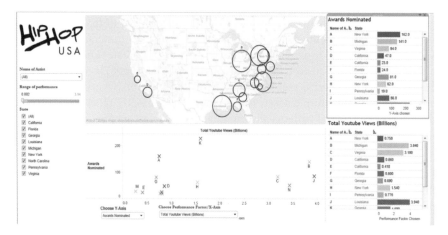

Figure 1.   User interface in this study.

For the non-expert group, we asked similar questions to a further 20 people and the 10 who scored the lowest were selected to participate in the study. All participants were in the age range 20–22 years. There were 12 male and 8 female participants (distributed evenly between the two groups).

Analytics and visualisation

The dimensions in the dataset were chosen to reflect ways in which success of Hip-Hop artists could be quantified and which could be discovered through search of the World Wide Web, i.e., (1) net worth of artist (2) awards nominated (3) awards won (4) Twitter followers (5) total YouTube views (6) studio albums produced. A total of 50 hip-hop artists were used in this dataset, generated from a number of web sources, including Twitter and YouTube. Data can be acquired from these sources by using an Application Program Interface (API). Twitter's API, for example, returns information in a JSON format which is easy to parse. YouTube provides a similar Data API through which we can access information about, for examples, 'hits' to a particular video or artist. Whilst it may be ideal for a visual analytics system to automatically parse the data into the system, for the purposes of this study the data were parsed by hand into an Excel file which was read into Tableau.

Tableau is a software tool that allows easy building of interactive data visual-izations. Tableau can be connected to data sources, such as Microsoft Excel. In addition, Tableau's dashboard feature allows developers to merge visualizations and interaction elements into a single interface through which any interaction that is performed affects the entire interface. Figure 2 shows the interface used for this study. The interface supports a range of different interactions, i.e., Filter: left of screen, allows the user to select information in terms of name of artist, metric used, US State where the artist comes from; Map: centre of screen, shows circles which indicate artists' location with the size of circles scaled to reflect metric chosen;

Correlation graph: bottom, select two metrics and show their relation; Bar charts: right of screen, top and bottom 5 artists on chosen metric.

Measures of performance

In order to create a task in which participants needed to search for and collate information, we used the Analysis of Competing Hypotheses (ACH) matrix (Heuer, 1999). Three hypotheses were provided to participants in the ACH matrix: the most successful artists perform well on social media (i.e. YouTube, Twitter); the most successful artists receive much recognition in their industry based on awards they have won and have been nominated for; the most successful artists are from the West-Coast and have produced many studio albums. Participants were required to find six pieces of evidence for and against hypotheses. We then classified the content of ACH responses in terms of a 'fact taxonomy' (Chen et al., 2009). In addition to this, we asked participants to rank a 'top 10' artists pre- and post-test (in order to see if there were any differences in ranking). Finally, the actions that participants performed when using the visualisation were categorised using Gotz and Zhou's (2008) action taxonomy of Explorations (Filter, Brush, Inspect, Merge, Pan, Query, Redo) and Meta-action (Revisit, Undo and Zoom). The focus on actions allows us to see whether the restructuring of information is important and whether interaction differs between experienced and less experienced participants.

Procedure

The design of the experiment was compliant with University of Birmingham ethics procedure, i.e., participants were asked to provide informed consent and were told that they were free to withdraw at any point in the study (including up to one week after the trial). Following an explanation of the user interface, participants were given a familiarisation session (using a different data set) in which they could practice the different forms of interaction. Once participants felt confident to use the interface, the experiment was explained to them. First, participants were given a list of 50 hip-hop artists and asked to indicate which they considered to be a top 10. Next, they were presented with the ACH matrix and told that their task was to find six pieces of evidence which they felt most appropriately related to the hypothesis. Participants were then allowed to complete the task at their own pace. Following completion of the ACH, participants were asked to provide rationale for their decisions. Finally, participants were asked to reselect a top 10 of hip-hop artists. Results were analysed using student t-tests. Initial f-tests were used to determine whether variance was equal or unequal, and appropriate adjustment was made to the test. Statistical significance was judged at an alpha level of 5%.

Results

The number of changes participants made in the rankings of whom they conceived to be the ten most successful out of the fifty artists pre- and post-test.

Table 1.   Differences in Actions.

	Merge	Inspect	Filter	Brush	Pan	Query	Revisit	Sort	Zoom
Expert	9.9	7.7	6	2.7	4.4	0	0.9	17.3	1.6
	(±5.8)	(±8.7)	(±4.7)	(±2.6)	(±3.5)	(0)	(±0.9)	(±4.5)	(±0.8)
Non-	2	28.9	21.5	8.1	15	1	3	38.1	4.9
expert	(±2.7)	(±10.7)	(±26)	(±7.3)	(±8.9)	(±1.3)	(±1.6)	(±11.5)	(±3.8)
t	2.89*	4.86*	1.85*	2.21*	3.5*	2.37*	3.71*	5.34*	2.7*

The mean numbers of changes were significantly larger amongst non-expert participants (4.5 ± 1.43) than expert participants (3.1 ± 1.29): ($t(18) = 2.3$, $p < 0.05$). The mean time taken to complete the task was significantly higher for non-expert ($46\,m \pm = 8$) than expert participants ($29\,m \pm 5$): ($t(18) = 5.52$, $p < 0.05$).

Using the Gotz and Zhou (2008) action taxonomy, there were significant differences between groups as shown in table 1. With the exception of 'undo', all actions show significant differences (* indicate $p < 0.05$). Non-experts show higher levels on all measures, except 'merge'.

Applying the fact taxonomy of Chen et al. (2009) to the ACH, it is possible to show two main differences between groups. The most prominently used fact type for both groups was Association (expert = 35; non-expert = 31). The fact type Difference was used more by experts than non-experts (expert = 16; non-expert = 7)., and the Cluster fact type, was used more by non-experts than experts (expert = 7; non-expert = 13).

Experts provided more structured responses and include more information from their own knowledge in support of the hypotheses, e.g., "Some artists, like *Lil Wayne don't have as a high net worth as someone like PDiddy*, but have a *higher following on social media*. More people would know of an artist like Lil Wayne due to his collaborations on many tracks, whereas an artist like Rick Ross has been delving into the hip hop genre". The *italics* indicate information which could be found in data set provided, and the rest of this material comes from the person's knowledge of artists and the domain. The non-experts conclusions were based on the facts they had discovered from the data set provided, e.g. "*The richer the artists, the more followers and views on social media. The more studio albums, the more awards won. Twitter followers aren't dependent on studio albums produced.*" In this example, the conclusions show comparison between two variables, i.e., net worth of an artist x Twitter followers, number of studio albums produced x awards won, Twitter followers x number of studio albums produced.

Discussion

This study demonstrates that domain experts exhibit a very different approach to interacting with visualisation in comparison with non-experts. The non-experts showed much higher levels of interaction with the visualisation than the experts.

The non-experts were also more likely to Cluster information. Presumably, to increase one's level of understanding data must be explored (in what Yi et al. (2008) term 'overview'). We also noted that non-experts also performed more of what Yi et al. (2008) term Adjust, in terms of exploring the dataset. However, if the dataset as whole was new, then all patterns would be equally new and it is not obvious how this could either result in insight or novel interpretation (as opposed to simply understanding the nature of the data set). While it is not surprising that non-experts spent so much of their time engaging in Overview and Adjust, and that they spent more time engaged in Clustering data (presumably in order to support Overview), it is interesting that the experts spent so little time in these activities. Experts used Difference fact types, presumably because they had more contextualised mental models. Indeed, expert participants commented after the trial that they had enjoyed the task and felt that they had something learned new.

Experts were significantly less likely to change their opinions pre- and post- test. This finding echoes previous work by Yi et al. (2008) who noted that experts tended not to change their opinion after interaction with a visual analytic system. For the experts, performance was more about seeking contrasts in the data (in terms of Differences), which relates to Klein's (2013) notions of coincidence and connection, and this implies that they were working with their own hypotheses which they then sought to map on to the ACH, i.e., in a Match Mental Model approach according to Yi et al. (2008). If the Mental Model is not matched, then insight might arise when an established hypothesis might become challenged. In this case, rather than feeling close to solving a problem, perhaps 'insight' is more likely to be an unsettled feeling that a current line of enquiry does not feel right (this is what Klein (2013) referred to as contradiction or creative desperation). This sense of dissonance might be a very different experience to the more positive emotion which is often associated with the term 'insight' but it might reflect the manner in which domain experts could benefit from the use of Visual Analytics.

While the suggestion that use of Visual Analytics is a matter of aligning the presented data to the person's prior knowledge might seem self-evident, this study suggests that it is the interaction between person and the visualisation which supports this alignment. In other words, knowledge is not simply a matter of either passively responding to the content of the display or simply drawing on prior knowledge, but is, rather, the active construction of knowledge by participants during this task. In order to explore the hypotheses presented in the ACH, participants need to make sense of these hypotheses and decide what sort of information is needed to support or refute them. This could imply that participants would create a list of required information, which they then search for in their interactions (or which they populate by recalling known information). In either case, it would be necessary for the participants to know (a) what information to look for and (b) where to find this information. This would imply a query-driven search of the space of information. Visual Analytics, in contrast, works on the assumption that the required information is discovered during the process of interacting with the space of information. In other words, the user is assumed to have a weakly specified notion of required information and to browse until they find something which might be relevant. On finding potentially relevant information, the person would seek to adapt the visualisation, perhaps by selecting similar information or by restricting the available information to a single type. This would allow an hypothesis to be raised and tested.

What is important to this suggestion is that the 'solution' will be found neither in the person's head nor in the data per se, but rather, following the notion of Distributed Cognition, during the interaction between person and data (or between person and the visualisations of those data). If this is the case, then there are two consequences which would be interesting to follow. The first is that the presentation of the data (i.e., the manner in which they are visualised and the manner in which the user is able to adapt them) will serve to constrain the type of solution that the person can find (Zhang and Norman, 1994). This means that it might be possible for different users to arrive at different solutions because they are able to produce different visualisations. The question is how best to maintain consistency (across possible solutions) while allowing freedom to explore the data. The second is that prior knowledge will influence the definition of the hypothesis to be tested and the recognition of plausible solutions. This means that the manner in which the expert seeks information in the information space will be very different from that of the non-expert (in our study, this difference is significant). The question is how best to recognise and respond to the level of 'expertise' that the user brings to the interaction. Developing this point further, one can see how combining data, i.e., the 'Adjust' activity identified by Yi et al. (2008), could be seen in the light of Kirsh and Maglio's (1998) notion of 'epistemic action'. In other words, the bringing together of different pieces of information need not be driven by information search or a specific hypothesis, but could be a means of shaping the information space in order to make particular types of search or decision easier to perform (e.g., by removing distracting information, by reducing the information set, by combining sets of information that one might usually associate with each other).

References

Bowden, E.M., Jung-Beeman, M., Fleck, J. and Kounios, J. 2005. New Approaches to Demystifying Insight, *Trends in Cognitive Sciences, 9*, 322–328.

Card, S.K., Mackinlay, J.D. and Shneiderman, B. 1999. *Readings in Information Visualization: Using Vision to Think*, United States of America: Morgan Kaufmann.

Chang, R., Ziemkiewicz, C., Green, T.M. and Ribarsky, W. 2009. Defining Insight for Visual Analytics, *IEEE Computer Graphics and Applications, 29*, 4–17.

Chen, Y., Yang, J. and Ribarsky, W. 2009, Toward Effective Insight Management in Visual Analytics Systems, *IEEE Pacific Visualization Symposium.* 49–56.

Gotz, D., Zhou, M.X. 2008. Characterizing Users' Visual Analytic Activity for Insight Provenance, *IEEE Symposium on Visual Analytics Science and Technology,* 123–130.

Heuer, R.J. 1999. *Psychology of Intelligence Analysis*, Washington: Center for Study of Intelligence.

Keim, D.A., Kohlhammer, J., Mansmann, F., May, T., Wanner, F. 2010. Visual Analytics, in: Keim, D., Kohlhammer, J., Ellis, G., Mansmann, F. (Eds.), Mastering the Information Age: Solving Problems with Visual Analytics. Eurographics Association, Goslar, Germany, 7–16.

Kirsh, D. and Maglio, P. 1994. On distinguishing epistemic from pragmatic action, *Cognitive Science 18,* 513–549.

Klein, G. 2013. *Seeing What Other Don't: the remarkable ways we gain insights,* New York: PublicAffairs.

Pousman, Z., Stasko, J.T. and Mateas, M. 2007. Casual information visualization: Depictions of data in everyday life, *IEEE Transactions on Visualization and Computer Graphics, 13,* 1145–1152.

Saraiya, P., North, C., Lam, V. and Duca, K.A. (2006) An Insight-Based Longitudinal Study of Visual Analytics, *IEEE Tranactions on Visualization and Computer Graphics, 12*, 1511–1522.

Thomas, J. and Cook 2005. *Illuminating the Path: The Research and Development Agenda for Visual Analytics,* United States of America: National Visualization and Analytics Center.

Yi, J.S., Kang, Y. and Jacko, J.A. 2008. Understanding and Characterizing Insights: How Do People Gain Insights Using Information Visualization, *Proc. 2008 Conf. beyond Time and Errors (BELIV 08)*, 1–6.

SPECIFICATIONS FOR INNOVATION

Daniel P. Jenkins

DCA Design International, UK

Design specifications are a critical part of most development processes. Well constructed specifications describe constraints, focusing the development process and providing efficiency. However, by their very nature, they can constrain creativity by encouraging, and even forcing, designers to adopt current solutions rather innovating to meet the overall need. The way a development process is constrained is of key importance. Too few constraints, and efficiency and direction may be lost. Conversely, if there are too many constraints, opportunities for innovation may be missed. Using the example of a network connected thermostat, an approach is discussed that supports innovative design by describing, modelling and managing system constraints.

Introduction

Regardless if you are designing a train or a toothbrush, specifications can be used to provide explicit requirements. This often come in the form of physical dimensions (e.g. maximum size, location and size of interfacing parts), environmental factors (e.g. temperature and humidity to withstand), ergonomic factors (e.g. description of the target population), aesthetic or sensory factors (e.g. requirement to represent brand values), cost (e.g. material, manufacture, purchase), maintenance that will be needed, quality, and safety.

In regulated industries, such as transportation or medical devices, regulations and widely adopted standards often form the cornerstone of a design specification. Taking the example of door controls within trains, they prescribe acceptable locations (e.g. heights above floor level), arrangements (open above close) and actuation forces, along with specifications for reliability and robustness.

Detailed specifications are something that most engineers are very comfortable with, particularly in the later stages of the design process. When described appropriately, they provide measurable and testable requirements for an artefact – creating a clear description of what the product is, and how it should perform. Conversely, the prescriptive nature of a specification can also be viewed as stifling, particularly by designers in the early concept generation stages of a project. Tightly defined physical features and functions can limit the scope for lateral thinking. As such, it could reasonably be argued that design processes that are reliant on a product specification are better placed for evolutionary, as opposed to truly innovative products.

One approach used to mitigate the constraining nature of specifications is to discount, or dramatically restrict, the specification at the initial stages of the design process – delaying its introduction until after a series of concepts has been created. The classic example of this would be a 'concept car' that explores a new design direction without being overly concerned by details such as the construction techniques required, the material costs, or its impact on fuel performance. While this approach of ignoring constraints in the early stages of the process can have clear advantages for creativity, it can reinforce an 'over the wall culture' between design and engineering, playing up to role stereotypes. Furthermore, it could also be argued that an examination of constraints can direct creativity to develop new ways of managing these constraints.

Few would argue with the notion that detailed specifications can stifle creativity; however, their value to the design process, in terms of efficiency and focus, is also clearly evident. The natural question then, becomes; how can the system constraints be managed to ensure that the process retains the advantages of a prescriptive specification, without constraining the process of innovation? In short, as this paper will explain, it is contended that the consideration of constraints and a specification can, and should, exist throughout the design process. However, the level of detail, and means of presentation, should change to meet the requirement of the design activity at hand.

The role of specifications in teams

The importance of a design specification is particularly evident in projects with sizeable development teams and supply chains. A clear and auditable development process is of paramount importance and the design specification plays a critical role in this. 'Ownership' for specific requirements can be assigned to individuals, regardless of whether they are in the core project team or cascaded to the supply chain. Likewise, where staged development processes are adopted, structured test plans can be employed to ensure that the design is compliant before project gateways are passed.

The reductional nature of a design specification is one of its great strengths in allowing the roles and responsibilities to be shared. However, without some form of systemic oversight, there is the very real danger that components, or constituent parts, may be designed to be compliant to the identified sections of the design specification, however, they may fail to adequately meet the purpose or values of the system. This is particularly relevant in cases where purposes are not cascaded with the requirements.

Specification generation

Where large organisations are ostensibly producing variants of the same core product (e.g. automotive companies, white goods producers), a highly structured process is particularly valued. The development process can remain common and be honed and refined based on previous practical experience (it is no coincidence that process improvement techniques such as 'six-sigma' have been widely adopted in these industries). At the start of the project, a template can be used to form the

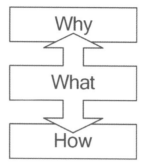

Figure 1.   Why-what-how triad.

base specification. Input from different roles in the company such as marketing, ergonomics and benchmarking teams can be used to set specific values within the specification. These can describe the target audience, along with their requirements. The resulting high-level of consistency between these specifications has clear advantages for working with ambitious time scales. The specification provides focus and direction across the wider team and substantially reduces the duplication of effort and excessive exploration. From a management perspective, this also has a number of advantages. According to Klein (2014), organisations value predictability because they like projects to run smoothly. Companies like to plan the steps that will take each project from start to finish, the resources for each step, and the schedule. That way, managers can quickly notice perturbations and make the necessary adjustments. The converse to this is that where time pressures are critical, it can be far quicker and less risky to develop a variant of a proven product than strive for true innovation.

Developing better specifications

At the risk of oversimplifying things, the design process can be captured by the relationship between the following three words.

The specification should sit central to this, it describes in detail exactly *what* the artefact, that is being developed, should do. As the development process progresses, the design team adds detail to the design to explain *how* each of these requirements can be met. What can be missing in this process is the explicit link up to *why* the requirement or even the product exists.

Based on the description thus far, there are a number of shortfalls with the classic design process that is designed around a specification. Many design specifications could be improved by:

1. Creating an explicit link that describes *why* a requirement is needed
2. Allowing specifications to be viewed at differing levels of abstraction (i.e. what should it achieve at a physical level, what impact should it have on the end users life)

3. Describing the inter-relationships between components in a system that influence how requirements are to be met.

One means of addressing these challenges is to adopt a systems representation called the abstraction hierarchy (Rasmussen, 1986; see Figure 2). As the title suggests, the technique and resulting diagram describes a system at a number of levels of abstraction (typically five). At each level, the model can be used to describe what the system should do at a different level of abstraction. Moreover, each of the nodes in the model can be used to explore the 'what-why-how' triad introduced in Figure 1. A given node answers the question of what is required, while the linked nodes below describe how the system can achieve this, whereas the linked nodes above can be used to answer the question why. These upwards links are particularly valuable as they provide a rationale for the system. Where there are multiple connections from a single node, the diagram also describes the interrelationships between components and how the same affordances can be achieved by different physical objects.

Example

The utility of the abstraction hierarchy is perhaps best explained with an example – in this case a network-connected thermostat. While the idea behind the internet of things (IoT) has been around for quite some time, it is recently receiving unprecedented levels of interest. Ostensibly at least, the IoT involves creating a connection with everyday house hold objects (e.g. heating systems, locks, doors and windows, fans, lights) to allow them to be controlled remotely or automatically in response to events (e.g. a time of day, a change in environmental conditions, a message from a human). The Nest learning thermostat is used by many as the de facto standard example when describing the IoT and connected devices. Due to its familiarity, it has been adopted here for the purpose of explaining the approach.

The physical functions of a product are described at the base of the abstraction hierarchy, while at the top, it purpose, or why it exists is captured (see Figure 2). The top row of the diagram, the domain purpose, describes the overall purpose of the product, in this case to reduce energy consumption while maximising the user experience. Unlike goals, these objectives do not change with time or as a result of different events, but remain fixed. The level below, domain values, provides high-order measures of performance used to determine whether or not the functional purposes are being achieved. In this case, to maximise thermal comfort and convenience, while minimising environmental impact and energy usage.

Moving to the very bottom of the hierarchy, the physical objects are described. Within the thermostat itself these include component such as temperature sensors, display screens, rotary dials, etc. In addition, other components in the wider system are also included such as smart phones and apps they may be required to interface with the product. Other artefacts that may have overlapping affordances can also be captured (such as smart meters and utility bills). The functions of each of these objects are described above in the physical functions row above. These physical

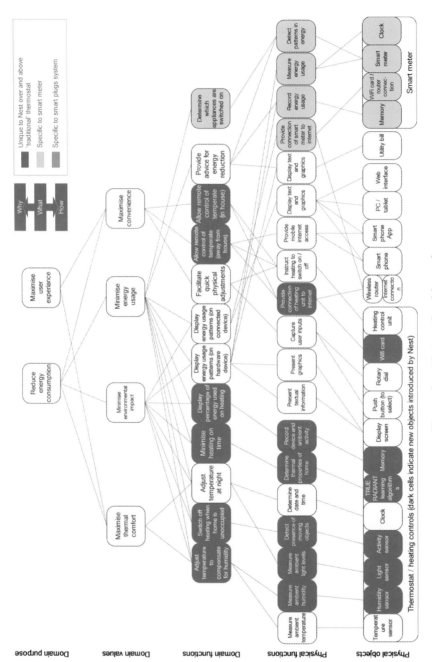

Figure 2. Abstraction hierarchy.

functions are the affordances or functions of the physical objects (what they do), independently of the purpose of the system. For example rotary dials capture user inputs, as opposed to capturing requests to change temperature. Describing function in this more generic way encourages the analyst to consider how functions and physical object can be used in different ways.

The row in the centre of Figure 2, the domain functions, links the diagram together. These are the functions that need to take place to meet the purpose of the system. For example 'switch off heating when home is unoccupied' or 'adjust temperature to compensate for humidity'.

Together the diagram creates an explicit link between the functions of the physical object at the base of the hierarchy and the user values and purposes at the top. The complexity of the linking provides an indication of how these different needs can be met by a combination of physical objects. The diagram has been coded to high-light what the NEST adds over and above a traditional heating system at different levels of abstraction. As such, the impact of the introduction of new components (physical objects) can be considered. For example, the introduction of humidity sensor allows ambient humidity to be measured, this allows the temperature to be adjusted to compensate for humidity which, in turn, maximises thermal comfort while minimising energy usage and environmental impact. This has the potential to positively impact both purposes of reducing energy consumption and maximising user experience. Explicitly considering these high-order impacts can help to reduce 'function-creep' where components are added as a result of technical ease rather than a user requirements.

At the base level of the hierarchy, each link can be taken in turn and questioned to establish if there are other components already in the systems that can perform the same functions. Similarly, other functions or affordances of the existing components can be considered to see if they have the ability to have a positive impact on the overall system goals. The boundaries of the system can also be expanded to consider other products and components and how functions can be shared between these. This is of particular interest in the IoT where numerous devices are to be connected.

The abstraction hierarchy provides a description of the system constraints at a high level. Furthermore, it encourages the analyst to consider the relationship between physical objects in the systems and the high-order purpose of the system. Accordingly, it lacks the detail of a traditional product specification. In most cases the design process would start with the abstraction hierarchy while the overall system architecture is being decided. The approach described is in no way intended to replace the standard specification, rather it aim to support it. Once a system architecture is agreed the abstraction hierarchy would inform the development of a more detailed design specification. This would include all of the traditional constraints such as relevant legislation, cost, size, safety, context restraints, and time. Explicit links can be made between these two representations (the abstraction hierarchy and the specification) allowing a clear audit trail and also allowing the specification to be updated in line with changes to the abstraction hierarchy should constraints or assumptions be modified.

Discussion

The challenge of generating a connected device can, of course, be viewed solely at a physical level by augmenting the current thermostat system with a means of communication. An existing detailed design specification from a legacy product could simply be modified to add a new requirement for connectivity. Accordingly, this connectivity can be viewed as a 'bolt-on' function – and the resultant focus would be on the decision of which type of communication protocol to adopt (e.g. WiFi or low energy radio communications) to reduce cost and increase reliability. Indeed an approach similar to this is likely to have been adopted by a number of manufacturers prior to the release of the Nest and deployed with varying level of success and adoption.

It is perhaps interesting though, that it was a small independent, albeit well-funded, company that designed and developed the product that is now viewed as the most innovative product. It was the Nest learning Thermostat, not a product from one of the large organisations that decided to move beyond a technical innovation, to one that seeks to actively engage the end user to meet the higher-order domain values. A number of the incumbent companies subsequently produced competitors to the Nest, offering products with similar components and functions at a physical level. However, by this point, benchmarking the competition would have allowed them to develop a detailed specification of what these products should do and the individual requirements for each of the components.

It would, of course, be naïve to think that the reason this level of innovation came from a start-up, and not a large organisation, was solely down to large organisations focusing on the physical and functional level of a project – brought about by over-reliance on design specifications. However, what is clear is that such a reliance on a 'traditional' product development process is not embracing opportunities for innovation.

The abstraction hierarchy is, again of course, not unique – there are a wide range of tools and techniques that encourage the design team to focus of user and stakeholder needs and seek to find innovative solutions to identified markets and problems (it is not clear what approach was used to develop the Nest product). Likewise, the approach described in the paper is certainly not proposed as a silver-bullet allowing a perfect balance to be struck between project efficiency and innovation. However, the described approach is proposed to help inform these tradeoffs, to encourage debate early on in the project of the purpose of the system.

References

Klein. G. 2014. No, your organisation really won't innovate'. Wired. Accessed 05/09/2014 http://www.wired.co.uk/magazine/archive/2014/05/ideas-bank/gary-klein

Rasmussen, J. 1986. Information processing and human-machine interaction: An approach to cognitive engineering. Amsterdam, The Netherlands: North-Holland.

ERGONOMIC GEAR KNOBS: A CASE STUDY IN TEACHING PLEASURE AND ATTACHMENT TO DESIGN STUDENTS

Elaine Mackie & Louise Moody

Department of Industrial Design, Coventry University

This paper presents a reflective case study of a student learning experience in design and ergonomics that focuses on user-centred research methods. The project involved the design of novel gear knobs by final year design students. This article discusses the range of methods taught to design students, with a particular focus on emotional design and product attachment. The case study considers how the students selected and construed these methods when exploring the brief, and the impact on their ability to apply emotional elements to their design work. The paper concludes by offering a model indicating how to link user-centred research methods, emotional design and product attachment.

Introduction

Teaching ergonomics to undergraduate design students can be challenging particularly imparting qualities such as empathy, and teaching approaches such as emotional design (Moody et al. 2011). It involves introducing subject material and methods which are essential to design, but not necessarily recognised by the student as a core requirement when their focus is on aesthetics, styling, functionality and technology (Oak 2000). There are numerous design-orientated user research methods that can be employed by designers to understand their users (Dong et al. 2008). Bruseberg and McDonagh-Philip (2000) argue however, that designers tend to rely on a more intuitive approach rather than formal methods. Within an educational context, students should be equipped with the knowledge of formal methods and how they can be employed, even if once experienced they source these skills from elsewhere, and rely more heavily on judgment and intuition.

One of the areas in which we have sought to develop a user-centred focus for students is in emotional design and attachment (Savaş 2004). The field of emotional design advocates that pleasure and emotion are key components in establishing user interaction and subsequent attachment (Jordan 2000). An understanding of the connection between aesthetics and emotional responses requires a focus on the cognitive and psychological processes involved in consumers' responses (Desmet 2007). Emotional design recognises the need to go beyond usability in order to understand customer satisfaction, with aesthetic quality also being essential to providing pleasure to the user (Norman 2004).

The Industrial Design course curriculum at Coventry University aims to teach pleasure principles and attachment within a user-centred design approach. This paper describes how these concepts have been taught and applied through the delivery of a design brief focused on the gear knob. The need to increase the perceived pleasure was recognised by Fiat through the Sensorial Quality Assessment (SEQUAM) (Bonapace 2002). The work investigated the impression of a car's 'feel' and initial perceived quality including the gear shift. Further automotive research has explored the level of attachment to cars and evaluated subjectively the pleasure offered by the interiors (Schifferstein & Zwartkruis-Pelgrim 2008, Karlsson et al. 2003, Benson et al. 2007).

The gear shift is one of the primary driving controls (Hultman and Larson 2005). Brands vary in the feel of the gear shift – some are stiff while others are more focused on comfort. Whilst new technologies (such as iDrive, Active Steering, Active Cruise Control and Dynamic Drive) have transformed our interactions, research has suggested users experience a sense of loss and uneasiness with the way new technology has changed gear changing (Benson et al. 2007). It is therefore, still an important part of the driving experience and something that our future designers need to consider. Here, a design brief for gear knobs is used as a way to explore the application of, and receptiveness to, user-centred, emotional design research methods by design students.

Methodology

This case study relates to a 5 month long teaching and learning design activity that ran during the second term of the academic year.

Participants

13 (12 male; 1 female) final year product design students from Coventry University took part in this design case study. This represents approximately just under half (43%) of the final year undergraduate product design cohort. They were UK students and qualified drivers with different levels of driving experience. They were a self-selected group, who chose to engage with this brief out of a choice of two other projects.

Procedure

Teaching and learning: A blended learning approach adopting a combination of both traditional and e-learning activities (Kahiigi 2008) was adopted. Face-to-face tutorial sessions once a week were supported by a digital repository of links to associated websites and anthropometry software. The user-centred methods taught were based on those identified in the SEQUAM and BMW study (Mao et al. 2005) and included benchmarking, observation, task analysis, questionnaires, prototyping and user trials involving interviews, focus groups and rating scales.

User-centred design exercise: The groups were expected to apply their learning on user-centred design methods to a design brief on emotional attachment. They were to use their research to generate design requirements regarding emotional attachment and address these through sketch ideas, concepts, technical drawings and 3D models. The students worked in 4 self-selected groups of 3–4 students in conjunction with an outside company. Each group (labelled A-D) had a different brief based on pleasure and attachment:

- Group A – a gear stick which connects a driver to the power of the engine to make the driving experience more exciting;
- Group B – a gear knob individual to the user and car with possibly a unique functional characteristic;
- Group C – a gear stick which evokes a sensual brand orientated pleasure;
- Group D – a gear stick which has an organic theme and flowing aesthetic.

The brief sought to address the themes of illumination, interior coherence, gear knob details, materials and production. The students were required to demonstrate all of the themes in their work, as well as considering pleasure and attachment. They were encouraged to document their working methods and user interaction through photographs and comments and publish via blogs or in design journals. Summative assessment of the work took place through a 15 minute interim 'work in progress' and final PowerPoint presentation of group work detailing the finished gear knob design.

Data collection and analysis

The design process and output was observed and assessed by two experienced tutors in design and ergonomics. The findings are based on their observations of the data collected by the students, tutorial discussions, the journal/blog entries, the design output, client feedback and the assessed presentations. The reflections are based on their perceptions of the students' willingness and approach to engaging with user-centred methods.

Results

The design work that emerged from the projects suggested that all of the groups considered the physical issues of their designs with their themes of emotional attachment. However, the consistency and quality of the application of user-centred research to the design process varied. Some variation is likely given the differing briefs, and the client's needs and interactions with them; however it is also argued that it is in part due to student willingness and ability to engage with certain methods. The students chose to use differing combinations of user-centred methods to inform their design.

Benchmarking

Two groups employed benchmarking. Group C tested 12 gear knobs that had been supplied by the client to assess ergonomic and comfort issues e.g. feel and grip, as well as aesthetics, materials maintenance and technology. They captured their responses by individually rating each of the designs from 0–10 based upon personal preferences and experiences. Those judged most comfortable were spherical in shape; but those judged most aesthetically pleasing were more of the appearance of an upturned hybrid (rounded head) golf club. Group D tested some of the client-supplied gear knobs as well as designs that the group had originated and modelled. They used radar diagrams to see how the gear knob designs performed in terms of comfort and fit, form and the likelihood of purchase. To prevent bias due to the 'finished' appearance of the client's gear knobs, all the products tested were sprayed with a matt grey paint. The rating scale categories chosen were based on research into the user market and the likelihood of purchase statistics of cars amongst UK adults. The group concluded that the client gear knob designs whilst comfortable were not of the form to encourage a likely purchase, whereas their own designs scored higher in design form language, but lower in comfort and fit.

Participant observation

Observation was chosen by 3 out of 4 groups. Group A used direct observation through trips to 4 different dealerships and a retail outlet to look at aftermarket gear knobs. Group B observed 5 users with different hand sizes changing gear. The interaction and various hand positions adopted at rest and pushing the gear knob forward and back through the gears were photographed. It was found that up to 6 different hand positions were employed to change gear, indicating a need to accommodate a variety of individual palmar grasp styles. The need for size optimisation was indicated through observations about the spacing of the fingers and a correlation with hand size. Group C simulated the gear changing task itself and observed a total of 12 users from different cultural backgrounds (6 male, 6 female) from which palm location diagrams were generated in order to conclude that males preferred to centre their palm on top of the gear stick whereas females palm positions were more variable.

Task analysis

Conducting a task analysis was recommended to all groups as it provides a structure for the description of the driving task and aids understanding of how users undertake different associated activities. Only Group C embraced this technique and used it to 'understand what a user interacts with or tasks they perform before switching on the engine of a stationary automobile and moving off.' Tasks were described in detail in terms of dynamic postures e.g. 'place hand on gear stick, move gear stick to first gear' as opposed to just 'change gear.' The students struggled with using it to create a meaningful evaluation of the driving task in terms of facilitating a change of gear, or how appropriate the design might be to the target market, or driving safety for example.

Questionnaires

Two groups designed and administered a questionnaire. Group B used a question-naire to capture the results of interaction with their mock-up rig of the drivers seating position and their proposed gear knob design and gearing action. The questionnaire elicited responses associated with visual appreciation, size and comfort of the gear knob, as well as capturing user observations about a unique turning feature. Group D rated the gear sticks produced by each group member and the client through a user trial and questionnaire. They determined which of the group member's designs was rated as the most comfortable and popular in terms of 'fitting' participant hand sizes.

Use of physical prototypes and user trials

All 4 groups enthusiastically developed prototypes and utilized static test rigs. The test rigs aimed to replicate relevant aspects of the relationship between the drivers' seating position and the location of the gear stick according to drivers' reach into specific zones. Left and right handed individuals participated to reflect different markets.

Conceptual prototypes were produced by all groups in the form of sketches, concept renderings and 3D appearance models. Concept renderings were often translated into Photoshop to better represent the materials and features of the design. Some student groups tried using automotive clay to conceptualise initial form ideas but this proved to be a poor medium for modelling smaller scale artefacts as it can be hard to control the surface finish.

Group A used polystyrene spheres and made a gated gear change mechanism for user testing to see how people held and used gear sticks. Participants hands were painted so that the surface of the sphere would be marked where contact was made with the fingers; colour being used to differentiate between the three users' grasp locations. This group also explored the aesthetic and tactile possibilities offered by use of negative spaces, creating grooves or gaps across parts of the surface to an interesting and pleasing visual effect.

Group B modelled their designs in blue foam before using Computer Aided Design (CAD) to realise the detail and to provide durable gear knobs for user trials. CAD files were created using Solidworks where full scale milled prototypes were made from polyurethane. Polyurethane was chosen to emulate the feel of existing gear knobs, and could be painted so that it was dimensionally stable in production and durable during user trials.

Form (in terms of physical elements such as dimensions, colour and aesthetics) was found to be analysed by all groups thoroughly and applied through prototypes and/or mock-ups. In terms of surface interaction with the products, some students used relatively simple methods to establish where contact was made by the users' hands on the gear knob. This interaction was appropriate for a focus on the quality of the touched experience as well as with factors relating to surface material and 'feel'.

Some of the designs presented included significant negative spaces formed as recesses within the enveloping exterior form, unusual in automotive gear knobs and more often found in devices such as games consoles, special purpose tools, or pistol-grips. This demonstrated that the students were taking into consideration the posture of the hand and associated potential pressure points.

Influencing design

Overall it was felt that although the research methods had been undertaken competently, the students did not always analyse their findings thoroughly against the emotions they wanted to embed within their design work. The design work was enthusiastically engaged in, but there were difficulties translating the findings into clear elements of attachment and emotion in the design output. All of the student groups failed to some extent to recognise the complexity of people and situations with their conclusions being very basic consisting of how the gear knob felt as a static object rather than appreciating its form and functionality.

Discussion

This case study involved student designers employing a range of different user-centred design methods with the aim of producing a design that better met the emotional and pleasure-driven needs of the users. It has been used as a way to explore the application of, and receptiveness to, user-centred, emotional design research methods by final year product design students, who are close to entering industry.

Two groups used questionnaires to gather data to some effect. Our broader experience with questionnaires suggests that students tend to focus on the number of questionnaires collected, rather than the quality of the questions and data collected. Therefore, where possible we encourage the use of interviews and focus groups so that user views and experiences can be explored in more depth. Despite the taught content, neither interviews nor focus groups were used. We suspect there is reluctance to commit to the organisation and time commitment that can be required to recruit, collect and analyse the data. Furthermore there is a tendency where possible to recruit participants internally to the University cohort, rather than reaching more varied participants. One of the four groups used task analysis. It is argued that the structured nature of task analysis might present a challenge for some design students and conflict with their natural creative approach.

The flexible approach adopted by the students in selecting and tailoring their methods to their projects supports research suggesting designers tend to employ methods in a fluid and intuitive way (Bruseberg and McDonagh-Philip 2000). Our teaching therefore, needs to recognise this and facilitate a more in depth understanding and application to ensure that students understand the important role of ergonomics and how it can lead to stronger product attachment. As the prototyping and creative activities were more enthusiastically engaged in by all groups, there is a need to further explore the inter-relationship between user-centred research methods, emotion and product attachment and help students to understand when and how to employ the range of available methods.

In order to address this issue the matrix in Figure 1 was developed. It seeks to map the research methods used by the automotive industry (and identified by SEQUAM), to the reasons for attachment (Savaş 2004), and the resulting positive and negative emotions (Desmet & Hekkert 2007).

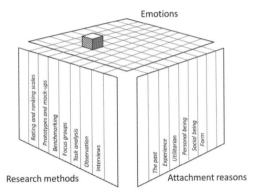

Figure 1.   Emotional design research methods matrix.

The matrix can be used as a visual model to aid students better understand possible emotional outcomes achieved through design. The visual structure offers a diagrammatic mapping of emotions to attachment reasons. The student would select a research method from the list based on their knowledge, time and resources. The type of attachment is then selected. This leads the user to a number of emotional outcomes. For example, suppose benchmarking has been chosen as the research method; the matrix could be used to capture 'positive' (e.g. joyfulness) (above the emotional plane of the matrix) and 'negative' (e.g. disgust) emotions (below the emotional plane) that might have been generated by users describing their 'experience' (attachment reason) with the range of gear sticks tested. As emotions are discussed with users, it encourages students to engage in further interrogation such as preferred hand positioning/surface location on the gear stick versus emotional experience ratings. The visual nature of the matrix should help explain the value and interrelationship between variables to design students. It is also anticipated that the matrix could support inter-disciplinary design teams by facilitating the communication of human factors requirements to engineers. The definition of user experience in both dimensional and form terms e.g. area of hand in surface contact, as well as consideration of the surface adherence/material on the user's perception may prove helpful to build a shared understanding of emotional requirements.

Conclusion

This paper has detailed a case study exploring the teaching and application of user-centred methods to elicit and apply emotional design principles in the development of a gear knob. It has been found that at final year level, design students will engage in research methods but struggle to apply the findings to the development of products that will have a strong emotional appeal. A matrix has therefore been developed that maps the research methods to attachment reason and emotions. In future work the effectiveness of the matrix in guiding method selection and application will be tested.

Acknowledgments

We would like to thank the final year Product Design students at Coventry University and Peter Atkinson who co-tutored this cohort.

References

Benson, R., MacRury, I. & Marsh, P. (2007) *The secret lives of cars and what they reveal about us*, London, BMW (UK) Ltd.

Bonapace, L. (2002) Chapter 15 – SEQUAM. In WS. Green PW. Jordan, *Pleasure with products: Beyond Usability*. London: Taylor & Francis. 189–217.

Bruseberg, A. & McDonagh-Philip, D. (2000) *User-centred Design Research Methods: The Designer's Perspective In*: Integrating Design Education Beyond 2000 Conference 4–6 University of Sussex, Brighton.

Desmet, P.M.A. & Hekkert, P. (2007) Framework of product experience. *International Journal of Design*, **1**(1), 57–66.

Dong, H. (2008) *Designer Orientated user research methods*. In Coleman, R., et al. Design for Inclusivity. pp. 131–147. Aldershot: Gower.

Hultman, L. & Larson, S. (2005) *Development of a Method for Subjective Expert Evaluation of the Human Driving Geometry*, MSc. Luleå University of Technology.

Jordan,P.W. (2000) *Designing Pleasurable Products*, New York: Taylor & Francis.

Kahiigi, E.K. Ekenberg, L., Hansson, H., Tusubira, F.R. & Danielson, M. (2008) Exploring the e-Learning State of Art, *The Electronic Journal of e-Learning*, **6**(2), 77–88.

Mao, J. Vredenburg, K. Smith, P. W. & Carey, T. (2005) *The State of User-Centred Design Practice*, Communications of ACM, **48**(3), 105–109.

Moody, L., Mackie, E. & Davies, S. (2011) 'Building Empathy with the User' in Karwowski W. Soares M. & Stanton N.A. (ed.) *Handbook of Human Factors and Ergonomics in Consumer Product Design*. CRC Press.

Norman, D. (2004) *Emotional Design: Why We Love (Or Hate) Everyday Things*. New York: Basic Books.

Oak, A. (2000) 'It's a nice idea, but its not actually real; Assessing the objects and activities of Design.' *Journal of Art and Design Education*, **19**(1), 86–95.

Savaş, Ö. (2004) A perspective on the person-product relationship: Attachment and detachment. In D. McDonagh et al. (eds) *Design & Emotion: The experience of everyday things*, 317–321. London: Taylor and Francis.

Schifferstein, H.N.J. & Zwartkruis-Pelgrim, E.P.H. (2008) Consumer product attachment: Measurement and design implications, *International Journal of Design*, **2**(3), 1–13.

THE ROLE OF CO-DESIGN IN WEARABLES ADOPTION

Sara Nevay & Christopher S.C. Lim

University of Dundee
BESiDE Research

Ageing and increases in longevity have become increasingly impact-ful on requirements for our built environments and designers need to create homes and spaces that support our changing needs. In the BESiDE project, care home residents were invited to use wear-able technology to explore how those living in care homes currently utilise their spaces. Though effective data collection tools, wearables designed specifically for older adults often fail to engage their wear-ers. This paper details two co-design workshops with older adults to identify design requirements for wearables that meet their spe-cific needs and capabilities. Our findings suggest that successful wearables prioritise comfort, utilise familiar materials and facilitate independent use.

Introduction

Coupling wearable objects with technology can be demanding when designing for older adults' health as there may be a tendency to trade aesthetic appeal for func-tionality (Newell et al. 2011). Dissatisfaction and abandonment of wearables can relate to weight and size as well as perceived social acceptability and whether it will attract unwanted attention (Hocking 1999). Indeed, as Dykes et al. state, "(wear-ables) need to avoid stigmas through medical styling and instead enforce a positive and familiar identity through the use of associated materials" (Dykes et al. 2013). Customisation, therefore, can be key to ensuring wearability, not only in terms of taste and style but also with regards to fit and comfort. Through a set of ongoing craft and co-design workshops, the researchers seek to achieve empathy with older users and their needs, and define design requirements for original wearables that are desirable, useful and usable.

Context and related work

The work reported in this paper contribute to BESiDE (Built Environment for Social Inclusion through the Digital Economy), a project from the University of Dundee that aims to inform the design of better, more enabling built care environments to promote greater mobility, physical activity and social connectedness and thus, wellbeing. By utilising wearable technology the project will determine how well older adults living in care homes currently navigate and utilise their spaces.

The adoption of wearables is an issue that our project acknowledges. Other studies, such as that conducted by, Endeavour Partners, a US based mobile and digital technologies think tank, reveals that users will quickly abandon devices that do not support them in making positive changes (Ledger & McCaffrey 2014). Furthermore, "adoption will ultimately be gated by the aesthetics and comfort of smart wearable devices". Their study states that a design-forward rather than a technology driven aesthetic may enable greater engagement with users; citing Withings' Activité watch (an activity and sleep monitoring device fashioned like a gentleman's standard watch) as a great example of the industry finding innovative user centric ways to 'disguise' technology. This approach to familiarise technology for older users has also been adopted by Care Predict, a US company specializing in assistive technology for older adults, who are developing wearable tracking systems for older adults using beaded bracelets as stylish base designs in response to users' design drawings (Rojahn 2014).

Methodology

In order to understand the users and their needs the researchers conducted empathic design and ideation workshops in conjunction with testing and the review of existing and emerging technology (Figure 1).

The empathic workshops were developed using a co-design methodology grounded in the understanding of real-life experiences, ideas and skills of the people who use need services (Szebeko & Tan 2010). Six participants aged 65 years or older with

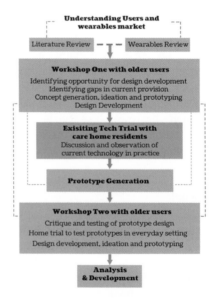

Figure 1.   Process map.

experience of care homes (e.g. as visitors or staff) were recruited via a project User Pool to pilot design methods and activities prior to working with potentially vulnerable residents. This comprised of two linked workshops; the first to ideate and produce an initial set of prototype wearables and the second engaged participants in a home trial to evaluate the prototypes.

The researchers also introduced GENEActiv watches to a small group of care home residents to test physical activity data collection and explore users' perception of current wearable technologies. The devices were specifically recommended by project colleagues – who are experts in gerontology and HCI-based upon their other recent and relevant activity studies.

Results

These workshops were held with older adults who have interest and insight into care home life. The aim of these sessions was to explore their perceptions as to apt wearables for residents and inform early prototype designs. With the aid of a visual card deck and mood boards, participants outlined their initial ideas for suitable wearables and critiqued physical mock-ups of their ideas, which were constructed by the designers in response to participant's ideas and discussion.

Workshop one

Participants were asked to sort through a bespoke card deck of wearables divided into four suits (Figure 2); Wearable Objects; Fastenings; Aesthetics; and Position of Wear. Some objects were everyday items and others more extraordinary, for

Figure 2. Card sorting in support of ideation.

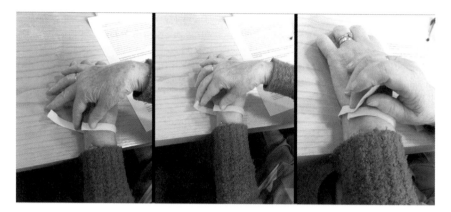

Figure 3.   Participant demonstrates difficulty in fastening narrow band.

instance, a t-shirt with an iPad incorporated into its front. The card sorting exercise enabled instant understanding of user preferences and acted as prompts for ideas that participants could note on blank cards provided for each suit. Participants commented that the incorporation of blank cards was useful to capture any 'missing' items or design components; one suggestion was a loose necklace with magnetic fasteners and a fabric pendant that could house a sensor. Participants also told us that wrist worn wearables would be less intrusive to manage and maintain and that familiar wearable items such as gilets would be 'less confusing' and promote comfort (both physically and emotionally) as the incorporation of sensor technology would be more subtle within this larger garment and not look 'out of place' when worn with other everyday clothing.

Participants identified appropriate forms, style and fastenings as those that facilitated independent use and simple physical prototypes were made in response. For example, Velcro should be used as this is simple to fix in place for older adults whose dexterity may be diminished. Being neutral and simple in form, the prototypes invited participants to freely manipulate them, imagine different capabilities and dismiss less suitable items without being precious. One participant commented whilst trying on an elasticated wristband, "This is just too fiddly. A wider strap would be better – just like putting a watch on" (Figure 3).

Favoured designs at this stage were wrist warmers with pockets as these could be put on/taken off independently with ease, are lightweight and could be worn underneath long sleeves (Figure 4). Participants' design suggestions included; Velcro sides should be contrasting colours to clearly indicate the position of the fastening; and bands should be wide as these are easier to use independently. Future designs should prioritise function and comfort over aesthetics and should be of acceptable weight and size, "if it's the size of a mobile phone (that's) too big and heavy and the people (residents) aren't likely to use them." Smaller, lighter objects – such as watch faces – are preferable as they are less obtrusive.

Figure 4.   Lightweight, pull-on wristband prototype.

Trial of existing sensor technology wearable

In conjunction with this workshop activity three local care home residents were asked to wear a GENEActiv for five consecutive days during waking hours. Residents volunteered as a result of informal visits to introduce the research. Observations and discussions with participants were conducted every second day by a researcher which revealed user experiences that are subjective relevant to comfort, functional conditions and social acceptability. One resident felt that the device "is a nuisance, too bulky and uncomfortable" and is difficult to wear along with her own watch. Another resident felt that the watch made her 'stand out', due to its' medical appearance and invited negative comments from fellow residents who asked "What's wrong with you?" These experiences enforced the researchers' intention to ideate original wearables for older adults with and for their users; a co-design focus – rather than a technology focus – driven by the users' needs aims to create useful, usable wearables and positive experiences.

Workshop two

Six User Pool participants were presented with 3 styles of wristbands further developed in response to the earlier workshop. Small sensors (43 mm × 40 mm × 13 mm) were incorporated into the designs at this stage via interior pockets though they were not charged or actively gathering data throughout the study – their presence was to solely test the effective function and comfort of the designs. Soft textiles and high performance fabrics were used to construct these wearables. The textile surfaces of the wearables were selected not only to take into account any diminished tactile sensitivity of older users but also facilitate good levels of hygiene, dryness and comfort. Each fabric was chosen to meet different factors indicative of comfort (Song 2011); Neoprene is thick, insulating, water repellent and a firm fit against the skin; Athletic Mesh is breathable, lightweight and soft; Bonded Mesh and Fleece offers a water repellent outer surface and a soft inner. All designs were plain in appearance and constructed without decorative embellishments as to be critiqued purely on comfort, fabric choice and effective function. Participants offered initial critique before choosing a preferred style to trial at home for 5 days; three chose the Athletic Mesh wristband as it looked 'least intrusive', soft and easy to put on and take off; two participants chose the Bonded Mesh and Fleece band as it fitted well in the initial session; and one chose the Neoprene band for its' bright colour.

Diaries were provided to record their experiences of wearing the prototypes and at the close of this 5 day home trial the participants returned to report on their wrist pieces in terms of comfort, ease of use and aesthetics whilst undertaking daily tasks and social activities.

Neoprene Wrist Band: The sole participant who chose to trial this wrist piece explained that she had done so because the colour "was fun" (a bright light blue) and that it was comfortable but that her experience hadn't been very positive otherwise; the neoprene, whilst supportive and ensured the correct placement of the sensor whilst worn, stretched too greatly when being pulled on and off which caused the sensor to fall or move out of position. The thickness of fabric required finger strength and dexterity to pull on and off independently and the synthetic foam structure of the fabric caused her wrist to perspire overly during wear.

Athletic Mesh Wrist Piece: All three participants who chose to wear this wearable at home described it as very easy to use (to pull on and off) and 'comfortable' throughout the day, at points "(they) forgot it was there". Two participants forgot to remove the wrist piece at the end of the day, as per the trial requirements but reported no resulting discomfort or improper function (all sensors remained in place). One participant reported that their spouse had expressed distaste at the appearance of the wrist piece as it reminded her of a medical bandage and wondered if it could be made "less medical in its' look". All reported that the fabric lost its' elasticity (and therefore compromised the positioning of the sensor) as the trial went on with one participant altering his to fit better on the third day by stitching the cuff with a fine wire thread. Another participant choose to 'roll up' her wrist piece so that it became half its original length and to accommodate for the lack of tension/stretch. These alterations, though indicative of the prototype's failings, infer that the participants felt a great degree of ownership in the design process, feeling free and being able to make adjustments to the designs as they saw fit. According to Fischer, (2002) when users participate in effective co-design, they flip between the roles of user and designer. The adaptations the participants made to their wristbands is indication of a successful co-design experience.

Bonded Mesh and Fleece Wrist Piece: Both participants expressed that the wrist pieces held the sensor in place effectively and that they enjoyed that the fit of the fabric was secure against their skin and underneath sleeves. However, the Velcro, covering the length of the wristband was, at times, sharp against the skin and could benefit from being shorter or having rounded edges. One participant also described that the comfort and fit of the design could be improved by altering the shape to be conical as opposed to a uniformly sized tube (Figure 5); his experience had shown that the rectangular tube shape could "go lumpy" and be restrictive whilst engaged in certain physical activity (in his case playing bowls). Indeed, as a result of the pressure of the fit whilst playing bowls and suffering from fragile capillaries,

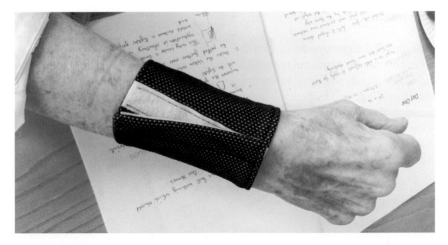

Figure 5.   Participant demonstrates altered position of Velcro.

the participant resulted in a bruise beneath the placement of the sensor. Design improvement must be made to minimise potential for this, especially in the case of future wearers who also bruise easily. As the participant states "Although it is not painful, this may prove very off putting for wearing again."

Discussion

From the workshops we learned from the group that the neoprene wrist piece seemed least suitable for future development and that the mesh wrist piece had most potential if its appearance could be improved upon along with retention of elasticity. The appearance of a future wearable should be "acceptable" but "not dull" and that further consideration of both age and non age related changes in capabilities should be explored further to ensure comfort, effective wear and fit. Although there are ergonomic guidelines available, (Smith, Norris & Peebles, 2000), it is through testing designs with users that issues such as fastening, tactility and elasticity are raised. A successful future wearable should also have the capacity to adapt to suit temporary physical changes; one participant had received wrist surgery some years earlier but dependent upon weather or health on a particular day, may still experience changes in the size or shape of her surgery wound. Wearable artefacts need to be appropriate within a care home setting; carriers must be easy to put on take off, engaging – to promote use and preserve dignity, comfortable to wear and facilitate good data capture.

We have learned that co-design approaches and methodologies may enable older adults to better engage with wearable technologies; shared activity, discussion and ideation, have generated motivations for wear as well as a sense of ownership for those involved in directing the design process, experiencing the wearing of prototypes and adapting the designs in a live development setting.

Conclusion

These initial design suggestions and requirements have set the scene for the discovery of not only the needs but the nuances of the intended user group. The card deck and simple prototypes, acted as prompts to explore design opportunities for potential wearables. Triggered by both the images and the tactile objects, participants shared and applied their personal understanding and experience. This inclusive experience of designing and developing enabled participants to feel a sense of ownership of the objects they were creating and adapting and the emerging prototypes have begun to address older users' needs in a design process driven by older user's priorities.

References

Dykes, T., Wallace, J., & Regan, T. (2013) Interactive Teaware: Sharing Experiences in Old Age. In *Proceedings of Praxis and Poetics, Research Through Design*, Newcastle upon Tyne and Gateshead, UK, Sept 3–5, 2013, 75–78.

Fischer, G. (2002) Beyond "Couch Potatoes": From Consumers to Designers and Active Contributors. *First Monday*, **7**(12) http://firstmonday.org.issues7_12/fischer/index.html

Hocking, C. (1999) "Function or feelings: factors in abandonment of assistive devices" *Technology and Disability* **11**, 3–11.

Ledger, D., & McCaffrey, D. (2014) *Inside Wearables: How the science of human behaviour change offers the secret to long-term engagement.* http://endeavourpartners.net/assets/Wearables-and-the-Science-of-Human-Behavior-Change-EP4.pdf

Newell, A., F., Gregor, P., Morgan, M., Pullin, G., & Macaulay, C. (2011) User-Sensitive Inclusive Design. *Universal Access in the Information Society* 10(3), pp. 235–243.

Rojahn, S. (2014) *An Activity Tracker for Seniors. MIT Technology Review*, URL: http://www.technologyreview.com/news/525016/an-activity-tracker-for-seniors/ Accessed 25.09.14.

Smith, S., Norris, B., & Peebles, L. (2000) *Older Adultdata: The Handbook of Measurements and Capabilities of the Older Adult, Data for Design Safety.* London: Department of Trade and Industry.

Song. G. (2011) *Improving comfort in clothing.* Cambridge: Woodhead.

Szebeko, D., & Tan, L. (2010), Co-designing for Society. *Australasian Medical Journal*, **3**(9): 580–590.

USABILITY OF VIRTUAL LEARNING ENVIRONMENTS THROUGH DESIGN PRINCIPLES

Rosamelia Parizotto-Ribeiro & Nick Hammond

Department of Industrial Design, Federal Technological University of Paraná
Curitiba, Brazil
Department of Psychology, University of York, York, UK

This article presents early results of two experiments on aesthetics applied to Virtual Learning Environments (VLEs). Previous research showed that aesthetics is correlated with perceived usability of ATM systems. The present research work proposed the use of five design principles as a model to measure aesthetics applied to VLE environments. An experiment was conducted in order to evaluate the role that the design principles play in interface aesthetics and usability. The results corroborate previous research findings. A second experiment was realized to validate results using a prototype of a VLE environment.

Introduction

Aesthetics has been object of many investigations over the past centuries. However, over the last decades, the subject has broadened to include the understanding of how art is related to what people feel and how they learn. The American Heritage Dictionary defines Aesthetics as *the study of the psychological responses to beauty and artistic experiences*. Aesthetics is an important part of the human experience and, far from being opposed to function, aesthetics is a complement to function (Norman, 2004). The literature shows evidence that aesthetics influence users' perception of usability. It also shows many theories about how to measure beauty and achieve aesthetic pleasure. The motivation for studying aesthetics applied to VLEs is that its acceptance by the user could be strongly related to its appearance and so influences the users' perception of usability. For the purposes of this experiment, we operationalize aesthetics as the use of design principles during the development of the interfaces of a VLE environment.

Related research

Recent research on the visual aesthetics of computer interfaces suggests that aesthetics is a strong determinant of users' satisfaction and pleasure (Lavie & Tracktinsky, undated) and that visual attractiveness of the site affects users' enjoyment as well as perceptions of ease of use (Heijden, 2003). This is supported by Ngo et al. (2003) who found that careful application of aesthetic concepts can aid acceptability, learnability, comprehensibility and productivity. Acceptability was investigated by Kurosu & Kashimura (1995) and Tracktinsky (1997) where the studies showed very high correlations between users' perceptions of interface aesthetics and usability. There are also several studies related to learnability. It was found that aesthetically pleasing layouts have a definite effect on the student's motivation to learn (Toh, 1998) and that good graphic design and attractive displays contribute to the transfer of information (Aspillage, 1991). In other words, good design helps the user to comprehend the information in a better, easier way. In a study conducted by Szabo & Kanuka (1998), subjects who used the lesson with good design principles completed the lesson in less time and had a higher completion rate than those who used the lesson with poor design principles. A study by Grabinger (1981) indicated that organization and visual interest are important criteria in judging the readability and studyability of the real screens. Screens that are plain, simple, unbalanced, and bare are perceived as undesirable.

Design principles

Ngo et al. (2003) have developed fourteen aesthetic measures based on Gestalt theory of visual aspects for assessing graphic displays' completeness. Their empirical studies have suggested that these measures may help gain users' attention and build their confidence in using computer systems.

In the present study, these aesthetic measures have been combined with relevant design principles that are most accepted by the designer community and widely used for the development of their practical work. The result was summarized in five design principles (unity, proportion, homogeneity, balance and rhythm) which seem to be the most relevant and suitable for the particular needs of a screen layout for a computer interface for Virtual Learning Environments.

An experiment using a static interface

This experimental study was conducted to verify the hypothesis (i) that the use of design principles can be an important factor to determine aesthetics of computer screen layouts and (ii) that aesthetics are related to users' perceived usability of the system. The participants had to evaluate static screen layouts based on their perceived attractiveness and usability.

Design

The experiment had a 3 × 4 within-subject design. The independent variables aesthetics and usability were manipulated using three different sets of screens, each one with four different layouts. The screens were shown in a Latin square design to counterbalance order effects.

Participants

The experiment involved 279 participants enrolled or working at Higher Education Institution in Brazil (CEFET-PR) and Siemens-Brazil, divided into seven groups as follows: (1) undergraduate students from the Design course; (2) undergraduate students from the Computer Science course; (3) undergraduate students from Engineering courses; (4) graduate students from a MSc course on Technology and Innovation; (5) lecturers of various courses; (6) members of staff; (7) employees of Siemens-Brazil.

The participants were chosen in order to create distinctive groups in relation to background, schooling, computer literacy, age and gender.

Materials

Three set of screens that simulated specific areas of a Virtual Learning Environment were created. Non-interactive software was used to allow participants to evaluate each one of the layouts. The first set (home) intends to give the participant a general idea about the environment. The content material set simulated an introduction to a Module of Photography. Finally, the e-mail set enables the participant to have access to the most commonly-used communication tool. Each set of screens used the same graphic elements, colours and typography on four different pages layout (Figure 1).

- Layout 1: followed the design principles (unity, proportion, homogeneity, balance and rhythm)
- Layout 2: violated all of them;
- Layout 3: complied with the first three design principles and violated the other two;
- Layout 4: violated the first three and obeyed the last two design principles.

The dynamic symmetry (Hambidge, 1926) technique was used to set up the grid to design the VLE environment used as experimental stimulus.

The graphical style chosen for this experiment was based on 'cartoon drawings', with the purpose of creating a more informal environment and clean interface using, as a composition space, the golden rectangle. When in use, sets 1, 2 and 3 were each prepared in four different layouts as shown in Figure 1.

The elements were simple, using complementary colours (mainly orange and blue) to create the necessary contrast. The typography used for the text was Verdana (black), which gives good legibility on this kind of media.

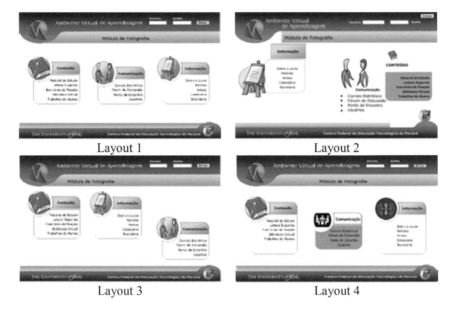

Layout 1   Layout 2

Layout 3   Layout 4

Figure 1.   Four different screen layouts for the set 'home'.

Procedure

The experiment tried to replicate the most usual conditions of use of a VLE. It was done in a controlled environment, where three 17″ PC were used to present the stimulus material allowing three participants to do the experiment at the same time. The experiment intended to collect data on users' perceptions of beauty and usability of the interface.

The first set of screens were shown so that the participants had a chance to see each set of four screens during four seconds, like a little *trailer preview*. This gave the participants the opportunity to see all four screen layouts before evaluating them. After that, the screens were presented again and the participants had to evaluate them for the aesthetic aspect using a five-point Likert scale, varying from Unattractive (1) to Attractive (5). At this point they were asked to evaluate how they perceived the organization of screen elements and how motivating the screen layouts seemed to be. This process was repeated for all three sets of screens (home, content and e-mail). Then, the participants were presented with the same set of screens were to evaluate the layouts' perceived usability using the same five-point Likert scale ranging from Difficult to use (1) to Easy to use (5). At this point they could not see the "trailer preview", but they still received instructions to evaluate each screen (twelve in total) taking into account their perception of how easy to use the environment was, and their satisfaction when navigating this particular VLE. They could spend as much time as they wanted to evaluate each screen and only when they clicked on the button submitting their rating would the screen change to the next one, registering the data in a separate file.

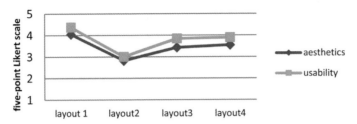

Figure 2.   Aesthetics and usability between four different layouts.

Results

The results analyzed the participants' perception of aesthetics and usability for the sets with different layouts. The results showed a positive correlation between aesthetics and usability for all three sets of screens as well as on four different layouts. The results demonstrate that the most important correlations were between applying and violating all the principles (Figure 2).

The experiment using an interactive interface

This experiment was conducted to (i) confirm the first experiment's results of a positive relationship between interface attractiveness and usability and (ii) verify if the results would have major differences after the user's interaction with the proposed VLE. Participants had to interact with the interface, performing two tasks, before evaluating the attractiveness and usability of the interface used. This study used the screen layouts from the previous experiment.

Design

The experiment had a between-subjects design with four levels. Each participant took part in just one condition and then evaluated the aesthetics and usability of the system. The tasks were performed using the Latin square design to counterbalance the order effect.

- Condition 1: high aesthetics with low usability (followed all the design principles with error messages or delays between screens);
- Condition 2: high aesthetics with high usability (followed all the design principles with no error messages and delays between screens);
- Condition 3: low aesthetic with low usability (violated all the design principles with error messages or delays between screens); and
- Condition 4: low aesthetics with high usability (violate the design principles with no error message or delay between screens).

Participants

The experiment was done with 97 participants enrolled or working at a Higher Education Institution in Brazil. The participants had different backgrounds, schooling, computer literacy and age; only gender was counterbalanced between groups.

Materials

This experiment was done using the same computers and environment layout as the previous one. It simulated part of an interactive VLE environment, with a homepage which the participant would see first, content material which they would navigate and the email where they would get and send messages. These different functionalities had the aim of taking the participants from one screen to another to perform the tasks and, at the same time, getting them involved with the environment.

Procedure

Once the participants started the experiment, there were intermediary screens with instructions conducting them from the beginning to the end. To perform the task the participant had to navigate through the environment interface. The VLE was developed to assure that they would follow the same path and just perform the task asked by this study, by enabling just one link at a time. On the first task the participant played the role of a student taking an online course who had received email from a friend asking his or her opinion about a particular topic of the content material. She or he was asked to go to the content material to find the answer and then reply to the email based on their aesthetic opinion. For the second task, the participants were told that they were tutors of an online course who had received an email from a group of students asking about the content of weeks 6 and 8. They were asked to go to the calendar and reply to all the students with the correct answer. The interface used *Previous* and *Next* buttons to advance and return the pages. They could navigate through the material and spend as much time as they needed

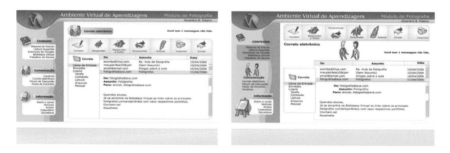

Figure 3.   Thumbnails of screens A and B.

to complete the task. When answering the emails, it was possible to return to the content material or calendar if they forgot the answer or wanted to check it again. After they finished the two tasks they were asked to evaluate the environments' aesthetics and usability using a five-point Likert scale. First, they evaluated the aesthetic aspects, varying from (1) Unattractive to (5) Attractive, and then the usability aspect, varying from (1) Difficult to use to (5) Easy to Use. Finally, they were shown one screen (Figure 3) with two thumbnail of the screens used in the study, screen A (applying all the principles) and screen B (violating all of them). They simply had to choose between screen A or B in terms of aesthetics and after that do the same in terms of usability.

Results

The between-subjects analysis of variance showed significant differences between ratings of aesthetics and usability on the four different conditions. Figure 4 shows that the better the aesthetics the better the usability was rated.

The overall sample also showed a positive correlation between aesthetics and usability of .678. Figure 5 shows a graph of the users' aesthetics and usability preferences

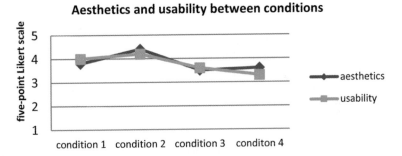

Figure 4.   Graph of the mean values for aesthetics and usability between conditions.

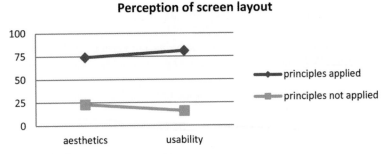

Figure 5.   Graph of the users' perception of screen layouts.

when comparing the layout that had applied all the principles and the other that had violated all the design principles. These final evaluations were done independently of the conditions that the participant had been submitted.

Conclusions

The analysis of the screen layouts applying the design principles found a positive correlation between aesthetics and perceived usability in all three different sets and all four different layouts. The outcomes showed that the difference between groups were not statistical significant giving indication that, at the visceral and behavioural levels, the perception of aesthetics and its relation to usability do not dependent on culture.

The results from the second experiment showed a positive correlation between aesthetics and usability on different conditions and they were statistically significant. The correlation between aesthetic and usability were much higher when the participants could compare between different layouts, showing strong evidence that, independent of background, schooling, computer literacy, age and gender, people can really distinguish good from bad layout design.

References

Aspillagae M. 1991. Screen design: a location of information and its effects on learning, *Journal of Computer-Based Instruction*, **18**, 3, 89–92.

Chang, D., Dooley, L. and Tuovinen, J. 2002. Gestalt Theory in Visual Screen Design – A New Look at an Old Subject, Australian Computer Society, Inc. [online] Retrieved 02 February 2004, from http://crpit.com/confpapers/ CRPITV8Chang.pdf

Grabinger, R.S. 1991. Computer screen designs: viewer judgements. *Educational Technology Research and Development*, **41**, 2, 35–73.

Heijden, van der H. 2003. Factors Influencing the Usage of Websites: The case of a Generic Portal in the Netherlands. *Information and Management*, **40**, 541–549.

Kurosu, M. and Kashimura, K. 1995. Apparent Usability versus Inherent Usability Experimental analysis of the determinants of the apparent usability, CHI'95 Mosaic of Creativity Proceedings, 292–293.

Lavie, T. and Tractinsky, N. undated. Assessing Dimensions of Perceived Visual Aesthetics of Web Sites http://burdacenter.bgu.ac.il/publications/LavieTractinsky .pdf

Ngo, D., Teo, L. and Byrne, J. 2003. Modelling interface aesthetics. *Information Sciences*, **152**, 25–46.

Norman, D. 2004. *Emotional Design: Why We Love or Hate Everyday Things*. Basic Books, New York, USA,.

Szabo, M. and Kanuka, H. 1998. Effects of Violation Screen Design Principles of Balance, Unity and Focus on Recall Learning, Study Time, and Completion

Rates. *ED-Media/ED-Telecom 98 Conference Proceedings*, Association for the Advancement of Computing in Education.

Toh, S.C. 1998. Cognitive and motivational effects of two multimedia simulation presentation modes on science learning. PhD dissertation, University of Science Malaysia.

Tractinsky, N. 1997. Aesthetic and Apparent Usability: Empirically Assessing Cultural and Methodological Issues, *CHI'97 Proceedings*, 22–27.

ERGONOMICS/HUMAN FACTORS AS A DISCIPLINE AND PROFESSION

TOWARDS A HUMAN PERFORMANCE
STANDARD OF EXCELLENCE

Barry Kirwan

EUROCONTROL, Brétigny-sur-Orge, France

Background

In Air Traffic Control it is recognised the human factors knowledge and techniques are not used to their full potential, and many air navigation service providers (ANSPs) have little or no human factors capability. In order to help rectify this, a human performance standard of excellence is proposed. This is a benchmarking system with thirteen scales that can each be rated 1–5, whereby organisations can see where they are in terms of their human factors capability maturity, and how they can improve. At the lower end of the scales, ANSPs can find ways to begin to address key HF issues relatively easily, whereas the upper ends of the scales are challenging even for organisations with an established human factors capability. Although the scales are in the context of air traffic, the overall framework and scale anchors may be of interest to other industries including aviation, rail, nuclear power, oil and gas, etc., since the issue of insufficient HF usage and uptake is an endemic and cross-industry issue.

Purpose

The purpose of the workshop is to gain critical feedback on the approach, the framework and the scales, so that a more robust and effective human performance standard of excellence can be delivered.

WORKSHOP: WORKING WITHIN CROSS-DISCIPLINARY TEAMS: HOW CAN WE HELP BRIDGE THE GAPS?

Becky Mallaband, Victoria Haines & Val Mitchell

User Centred Design Research Group, Loughborough Design School, Loughborough University

Workshop Background

Historically within academia, there have been distinct boundaries between subject areas and sometimes even between specialities within those subject areas. Conversely, the need for research between and across disciplines has been acknowledged as necessary in order to address complex problems and this has been reflected in recent funding calls. However, working across disciplines encounters a number of difficulties. Some of these difficulties are more practical, such as identifying an appropriate place to publish research outcomes, whilst others are more ingrained, such as a difference in the theoretical approach of team members or where the use of complex discipline-specific terminology alienates other members of the team. Regardless of the challenges, this type of working is important and can provide very rewarding outcomes. This workshop aims to address some of the challenges encountered, encouraging discussion and sharing between participants, before finding ways to address these challenges and identifying key skills to do so.

The workshop organisers have substantial combined experience of working within cross-disciplinary research teams, particularly across social and engineering disciplines and through these experiences have developed a bridge building concept for the role of the user centred specialist within these teams. In addition they have developed a set of principles which aim to aid the process of cross-disciplinary research (Mallaband, 2013; Mallaband & Haines, 2014). The workshop will build on the work already carried out and will explore participants' experiences of cross-disciplinary work, aiming to further conceptualise the process and review, expand on and evaluate the principles of cross-disciplinary working. The workshop will be interactive and creative, seeking to encourage participants to conceptualise their own cross-disciplinary teams and experience in order to better understand the process dynamics and to improve future practice.

The workshop intends to explore the challenges which participants have experienced with cross-disciplinary working, instigating in-depth discussion to identify shared experiences and potential solutions. The workshop will then focus on the skills set of a user centred specialist (Human Factors/Ergonomics practitioner) and identify ways in which these skills lend themselves to the role of bridge building in a cross-disciplinary working context.

Workshop purpose

The workshop aims to identify the roles within cross-disciplinary teams through the exploration of participants' past experiences and identify approaches that can facilitate effective involvement of a user centred specialist within a cross-disciplinary team.

Workshop audience

Anyone involved in cross-disciplinary research teams. Participants should leave the workshop with an overview of the potential role which they can take within cross-disciplinary teams, identification of the skills needed for this role and practical applications of how to develop and put into practice these skills.

Workshop activities

Interactive activities will be used to engage participants and encourage them to think of their own experiences and consider a more theoretical perspective of the cross-disciplinary working process.

Practical bridge building activities with the use of Lego (or similar) will be used to help participants visualise the process which needs to take place between different members of the research team and other stakeholders (including participants).

A 'talking heads' video will also be created in which participants will be asked to describe their thoughts and challenges in relation to cross-disciplinary working which will be edited together into a short film to show the range of barriers encountered.

References

Mallaband, B. (2013) *Integrating User Centred Design into the development of energy saving technologies*. PhD thesis, Loughborough University, UK.

Mallaband, B. & Haines, V.J. (2014) Blurred Lines: How Does Cross-Disciplinary Research Work In Practice. In proceedings of *ACM UbiComp 2014* Workshop: HomeSys, 13 September, 2014.

ERGONOMICS/HUMAN FACTORS – ART, CRAFT OR SCIENCE?: A WORKSHOP AND DEBATE INSPIRED BY THE THINKING OF PROFESSOR JOHN WILSON

Sarah Sharples[1] & Pete Buckle[2]

[1] *Human Factors Research Group, Faculty of Engineering,*
University of Nottingham,
[2] *Royal College of Art, London*

The new (soon to be launched in 2015) edition of "Evaluation of Human Work" has afforded a contemporary overview of what methods leading researchers in E/HF are currently adopting and advocating. It has also thrown out a challenge regarding how E/HF professionals should be able to evolve as individuals, reflect on what tools work and do not work, in which circumstances, and why. By acknowledging the notion of E/HF as reflective practice we can continue to develop appropriate methods and tools, along with the underlying theoretical frameworks that support the translation of our work to different domains. E/HF may not always be about delivering the 'right' solution(s), but is always about delivering the right approach.

This has lead, in the past, to a sometimes uncomfortable debate regarding the 'hard' scientific approaches versus the merits of a 'softer' qualitative methods. This debate however, will be even more contentious as it explores the notion of ergonomics as a reflective, creative craft v an evidence driven, research based science.

Reference

Wilson, J.R. & Sharples, S. (2015) *Evaluation of Human Work: 4th Edition.* Boca Raton: Taylor & Francis (CRC Press).

HEALTH AND SAFETY

HISTORY REPEATING – OR CAN BETTER LEARNING ALSO IMPROVE RESILIENCE?

Jonathan Berman

Greenstreet Berman Ltd, UK

Much has been written about the importance of organisational learning yet failures continue. This paper offers a view on both the importance of effective organisational learning, and some challenges and barriers to achieving it. It considers why it seems so difficult to achieve, and what arrangements might be effective. The paper presents ideas for creating a robust and sustainable learning organisation, based on experience over many years across a range of high-hazard industries. It seeks to encourage debate about learning, resilience and performance improvement.

Introduction

The importance of learning from experience is well recognised. Many aphorisms and quotations show that the requirement seems embedded in popular culture: "Those who cannot remember the past are condemned to repeat it" (Santayana) and "History repeats itself, first as tragedy and then as farce" (Marx). The first quotation touches on the importance of being able to recall past experience – the knowledge and understanding that can come from previous incidents and events is often gained at significant expense, and should not be discarded lightly. The second quotation presents a slightly bleaker picture, suggesting that the repetition of previous failure has an element of inevitability about it. Are either, or both, of these positions correct?

We can review the litany of major accidents and see repetition and lost opportunities for prevention. Whilst there may be subtle differences from one accident to the next, there are also painful similarities. Ladbroke Grove and Southall train crashes, Hatfield and Grayrigg derailments, Grangemouth and Texas City, Flixborough and Piper Alpha – the list goes on. Not only are there challenges associated with learning from the experience of others, but also there appears to be real difficulty in learning from events within a single organisation.

Many if not most organisations have established processes for learning from experience – they seek to understand the causes of performance shortfalls and failures, and to take from that experience the lessons that could be used to prevent their recurrence. Increasingly, organisations use the same or similar processes also to understand positive contributors to good performance, in order to promote them more widely. Surely such organisations will then be able smugly to reflect on Santayana's quotation and bask in the confident knowledge that they are immune to repeated events?

We know this is not the case and that many if not most organisations are vulnerable to Organisational Drift – the inexorable but imperceptible decline in performance that is eventually revealed in the next major accident or failure. Whilst trying to avoid generalising, one can look at BP's experience with Grangemouth, Texas City and Deepwater Horizon and reflect on what might be a 7–8 year cycle of drift, incident, response, improvement, drift ... Berman and Ackroyd (2006) noted some of the precursors to drift, such as the apparent stability of previous good performance and gradual changes within the organisation, and also noted some of the potential defences, including effective oversight and external benchmarking. Dekker (2011) notes that system complexity might make recognition of drift difficult, as each failure is ascribed simply to the (poor) reliability of individual components. We need to be better at understanding the systemic implications of individual failures.

A constant challenge appears to be associated not with the ability to understand incidents and events, but rather with the ability to extract the key learning points from that understanding and, more importantly, to embed them within normal operations. This paper reflects on mechanisms to enhance that process of embedding change, and hence of achieving real learning.

Investigation

Whilst it is easy to assume that incident and accident investigation processes are now well understood and well-established, in practice there remains work to do. Few organisations now can be found that do not have trained investigators, equipped with such analysis tools as change and barrier analysis, root-cause investigation, causal factor charting and so forth. These are necessary tools, and skills, but not sufficient to ensure learning.

Whether it is formal investigation following an occurrence, or more embedded processes for learning from experience, 'investigation' provides the data to input into the learning process, and should also provide a level of analysis that ensures that those data are retained in a form that enhances learning and accessibility. Most investigation processes include a formal taxonomy of root causes, if only to aid recording of data in a database, that influence the course of the investigation and analysis. There is a need to consider carefully how the root-cause taxonomy might influence the information that emerges from the investigations, and how it sits alongside data that can be gathered from examples of positive performance. Many taxonomies are, understandably, oriented around Ergonomics and Human Factors (E/HF) principles – the factors that affect performance.

For example, NUREG CR-6751 (USNRC, 2002) presents a Human Performance Evaluation Process which is structured around problem identification and resolution, but focuses mainly on factors at the individual level: fitness for duty; knowledge, skills and abilities; attention and motivation; procedures; tools and equipment; etc. Whilst it also notes systemic causes for each of these factors, it does not explicitly address management systems.

There may be merit in taxonomies oriented towards organisational structures and constraints and those facets of organisational performance amenable to change, as well as the more traditional E/HF topics. Whilst it is essential that the underlying performance shortfalls are understood in E/HF terms, it is also important that the organisational response can be categorised and understood. Such taxonomies might include the factors more associated with individual performance such as training, procedures, interfaces, process, etc. as set out in NUREG CR-6751, but also usefully include topics related to learning and behaviour change, such as previous experience, formalisation of approach, organisational learning, knowledge management, succession planning, prioritisation, resource allocation; organisational performance management (KPI setting and monitoring), etc. There is a need to consider the extent to which collection of information should be oriented also towards addressing shortfalls in learning, rather than only shortfalls in performance.

Furthermore, this orthogonal approach may help to avoid the contrasting pitfalls of searching for *the* root cause (i.e. a notional single shortfall that caused the event – unlikely in a highly defended system), or of creating such a large list of shortfalls that there is no hope of resourcing implementation of even a small fraction of them, or of failing to recognise the generic learning that should ensue.

However, investigation/data collection is only the initial stage in a learning and change management process. There is also a need to understand an appropriate change process and how it can be embedded within the organisation. Whilst the investigation should highlight potential changes, the organisation needs to be competent to take those recommendations and prioritise and implement them, taking wider account of the organisational implications. Not all potential solutions are equally palatable.

Change process

Where do many organisations start when considering change following an occurrence? A frequent complaint is that recommendations will inevitably focus on training and/or procedures, with a cynical rider that these two tend to be the lowest cost options (hardware and process change are far more painful).

This cynicism is somewhat unfair on two powerful methods for achieving rapid improvement. Designing and procuring a change in hardware to prevent a particular failure is time-consuming and expensive. It also has a potentially limited benefit – it will (if implemented correctly) prevent the recurrence of the specific incident under investigation, but may have minimal effect on other similar activities. In contrast, training and procedures are readily transferable. This then highlights a critical challenge for high-reliability organisations – how can they achieve cost-effective change that influences the way they operate, rather than merely affecting certain activities within the organisation?

There is recognition of the importance of achieving this – the nuclear sector has a vigorous Human Performance theme that seeks to understand a range of factors that influence performance, and recognises that no single approach will yield comprehensive and robust solutions. Human Performance touches on many aspects of E/HF, and illustrates the importance of a holistic approach, although as implemented it can be criticised for being more person-centred than system-centred. It might be argued that the focus remains more on task performance rather than organisational performance, and it is perhaps here that the greatest opportunity for achieving enduring change resides.

Organisational performance and change

Most organisations are good at identifying improvement opportunities arising from incidents and events (training, warnings and cautions in procedures, signs and barriers, interfaces, etc.). They might consider the contributors to decision-making failures, and seek to prevent their recurrence through awareness-raising of human performance characteristics. In aviation there is a requirement for pilots to undergo human factors training such that they understand the key factors affecting their performance. Increasingly in other domains as diverse as nuclear power and healthcare there is a focus on understanding crew-resource management (albeit sometimes using other labels) in order to improve the behaviours and decisions of personnel. The concept of situation awareness has become widespread and provides useful shorthand for considering the demands of complex systems with multiple information sources and activities.

In such cases, the investigation and change management processes demonstrate some ability to address the causes of incidents and, within limits, to extend the remit of the improvements to cover similar systems or incidents that can be reasonably foreseen. However, this 'traditional' process focuses on analysis of performance to identify shortfalls, and on opportunities to address and remove them. It takes the output from the investigation process and seeks to embed it in the organisation, but in a manner linked to the previous event. It may not address the way the organisation considers its performance, or the way that the organisation considers uncertainty and change.

Learning

There is merit in reflecting on the meaning of learning. It's not about being able to repeat (different) arrangements 'by rote' – it's about ensuring that the organisation changes beneficially as a consequence of a positive or negative experience. Often 'learning' is considered to be a change in arrangements or procedures. However, the underlying behaviour may be similar – 'follow whatever are the current procedures'. This does not equip the organisation to accommodate the inevitable differences that underlie the 'next' accident.

For example, a driver lost control of a car on a bend, due to excessive speed. The outcome: a lower speed limit on that section of road. The required 'organisational behaviour' remains compliance with the posted speed limit. A more adaptable (but more costly) solution would be to enhance driver competence so that they are better able to assess an appropriate speed. But it is also necessary to consider extrinsic factors that affected the ability of the driver to assess the hazard. This in turn requires an assessment of what the factors contributed to the driver's failure to do so originally. Did they misperceive the hazard? Did they misjudge their speed? Did they fail to understand that speed was relevant? What would those factors tell us about the 'organisation'?

There appears to be a tendency to reduce speed limits on rural roads in the UK. A consequence is that drivers may inadvertently abdicate responsibility for determining a safe speed and base their decisions on an assumption that the posted speed is 'safe' – even though sometimes (poor weather etc) it may be far from safe. The simplistic approach has potentially reduced safety. An alternative solution is to improve driver competence and although less immediate is likely to increase the resilience of the system by enabling drivers to cope with other roads and bends, and with changes in the environment – the inevitable uncertainty in complex systems.

The same approach is embedded in Resilience thinking, and in Safety II (e.g. Hollnagel 2014), with the latter addressing the importance of considering successful performance rather than only failure. To be successful in these terms, the organisation needs both to look at good practice, which many organisations currently do, and also to understand the reasons why those practices were implemented at the time – how did they arise and why were they considered appropriate? It's important to recognise that the behaviours were, at least initially, 'unusual' and maybe even 'non-compliant'. The requirement is to be able to look in behavioural terms at the process rather than merely the output and to understand what initially triggered the behaviours, and what embedded them.

Resilience is about understanding uncertainty rather than merely coping with it. This in turn implies that the organisation is able to accept uncertainty, and to have in place processes that allow it to adapt within accepted constraints. A learning organisation needs to accommodate this requirement for flexibility and understanding uncertainty – to be able to identify and introduce managed change.

A learning organisation

Being a learning organisation is not only about reviewing past experience and identifying changes that might add value. It's also about controlling drift and erosion of knowledge. Such loss of knowledge frequently underpins adverse behaviours, either because the required behaviours are no longer understood, or because their relevance to specific situations is no longer recognised.

Weick (2004) discusses abduction as the process that allows a person to create a hypothesis that will explain new information, drawing on past experience and knowledge to allow the person to make sense of what they are experiencing. Abduction is important when understanding operator behaviour in complex systems, and understanding why people might choose to 'interpret' procedures rather than follow them unquestioningly. The notion of interpretation is often resisted by high-hazard organisations as it introduces a perceived loss of control. Organisational learning might be better served by accommodating this process and enabling arrangements that support and contextualise abduction. Pettersen (2013) notes that acceptance of abductive reasoning allows an organisation to take account of the accumulation of shared memories, lessons learned, and heuristics. It is by supporting such processes that an organisation becomes capable of learning not only from formal incidents and their investigation but, more importantly, from the accumulation of small pieces of positive experience.

To support abductive reasoning, a number of arrangements are required. One is to ensure that there are mechanisms not only for sharing information and experience, but also for understanding how that experience affected decisions and outcomes. This in turn places a requirement on investigation and analysis to provide that understanding, rather than merely describing inputs and outputs.

But abductive reasoning can seem an anathema to high-hazard organisations, as the implication of a degree of 'uncontrolled' decision-making is worrying where a process needs to be highly controlled. This opens discussion around resilience and how organisations enable a flexible response to the unexpected. The resilience debate needs to acknowledge the need for a degree of control allied to flexibility – the notion of 'resilient procedures' is considered by some to be an oxymoron, whereas in practice it may be a solution to the competing demands of controlled performance and flexible response. The deviations from 'expected' behaviours that often are the precursors to incidents also, in other circumstances, are the flexible recoveries that we applaud. We need to be better at understanding the factors that influence them – how they arise, and why they were considered appropriate in the circumstances.

There is therefore a parallel requirement, alongside the conventional approach to investigation, to consider how information is encoded, stored and accessed within the organisation. This touches on the concept of corporate knowledge. Without such knowledge in an accessible form, the sharing of information will be impeded. But supporting knowledge sharing is critical to effective learning (and resilience). The investigation process must go beyond 'root cause'. Why did the organisation fail to detect early signs of being unable to respond to the occurrence? Why did it fail to detect the inherent weaknesses? Why was the learning process unable to identify and offer solutions?

A further challenge arises from loss of staff, perhaps due to aging workforces or organisational change. Organisations typically attempt to capture and codify formal knowledge before it is lost (such as to do with the maintenance of legacy systems with insufficient documentation). It is less clear how to capture the informal knowledge that supports abductive reasoning. This informal knowledge may be of particular importance in the control of drift and enhancement of resilience. Investigation needs to provide clarity concerning the management and use of such knowledge, and to improve understanding of the risk perception associated with, and derived from, that informal knowledge.

Effective learning may therefore be about changing processes to accommodate greater flexibility aligned with greater risk awareness. Many organisations have huge appetites for learning – why can they not translate this into effective arrangements? One reason might be a focus on codifying learning content in a form that is inconsistent with how people tend to try to use such information informally. For example, it was found that a slinging and lifting process in a constrained workspace could not be undertaken in full accordance with the handbook yet, despite this fact, near-misses and dropped loads were assessed in terms of (non)-compliance with that handbook. Much informal knowledge was applied by the slingers, but the organisation took no account of it.

A way forward

Having worked with a wide range of organisations (all of which implement forms of 'learning from experience' but which still sometimes demonstrate failure to learn) a number of ways that could improve the learning process can be identified. These are not 'magic bullets' that will immediately solve the learning challenge. They are, instead, a prompt for further discussion about how we can create resilient organisations that are mindful and responsive, and which can internalise learning in a more embedded and enduring manner.

Knowledge capture (from incidents and other sources) needs to acknowledge the importance of understanding how the organisation has learned (or not) from previous events, and how it can be best supported in identifying changes to process and organisational behaviour. The investigation should explicitly seek out the missed learning opportunities and understand why they were missed.

Individual behaviour needs to be understood in the context not only of 'conventional E/HF' (the factors that affect performance) but also of 'the organisation'; this is very much the province of E/HF and acknowledges the centrality of the complex systems approach. Explicitly seek out the emergent properties of the system that might have affected behaviours.

Organisational behaviour needs to be understood in the context of shared and informal knowledge and experience, and sensemaking. Explicitly consider the nature of shared knowledge and its role in influencing behaviours, why the incident occurred, and how that knowledge affected perceptions at the time.

The process rather than the outcome needs to be a greater part of the focus of exploration of incidents/occurrences. The event will have been, in part, an accident of timing, but it is likely that the relevant behaviours have been previously exhibited. There is a need explicitly to consider how the process encouraged the observed behaviours prior to the event under investigation. All investigations should include outputs that reflect on the learning processes within the organisation, to understand better how those processes can be strengthened without inappropriately constraining challenge and exploration.

In summary, the challenge for organisations operating complex systems includes the need to understand why their learning processes sometimes fall short of what is needed. Few incidents arise from an unforeseeable scenario. Enhancing organisational learning should be a key goal in complex systems with complex emergent properties. Those very incidents that highlight shortfalls in learning or organisational memory are opportunities not only to improve learning but also to oppose drift. Explicit consideration of learning when examining incidents is one element in support of this. This approach also might lead towards better accommodation of flexibility and resilience within a proceduralised process, as it starts to formalise how the organisation understands and manages flexibility.

We need to understand better what resilient processes look like, and how to foster the flexibility that needs to underpin them. We need to consider how best to utilise learning processes to support that improved understanding. We need to consider what we can learn about preventing history from repeating.

References

Berman, J. and Ackroyd, P. 2006, Organisational drift: a challenge for enduring safety performance. In *Hazards XIX: Process Safety and Environmental Protection*, (IChemE, Rugby), 138–151.

Dekker, S. 2011, *Drift into Failure*, (Ashgate, Farnham).

Hollnagel, E. 2014, *Safety I and Safety II*, (Ashgate, Farnham).

Pettersen, K. 2013, Acknowledging the role of abductive thinking: a way out of proceduralisation for safety management and oversight? In C. Bieder and M. Bourrier, *Trapping Safety Into Rules* (Ashgate, Farnham), 107–120.

USNRC. 2002, The human performance evaluation process: A resource for reviewing the identification and resolution of human performance problems. NUREG CR-6751. Washington, DC.

Weick, K.E. 2004, Faith, evidence and action: better guesses in an unknowable world. *Organization Studies*, 27(11).

CONFIRMATION BIAS IN A ROUTINE DRILLING OPERATION: A CASE STUDY

Margaret T. Crichton & John L. Thorogood

*People Factor Consultants Ltd,
Drilling Global Consultants LLC*

Confirmation bias is a pervasive cognitive bias, with examples being cited in various settings such as economics, medicine, and law. In terms of process safety, cognitive biases tend to be identified following severe incidents. However, as this recent case study illustrates, confirmation bias also emerges during routine operations, and, while no harm to people or the environment occurred in this example, the potential always exists for situations to escalate leading to undesired outcomes. This paper presents brief definitions of confirmation bias and process safety, the drilling context, and the case study describing the emergence of confirmation bias. Recommendations are proposed to minimise the effects of cognitive biases, especially confirmation bias.

Introduction

The issue of cognitive biases is increasingly being addressed in relation to process safety (OGP 460, 2012; OGP 2012p, 2013). The importance of process safety, particularly in the oil and gas industry, was highlighted by the recent tragedy of the Macondo blowout (2010; Chief Counsel, 2011), as well as previous incidents such as the Texas City Refinery (2005; Hopkins, 2009), Longford Refinery (1998; Hopkins, 2000), Petrobras semi-submersible rig (2001; ANC/DPC 2001) and Piper Alpha oil platform (1988; Cullen, 1990). Improved understanding of human factors, and especially the impact of cognitive biases affecting real-time decision making, is a topic which has the potential to reduce human error, and minimise the possibility of incidents.

Human error, according to Dekker (2006), is not a cause of failure, but rather is the effect of deeper and more widespread trouble in organisations. The UK HSE (HSE, 1999) has identified a number of Performance Influencing Factors (PIFs), including job, person, and organization factors, which can affect the likelihood of error. Recognising the impact of these factors on workplace performance, errors are often a result of decision errors (Orasanu, Martin & Davison, 2001) whereby decision makers are cognitively economical, in that they adopt heuristics and biases in the decision making process. Cognitive biases refer to the shortcuts that people unwittingly use when interpreting situations and making decisions, such that inferences about other people and situations may be drawn in an illogical fashion (Hasleton,

Nettle & Andrews, 2005). A number of cognitive biases exist, but one of the most pervasive is confirmation bias, leading to unexpected outcomes and adverse events. Examples of confirmation bias have been presented from economics (McMillan & White, 1993), medicine (Pines, 2006), and law (Hill, Memon & McGeorge, 2008). Hopkins (2012) cites confirmation bias as one of the contributing factors to the Macondo disaster. However, confirmation bias may be even more ubiquitous than often thought, and may impact hugely on operations in high hazard industries without necessarily resulting in serious harm. Cognitive biases may be considered to arise only in time-pressured complex situations, however, an example is provided here where confirmation bias occurred during routine oil and gas drilling operations. The purpose of the paper is to define the bias, to illustrate how easily this particular bias can occur, and to discuss how it can be identified and managed.

Confirmation bias

Confirmation bias can be defined as the seeking, or interpreting of evidence in ways that are partial to existing beliefs, expectations, or a hypothesis in hand (Nickerson, 1998). Wason (1960) originally proposed the concept of confirmation bias when he noted that participants in a research study indicated a tendency to confirm rather than disconfirm their hypotheses. Subsequently, confirmation bias has been identified as a ubiquitous phenomenon, according to Nickerson (1998), and the way that people search, interpret, and remember information, to support pre-existing beliefs or assumptions, has been replicated in many situations (Oswald & Grosjean, 2004; Weick & Sutcliffe, 2007).

Nickerson (1998) proposes that confirmation bias occurs due to an unwitting selectivity in the acquisition and use of evidence. In relation to System 1 and System 2 thinking, Kahneman (2011) comments, that System 1 is gullible and biased to believe, thus without the capacity or time to deliberately search for disconfirming evidence, suggestions can often be accepted uncritically. Confirmation bias is particularly evident during attempts to diagnose what has gone wrong in a malfunctioning system (Reason, 1995), whereby investigators can frequently, unintentionally, match possible causes to available signs and symptoms and then look for evidence supporting this assumption, whilst explaining away contradictory information.

Cognitive biases and process safety

In terms of process safety, the International Association of Oil and Gas Producers (IOGP) recently published a report outlining the cognitive issues associated with process safety and environmental incidents. The report discusses safety-critical tasks and non-technical skills, in particular the influence of cognitive biases on performance, with the aim of creating a better understanding of the psychological basis of performance and future improvements (OGP 460, 2012). Without a clear understanding of the cognitive processes involved in carrying out safety-critical tasks, then incidents will continue to occur, and the solutions often proposed, such as providing more data to operators, will not improve the situation. Investigations following major incidents do not always examine the human factor or cognitive processes that contributed to the event, meaning that lessons are seldom identified or acted upon (Woods, Dekker, Cook, Johannesen & Sarter, 2010).

This case study illustrates the effects of confirmation bias during routine drilling operations. The members of the drilling team were faced with a number of data points that did not match their expectations, but this incongruous information did not influence their decision making, such that operations continued, and anomalies were explained away, rather than stopping and re-assessing the situation.

Overview of drilling operations

In order to produce hydrocarbons, a well will undergo a process from exploration to production. An exploration well is the first well drilled in a new location, and is often drilled for information gathering purposes, and to locate a hydrocarbon reservoir. An appraisal well is then drilled to determine the extent of hydrocarbons in the reservoir. Next, the well will undergo completions, where the well is made ready for production, typically by the insertion of tubing and casing. Finally, the well will be readied for production and valves will be fitted to the wellhead. The flow of hydrocarbons is controlled through these valves, which are then connected to pipelines leading to refineries or storage facilities.

Prior to well operations commencing, a drilling plan will be created. The drilling plan details the design of the well and specifies the procedures to be used to drill the well as safely and effectively as possible. Calculations of expected drilling pressures, casing sizes, and final depth of the well, will be used to provide a design which will constructed by the drilling team during the drilling operation. During the exploration well phase, the team comprises personnel both onshore and at the rigsite. Key disciplines in the onshore team include: Operations Management, Drilling Engineering, Logistics, Cost Control, and Geological supervision. At the rigsite, the team comprises personnel from the operator (e.g. Drilling Supervisor), the rig contractor (e.g. Driller, Assistant Driller, Toolpusher), and service companies (e.g. Mud Logger, Cementer). Many of these team members will be working together for the first time, and some may only join the team for limited durations throughout the lifespan of the drilling programme. This raises the possibility of lack of clarity of and assumptions about roles and responsibilities, and of poor communication between team members.

The case study

The drilling team in question was tasked with drilling a 4500 m deep exploration well in Southern Chile during the southern hemisphere winter. Daylight hours were short, and the weather typically comprised high winds, blizzards, and sub-zero temperatures. The section drilling plan in question was for a vertical hole to be drilled to a depth of 1563 m over a period of eight days. The interval was not

Table 1.   Inclinometer readings taken over the period of drilling.

Run	Depth (m)	Readings obtained (degrees)	Actual reading[*] (degrees)	Comment
1	578	0.25	1.05	Completed apparently successfully
2	805	0.80	1.70	Problems with equipment; modifications to equipment made
3	1108	4.75	6.15	Problems with equipment; reading discounted due to running out of time to complete task
4	1108	1.50	6.15	Second attempt using modified equipment
5	1431	No data	12.10	No result due to running out of time to complete task; modified equipment used
6	1508	No data	13.75	Second attempt failed; data considered to be "crazy"
7	1535	8.00	14.19	Initial growing awareness that situation was not as expected
8	1583	14.00	15.02	Confirmation that well had deviated from vertical

[*]Note: The actual readings were acquired during a correction run after drilling was halted and using a different type of equipment.

expected to be particularly challenging, and no problems had been noted in other wells in the area. While drilling the well, the plan called for inclinometer readings to be taken at set intervals of depth, to indicate whether the well was true to vertical or was drifting and might have to be corrected back to vertical.

Between Day One and Day Eight of drilling, eight inclinometer readings were taken. As Table 1 illustrates, indications of deviation from vertical were evident over the eight day period, but these unexpected data were either explained away due to, for example, problems with the inclinometer equipment, or were ignored as being flawed. The data being collected did not match the team's expectations and were readily disregarded, therefore drilling continued according to the plan.

Even as the unexpected and surprising data were being collected, the rig-site team did not question the results. As far as they were concerned, the well was being drilled vertically, and minor deviations of up to 0.5° could be accepted. Any concerns about the increased deviations were not raised, as there were mitigating circumstances that could be taken into consideration in interpreting the situation, and making the reasonable decision, as far as they were concerned, to continue drilling. Such circumstances included the extreme weather conditions and difficulties with the equipment used to deploy the instrument which could cause the inclinometer task to take up to six hours to complete, meaning that a team member had to stand outside in severe weather conditions for that length of time and carry out a challenging activity. This often resulted in the job being rushed. In addition, replacement inclinometer equipment was not readily available, and would take days to source, leading team members to continue to use modified equipment. As Kahneman (2011) proposes, a plausible scenario was used to explain the situation and to continue drilling. Drilling teams also suffer from equipment usability issues in that designers often do not adequately consider the end-user when offering their products. Even recognizing that the team members were using inaccurate equipment and modifying equipment to gain data, they still continued to believe the data they were receiving – as this seemed more acceptable than having to re-assess and re-evaluate their actions.

Discussion

That a supposedly vertical well was drilled to a deviation in excess of 15° came as a surprise to the team. The question is, how did this occur? The well was planned ahead of drilling commencing, and the team executing the plan were highly experienced, and had extensive experience in the area and internationally. However, during the planning phase, inadequate technical expertise appears to have been sought leading to the characteristics of the drilling equipment being unknown. The team executing the drilling plan assumed that the planning team had produced a rigorous plan, along with contingencies, that defined how the well was to be drilled. As far as the team was concerned, this was a routine drilling operation. The prevailing culture was a sense of "time is money" albeit that this was frequently self-induced and rarely explicitly promoted by management. Moreover, the organizational culture was influenced by tensions existing between management and the unions, which led to a tendency for blame. As operations were underway, the original drilling plan was altered without sufficient attention being paid to the implications of the change. No formal Management of Change process was practised in this team. In this sense, a situation characterised by uncertainty, complexity, and ambiguity arose.

As the unexpected data were being collected, a tendency towards plan continuation (Orasanu & Martin, 1998; Reason, 2013) also emerged. Plan continuation occurs when those involved are focused on achieving the goal, for example, of drilling a well to the required depth in the expected timeframe, even though data was present indicating that a change of plan was required. The team continued with the drilling plan in the face of the unanticipated data. When a plan is being executed, the plan is considered to be sound and reliable, and there is a strong resistance or unwillingness to change. A further contribution could be that the team experienced an 'error of decision making', resulting in a mistake, however confirmation bias appears to have contributed to this error.

The pervasive nature of confirmation bias affects operations in that people rationalise information to make it fit what they want to believe (Dekker, 2006; Weick & Sutcliffe, 2007). This also means that it can arise in both routine and non-routine operations. Typically, confirmation bias is identified in hindsight after an incident has occurred. Data that could be described as 'weak signals', in other words, a small anomaly observed during an activity considered at the time to be innocuous and readily dismissed (Weick, Sutcliffe & Obstfeld, 1999), is later identified as significant during incident investigations. Such in-depth investigations are seldom conducted after a routine operation has had to be modified, meaning that the weak signals are not noted, nor is the impact of confirmation bias on the operations picked up. Nonetheless, the effect of confirmation bias can lead to operations continuing in the mistaken belief that everything is acceptable. In this case study, no harm to people or the environment occurred. Once it was recognised that the well had been drilled off vertical, operations were halted, and new plans made to re-drill the well. The ensuing increased costs were in terms of time, resources, budget, and, for many concerned, frustration. Continual checking by an engaged management, even though geographically distant, could have picked up on these weak signals, however, there was little focused discussion about these anomalous readings and management viewed them as routine 'noise'. A review, or debrief, of the operations highlighted both the non-technical and technical aspects of the well, and identified lessons for future operations. Whether those lessons are actually learned and implemented, remains to be seen.

Whilst acknowledging that a number of performance shaping factors contributed to this event, such as the use of measurement equipment in which the team had little trust, the failure to follow the management of change process, and the environmental conditions, this case study is also interesting in how it highlights the effects of cognitive biases on workplace performance. In order to minimise the occurrence of cognitive biases, various solutions can be considered (Thorogood & Crichton, 2014). Referring back to the case study example, recommendations include:

Recommendation I: Awareness of cognitive biases: Training should be provided to all members of drilling teams in the non-technical skills necessary for safe and effective performance. Such training could be provided through an organisation's commitment to developing and providing a Crew Resource Management (CRM) form of training (Energy Institute, 2014). Such training should specifically address the cognitive biases associated with process safety and environmental incidents (OGP 460, 2012).

Recommendation II: Management support for "Stop the Job": Management should support teams to stop operations if the situation appears to be deviating from expectations. Although tools such as Stop the Job exist in the oil and gas industry, where teams are encouraged to halt operations while the situation is re-assessed, team members appear to be hesitant to stop operations usually due to *perceived* production or project deadline concerns. Effective checks by offsite management should also help to raise awareness when 'weak signals' are increasing, and encourage timeouts to re-assess situations. Such tools need to be practised and reinforced.

Recommendation III: Debrief routine as well as non-routine events both during and at the end of operations: A debrief may occasionally place at the end of the operation, but this would typically address technical issues that arose during the operation to identify what went well and what could have been done better. It would appear to be extremely rare that human factors are included in the debrief. Yet, as the aviation sector has demonstrated, debriefs provide an excellent opportunity to expand the "what went well/what didn't go well" discussion to include non-technical skills, such as: How were communications? What issues with teamwork arose? Who made decisions and who was involved in the decision making process? If decisions were complex or challenging, why was that? When did the team members call a time-out to re-assess the situation? In this way, human factors, non-technical skills, and cognitive biases can be identified and their effects on performance can be recorded and shared with other teams.

References

ANC/DPC Inquiry Commission Report (2001). *P36 accident analysis*. Report by Agencia Nacional do Petroleo/Diretoria de Portos e Costas.

Chief Counsel's Report (2011). *Macondo: the Gulf oil disaster*. National Commission on the BP Deepwater Horizon Oil Spill and Offshore Drilling

Cullen (1990). *The Public Inquiry into the Piper Alpha Disaster* (Volumes I and III (Cm 1310)). (HMSO, London).

Dekker, W. (2006). *The field guide to understanding human error*. (Ashgate, Aldershot, UK).

Energy Institute (2014). *Guidance on crew resource management (CRM) and non-technical skills,* Report by the Energy Institute, London

Haselton, M.G., Nettle, D., & Andrews, P.W. (2005). The evolution of cognitive bias. In D.M. Buss (Ed.), *The Handbook of Evolutionary Psychology*: Hoboken, NJ: Wiley.

Hill, C., Memon, A. & McGeorge, P. (2008). The role of confirmation bias in suspect interviews: A systematic evaluation. *Legal and Criminological Psychology*. 13(2), 357–371.

Hopkins, A. (2000). *Lessons From Longford: The Esso Gas Plant Explosion*, Australia: CCH.

Hopkins, A. (2009). *Failure to learn: The BP Texas City refinery disaster*. Australia: CCH.

Hopkins, A. (2012). *Disastrous decisions. The human and organisational causes of the Gulf of Mexico blowout*. Australia: CCH.

HSE. (1999). *HSG48 Reducing error and influencing behaviour: Human factors and occupational health and safety*. Norwich: HMSO.

Kahneman, D. (2011). *Thinking, fast and slow*. London: Penguin Books.

McMillan, J.J. & White, R.A. (1993). Auditors' belief revisions and evidence search: The effect of hypothesis frame, confirmation bias, and professional skepticism. *The Accounting Review*. 68(3), 443–465.

Nickerson, R.W. (1998). Confirmation bias: A ubiquitous phenomenon in many guises. *Review of General Psychology*, 2(2), 175–220.

OGP (2012). *Cognitive issues associated with process safety and environmental incidents*. Report No 460. International Association of Oil and Gas Producers.

OGP (2013). *Process Safety Events — 2011 & 2012* Data Report No. 2012p (December). Retrieved 1 August from: http://www.ogp.org.uk/pubs/2012p.pdf.

Orasanu, J.M & Martin, L. (1998). *Errors in aviation decision making: A factor in accidents and incidents*. Human Error, Safety and Systems Development Workshop (HESSD). Retrieved August 2014 from: http://www.dcs.gla.ac.uk/~johnson/papers/seattle_hessd/judithlynnep.

Oswald, M.E. & Grosjean, S. (2004). Confirmation bias. In R.F. Pohl (Ed.). *Cognitive illusions. A handbook on fallacies and biases in thinking, judgement and memory*. Hove: Psychology Press.

Pines, J.M. (2006). Profiles in patient safety: Confirmation bias in emergency medicine, *Academic Emergency Medicine*, 13(1), 90–94.

Reason, J. (1995). Understanding adverse events: Human factors. *Quality in Health Care*, 4, 80–89.

Reason, J. (2013). *A life in error: From little slips to big disasters*. Farnham, UK: Ashgate.

Thorogood, J.L. & Crichton, M. (2014). *Threat and error management: The connection between process safety and practical action at the worksite*. Proceedings of the SPE/IADC conference, Fort Worth, TX.

Wason, Peter C. (1968). Reasoning about a rule. *Quarterly Journal of Experimental Psychology*, 20 (3), 273–28.

Weick, K.E. & Sutcliffe, K.M. (2007). *Managing the unexpected: Resilient performance in an age of uncertainty*, (2nd ed). San Francisco: Jossey-Bass.

Weick, K.E., Sutcliffe, K.M. & Obstfeld, D. (1999). Organizing for high reliability. Processes of collective mindfulness. In B.M. Staw & R. Sutton (Eds). *Research in organizational behavior* (81–123). Greenwich: JAI Press.

Woods, D.D., Dekker, S., Cook, R., Johannesen, L. & Sarter, N. (2010). *Behind human error*. (2nd ed). Farnham, UK: Ashgate.

HAZARD PERCEPTION AND REPORTING

Ewan Douglas, Sam Cromie & Chiara Leva

Centre for Innovative Human Systems, Trinity College Dublin, Ireland

Reporting of hazards is a key aspect of safety management in industry, but relatively little empirical investigation of reporting has been undertaken. This research reports on a study to explore the detection and reporting of hazards by members of the public. Three simulated hazards were developed. The experiment aimed to assess the capacity to recall recognise and report hazards of the participants by means of an exit survey. Participants performed better at recognition than recollection with no actual reporting of hazards recorded. The results validated some of the findings suggested by the literature and can assist in the development of a new experimental methodology as training within organisations to improve awareness of hazards and reporting practices.

Introduction

Across many industries there is now a requirement for the implementation of a Safety Management System (SMS) including proactive risk assessment (Leveson 2011). For example, within the aerospace industry there is the European policy for aeronautical repair stations (EASA 145) which specifies a requirement for collecting proactive information on risks and hazards as they are encountered within the life cycle of the organisation (Pérezgonzález, McDonald, & Smith, 2005). A reporting system is an effective way of addressing these requirements and collecting information on hazards from the workforce. Many reporting approaches use reports submitted from the "shop floor" as one of the inputs on risk and hazards that will be managed by the SMS. There has been significant literature on the factors that can influence the level of reporting within the organisation from the design of the data collection forms, to the procedure, to cultural considerations of the SMS system (Johnson, 2003; Leveson, 2011). However, before reporting a hazard, the reporter has to successfully notice and identify the hazard. The literature in the area to date has not investigated this aspect of reporting. The study reported here took advantage of an exhibition called the "Risk Lab" in the Science Gallery in Trinity College Dublin to explore the rate at which the general public will notice, identify and report hazards. Although outside an industrial setting, the hazards used in the study represented a clear and recognisable danger to the public.

Existing efforts to stimulate hazard reporting within industry

Several studies have looked at developing a proactive approach to risk management; a recent example can be found in Leva, et al. (2010) where a proactive "daily journal" was developed and implemented in a small Italian regional airport. The new methodology rolled out was a web-based tool consisting of an anomaly log that should be completed by ground staff after each aircraft "turnaround". The anomaly log, while providing immediate benefits to the ground staff by assisting in the shift handover procedures, also collects proactive data on anomalies that are encountered during each turnaround allowing these anomalies to be captured immediately after they occur (Leva, Mcdonald, et al., 2010). Literature such as Wiegmann & von Thaden, (2003) highlight the importance of collecting data on anomalies and incidents immediately after the actual occurrence of an incident. Leveson, (2011) highlights the importance of designing a suitable reporting form to allow the reporters to provide the information they want to provide without being an extra burden into their day-to-day workload. Reporting approaches that are cumbersome or add more paperwork to already overburdened staff can act as a barrier to reporting. Furthermore, the reporting system should be designed with the objective of delivering benefits to the day-to-day operations of the staff expected to use them (Kongsvik, Fenstad, & Wendelborg, 2012; Leva, Cahill, Kay, Losa, & McDonald, 2010)Therefore, communication and training about the benefits of the reporting system should also be provided to the staff. Industrial initiatives in this area all rely on staff seeing hazards, and then reporting them into some form of a system. The majority of the research so far has been on the procedure behind compiling a reporting form and its follow up and considerable effort is made to raise the awareness of hazards through, for example, posters or training, with the assumption that these will increase the detection and reporting of hazards. However there is a crucial question before this process can begin: Are the reporters able to notice and identify all relevant hazards?

From hazard detection to reporting

This paper assumes that there are four steps in a reporting process: 1) witnessing a hazardous scenario, 2) identification of the scenario as hazardous, 3) risk assessment of the scenario, and 4) reporting. First, the individual needs to pay attention to the visual/auditory/olfactory/tactile stimuli that represent the hazard. There are several factors that could influence this process ranging from the salience of the hazard to the "unexpectedness". Wogalter et al. (1999) suggest that hazard perception is affected by the likelihood of the hazard introducing a risk of injury to the reporter. There are also environmental considerations to be taken into account: a busy loud environment can make some hazards harder to distinguish from the background noise. Personal experience has been found to have a significant role in hazard detection. There have been studies on the hazard perception habits of new drivers (Deery, 1999) that show how inexperienced drivers will treat all hazards with a similar priority, while more experienced drivers will tend to assess hazards more proactively and prioritize them accordingly (Wiegmann & von Thaden, 2003). (Deery, 1999), in a study of hazard perception with regards to driving age, found that age groups are non-homogenous with regards to the level of risk perception and factors such as personality and task attitude can have a significant influence on how an individual perceives a hazard.

Second, scenarios have to be sufficiently processed by the individual to be identified as hazardous. This study explores the level to which hazards are processed by asking participants to recall hazards. This was explored in two ways using a computer-based survey at the end of the exhibition. Participants were first asked to recall and note any hazards they had seen during the exhibition. They were then presented with a set of "hazards" on the computer screen, some of which had been present during the exhibition and some which were not, to investigate if participants can recognize hazards that they may not have identified and recalled in the first place. The study draws on the levels of processing theory, which posits that the more cognitive processing are applied to stimuli the more readily accessible they will be (Craik & Lockhart, 1972). The hazard recall (i.e. being able to remember the hazard unprompted) is taken as an index of a deeper level of processing than recognition.

Third, the hazard needs to be assessed as being of significant risk to be worthy of reporting, and then, fourth, reported. Steps three and four were measured together in terms of whether or not the hazards were reported.

The objectives of the study were to:

1. Examine the levels of reporting of reporting hazards from the general population
2. Examine the difference between levels of recall and recollection of hazards from the general populations

Method

Design

The investigation consisted of three hazards that varied in salience, size and "unexpectedness" located within the science gallery. Participants explored the Science Gallery exhibit where the hazards were located; however the hazards were not the focus of the exhibition. They were part of the environment as would be expected within an industrial setting.

Reporting, recall and recognition were assessed by exit survey as participants were leaving the gallery. *Recall* was measured by means of a free text box that asked participants to report hazards that they had encountered in the risk lab. Then they were asked if they had reported any of these hazards to the Gallery staff (*Reporting*). Subsequently they were presented with pictures of the hazards and asked if they had seen that specific hazard during their visit (*Recognition*). False hazards, which were not present in the gallery, were included in this section to ensure genuine recognition was being measured.

Figure 1.   Switchbox.

Participants

The investigation was hosted within the "Risk Lab" exhibition at the "Science Gallery" at Trinity College Dublin, which was a free exhibition open to the public. 153 participants completed the exit survey, with a mean age of 26.6 years and a range of 37 years.

Equipment and materials

A professional prop company produced three realistic (but not hazardous) hazards for the experiment. The three hazards were: a leaking chemical cupboard, a faulty switchbox and a leaking pipe. The "hazards" were placed around the risk lab exhibition before it was open to the public. The fuse box was placed near the entrance lobby, the pipe was placed in a busy corridor and the chemical cupboard was placed at the top of the main staircase in the exhibition.

Figure 1 is the switchbox hazard used. As shown there is an inadequately secured panel on the side of the unit, several "tripped" circuit breakers and there was a device installed that simulated a blue light and sounds indicating electrical arcing occurring. The unit was designed to look as if it belonged in the science gallery owing to electrical piping joining the real piping in the environment.

The second hazard was the chemical cupboard shown in Figure 2. This consisted of a cupboard with a "staff only" sign, an open lock with the key left in, several containers with evident hazardous substance symbols being visible and a simulated leak onto the floor.

The final hazard was a leaking pipe (shown in Figure 3). The pipe was placed beside identical real pipes, and held a pressure gauge showing a pressure reading in the red area and making occasional hissing noises.

Figure 2.   Chemical Cupboard.

Figure 3.   Leaking Pipe.

Data collection

The participants were asked to complete a short (approx. 1 min) survey before leaving the exhibition; the survey was hosted on an apple iPad running Survey Gizmo software.

Results

Figure 4 shows the results for the study in terms of hazard recall, recognition, and reporting rates to exhibition staff. The chemical cupboard had the highest rates of recognition, recall and was the only hazard that was reported to staff. 51% of participants could recall having seen this hazard, as compared to 37% for the

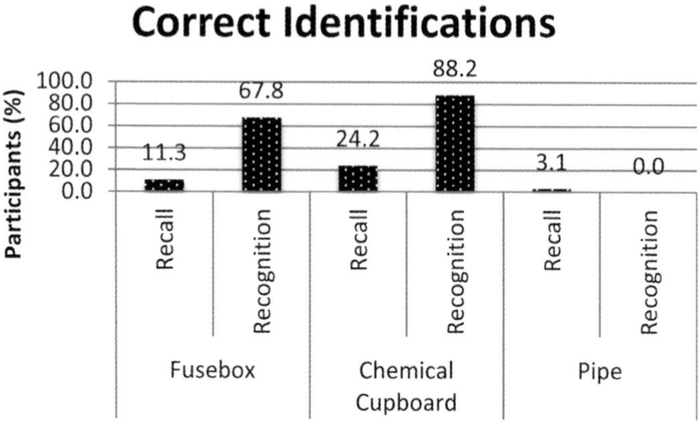

Correct Identifications

Figure 4.   Frequency of Recall and Recognition.

fusebox and 0% for the pipe. Recall rates for both the chemical cupboard and fusebox were lower than recognition rates, as would be expected. In total, only three people (approx. 2%) reported any hazard to the exhibition staff.

Discussion

The objective of this study was to investigate the levels of reporting, recall and recognition of hazards by participants in the general population. Each of these will be discussed in this section.

In the exit survey, participants were first asked to recall any hazards they had encountered during their visit. 33% of participants were able to recall a hazard from their time in the exhibition, suggesting that they had both witnessed the scenario and identified it as hazardous (steps 1 and 2 of a reporting process). Further, 51% of participants recognized the chemical cupboard hazard when presented with it, showing that they had witnessed the hazard (step 1) but had not necessarily identified it as hazardous (step 2). However, only three participants went on to report the hazard, suggesting that steps 3 and 4 of the reporting process were a blocker to reporting in this study.

The main result from this study is the extremely low reporting of hazards to the exhibition staff. Of 153 participants who completed the exit survey, and the several hundred overall visitors to the exhibition, only three proactively reported a hazard. The participants may not have noticed the hazard to begin with (i.e. they did not witness a hazard). However, the recognition data shows that this was not the case for a large section of the participants. Secondly, the participants may not have identified that hazard, but again the recall data shows that this was not the case for a substantial section of the group. The data appears to show unwillingness among the participants to proactively report hazards and a "filtering" of hazards through perceptual, cognitive and social processes. The hazards could be seen/heard by participants but were not identified as hazards until presented as such. Hazards were recalled that were not reported. As discussed earlier in the paper, a number of reasons are outlined in the literature to explain poor reporting. Wogalter et al (1999) suggested that the likelihood of injury may drive reporting and in this experiment the exhibition environment may have been perceived as a 'safe' area, meaning that participants did not expect to be injured during their visit. The literature also suggests that unexpected hazards are more likely to be reported than expected hazards which explains why the chemical cupboard was reported the highest as the fuse box and the pipe are features one would expect to find in this environment. The environment may also influence reporting; in this case, the busy and noisy environment of the Science Gallery may have reducing the ability of participants to notice the hazards, particularly the pipe which make a hissing noise but may not have been sufficiently loud to overcome background noise. The other exhibitions around the hazards were specifically designed to engage visitors and may have reduced their attention to their environment.

The study was limited by the lack of control over the sample that was drawn from the general public meaning there may have been a mix of people from backgrounds with different levels of risk and hazard awareness. However, the size of the sample should provide some balance for this. The study also lacked control over the experimental conditions with varying levels of noise and busyness as well as a lack of ability to monitor participants during their visit. Future studies may replicate the experiment in a more controlled, industrial environment to collect results in a setting with higher face validity.

The outcomes of this investigation provide lessons for the process of risk reporting within industry. In particular, the suggestion that a 'safe' environment generates less reports is also applicable in an industrial context and reinforces the messages from Safety Culture research around engaging staff and taking personal responsibility for safety, not expecting safety to be provided for them. Secondly, busy or noisy environments containing highly engaging activities may also reduce reporting as individuals focus on the specific aspect of the environment of interest. Staff within an organization will have to be more engaged and trained on hazard perception if good quality data is to be derived from a reporting system as the results are suggesting staff needs to be made aware of hazards and be involved in actively assessing them themselves if they are expected to remember and recall the hazards accurately. There is scope to use this experimental methodology as training within organizations to improve awareness of hazards and reporting practices. Furthermore there may be benefits to study possible alternatives in the actual physical methods of reporting to see which method may produce the best reports.

References

Craik, F. I. M., & Lockhart, R. S. (1972). Levels of processing: A framework for memory research. *Journal of Verbal Learning and Verbal Behavior*, **11**(6), 671–684. doi:10.1016/S0022-5371(72)80001-X

Deery, H. A. (1999). Hazard and Risk Perception among Young Novice Drivers. *Journal of Safety Research*, **30**(4), 225–236. doi:10.1016/S0022-4375(99)00018-3

Johnson, C. (2003). *A Handbook of Incident and Accident Reporting* (1st ed.). Glasgow: University of Glasgow Press.

Kongsvik, T., Fenstad, J., & Wendelborg, C. (2012). Between a rock and a hard place: Accident and near-miss reporting on offshore service vessels. *Safety Science*, **50**(9), 1839–1846. doi:10.1016/j.ssci.2012.02.003

Leva, M. C., Cahill, J., Kay, a M., Losa, G., & McDonald, N. (2010). The advancement of a new human factors report–'The Unique Report'–facilitating flight crew auditing of performance/operations as part of an airline's safety management system. *Ergonomics*, **53**(2), 164–83. doi:10.1080/00140130903437131

Leva, M. C., Mcdonald, N., Del Sordo, D., Righi, P., & Mattei, F. (2010). Performance Management in a Small Regional Airport, from Day to Day Data Collection to Resilience. In *EUROPEAN SAFETY AND RELIABILITY CONFERENCE*. Helsinki.

Leveson, N. G. (2011). *Engineering a Safer World*. Cambridge, MA: Massachusetts Institute of Technology.

Pérezgonzález, J. D., McDonald, N., & Smith, E. (2005). A review of the occurrence reporting system proposed by EASA Part-145. *Safety Science*, **43**(8), 559–570. Retrieved from http://www.sciencedirect.com/science/article/pii/S0925753505000640

Wiegmann, D. A., & von Thaden, T. L. (2003). Using Schematic Aids to Improve Recall in Incident Reporting: The Critical Event Reporting Tool (CERT). *The International Journal of Aviation Psychology*, **13**(2), 153–171. doi:10.1207/S15327108IJAP1302_04

Wogalter, M. S., Young, S. L., Brelsford, J. W., & Barlow, T. (1999). The Relative Contributions of Injury Severity and Likelihood Information on Hazard-Risk Judgments and Warning Compliance. *Journal of Safety Research*, **30**(3), 151–162. doi:10.1016/S0022-4375(99)00010-9

STAKEHOLDERS' VIEWS ON THE AGEING CONSTRUCTION WORKFORCE: PRELIMINARY FINDINGS

S.J. Eaves, D.E. Gyi & A.G.F. Gibb

Loughborough University Design School, Loughborough University
School of Civil and Building Engineering, Loughborough University

The UK population is rapidly ageing, resulting in an older workforce. In construction, older workers face particular challenges due to the harsh conditions of the workplace and the heavy manual nature of tasks. Although perceptions of older workers include them being slow and averse to health and safety training, the construction industry needs to consider the ways in which healthy ageing can be encouraged in the workplace. This is an important issue for both managers and the workers themselves. Focus groups were held with construction stakeholders in three companies to investigate their views on older workers, healthy ageing and opportunities and barriers. This paper presents the preliminary findings from these focus groups.

Introduction

The UK is experiencing an ageing population due to a decrease in birth rates and an increase in life expectancy (Jacobzone, 2000). This has coincided with the abolishment of an official retirement age and an increase in state pension age to 68 by the year 2046 (EUOSHA, 2014). In construction, ageing can be tough. For example, harsh weather, poor natural light and ventilation and high levels of noise have the ability to exacerbate natural declines in ageing such as vision and hearing difficulties. Heavy manual tasks and repetitive bending and twisting have been shown to be important predictors in early retirement from the construction industry, as well as increasing the likelihood of musculoskeletal symptoms (Hengel et al., 2012).

Older workers have been shown to be negatively viewed by some managers in construction. For example they are considered to be slower, more difficult to train than younger workers and averse to health and safety regulations such as wearing personal protective equipment. However, positive perceptions of older workers include their dedication, reliability, knowledge and experience and they are often respected by younger workers, who appreciate their experience and advice (Leaviss et al., 2008).

In the light of an ageing workforce, research is needed to investigate healthy ageing at work. Research has shown that changes to the design of the workplace and working behaviours can influence the health and well-being of employees, leading to an increase in productivity and a decrease in absenteeism (Loch et al.,

Table 1.   Structure and content of focus groups.

Theme	Topic	Description
Introduction	Welcome and introduction	Context & background of the older workforce Experience and knowledge of ageing in construction
Views	Perceptions of older workers	Advantages, disadvantages, barriers and issues faced with an older/ageing workforce
Design	Ideas from the workforce	Company specific ideas presented Do you think these would work?
	Discussion	Opportunities/barriers
Future	Moving forward	How will you keep your workers involved? How will you continue to capture their ideas?

2010). Healthy ageing can be encouraged by workers of all ages sharing ideas about good working practice and healthy behaviours. Intervention and change can have a strong impact when the end-user is involved (Hignett, 2005) which provides sound evidence for participatory ergonomics in the construction industry.

Previous research by the authors has found that construction workers of all ages have good ideas about how to improve their health and well-being at work (Eaves et al., 2014). The findings suggest that involving the workforce in improving the workplace is beneficial for all workers. For change to be effective, managers must be supportive of both the initiatives and of the workers. Based on this, focus groups were held to investigate stakeholders' perceptions of older workers and their views on the opportunities and barriers to change in the industry.

Method

Stakeholders (such as site managers, engineers, health and safety professionals and client liaison officers) of a maintenance facility, domestic build company and a civil engineering company were asked to take part in focus groups. Table 1 summarises the structure and content of these focus groups.

Stakeholders were given a short presentation providing background to the research and were then asked to discuss their opinions and perceptions of older workers both in general and specifically in the construction industry. They were then presented with ideas on potential changes to improve health, well-being and working behaviours from their own workforce, these were extracted from in-depth interviews from previous research (Eaves et al., 2014). Preliminary findings from the focus groups are presented in this paper.

Results

Three focus groups were held with stakeholders (17 participants). Audio recordings of the groups were uploaded, transcribed and thematically analysed in NVivo10 by the researcher under the themes shown in Table 1.

There were similar numbers of positive and negative perceptions of older workers, with 22 negative and 24 positive references in total. Negative perceptions included older workers having 'old school' habits; not being very open to change; older workers thinking they know better and not complying with health and safety regulations. Older workers were positively perceived to be more experienced, loyal, knowledgeable of their trade, more respectful and less lazy than young workers. Managers felt that they worked at a slower pace but in doing so were more consistent and productive over long periods of time.

Stakeholders were positive about the ideas to improve health and well-being from their own workforce. They agreed changes could be easily made for low cost solutions such as improving facilities and increasing engagement in toolbox talks. There were lengthy discussions held about how these improvements could be made, including revising previous techniques used within the companies such as incentives of certificates, awards evenings or free meals in the canteen. However it was felt that incentives such as free fried breakfasts undermined the purpose of health and well-being initiatives. Ideas less favourably perceived included physiotherapists providing massages for workers and having on-site occupational health professionals; the biggest barrier to these suggestions was money. Another barrier was the engagement of the workforce; whilst they recognised that the ideas were good, they felt that in reality, long-term engagement was impractical due to the workforce losing interest.

Stakeholders were positive about moving forward and there was much discussion on how the workforce could continue to be involved. Ideas included noticeboards to present posters on health and well-being and feedback on initiatives and ideas. They felt it was important to place less emphasis on 'safety' and more on 'health and well-being'. They were wary of making big changes as previous initiatives had not been followed through very well on sites. More importantly they were prepared to listen to the workers' ideas and seemed keen to make improvements.

Discussion

Stakeholders appear to hold mixed perceptions of older workers, echoing findings of previous research. As older workers are perceived to be knowledgeable and experienced, industry stakeholders should consult them when making changes to the workplace: previous research has shown that workers of all ages can make valuable contributions to change which can encourage healthy ageing in the workplace and potentially help combat issues arising with the ageing workforce (Loch et al., 2010).

Previous research has shown that including workers in decision making can lead to them feeling valued with a higher sense of worth in the workplace (Wilson, 1995). This may reduce negative perceptions about older workers such as their aversion to health and safety regulations and personal protective equipment, and instead improve communications between the workforce and management.

Conclusion

Older workers are perceived to be an asset to the construction workforce but stake-holders also have concerns regarding their attitudes to change and health and safety. Changes are necessary to ensure workers are able to age healthily in the work-place: it is essential that the benefits of older workers are exploited to ensure their knowledge, experience and loyalty is not lost.

Statement of relevance

Good practice and working behaviours can be facilitated to encourage healthy ageing in the light of an ageing workforce. Managers and supervisors have a responsibility to listen to the ideas of their workers in order to improve health and well-being in the construction industry.

References

Eaves, S., Gyi, D. and Gibb, A. 2014. Construction workers' views on work-place design and healthy ageing. In A. Raiden (ed.) and E. Aboagye-Nimo (ed.) *Proceedings 20th Annual ARCOM Conference,* 311–320.

EUOSHA, 2014. Safer and healthier work at any age: Country inventory: UK No. EUOSHA-PRU/2013/C/02 Draft V3.

Hengel, K. M. O., Blatter, B. M., Geuskense, G. A., Koppes, L. J. L. and Bongers, P. M. 2012. Factors associated with the ability and willingness to continue working until the age of 65 in construction workers. *International Archives of Occupational and Environmental Health,* 85(7), 783–790.

Hignett, S., Wilson, J. R. and Morris, W. 2005. Fiding ergonomic solutions – participatory approaches. *Occupational Medicine,* 55, 200–207.

Jacobzone, S. 2000. Coping with aging: International challenges. *Health Affairs,* 19(3), 213–225.

Leaviss, J., Gibb, A. and Bust, P. 2008. [Online] Available at: http://www.sparc.ac .uk/media/downloads/executivesummaries/exec_summary_gibb.pdf [Accessed January 12th 2013].

Loch, C., Sting, F., Bauer, N. and Mauermann, H. 2010. How BMW is defusing the demographic time bomb. *Harvard Business Review,* 88(3), 99–102.

Wilson, J. R. 1995. Solution ownership in participative work redesign: the case of a crane control room. *International Journal of Industrial Ergonomics,* 15, 329–344.

HUMAN FACTORS AT THE CORE OF TOTAL SAFETY MANAGEMENT: THE NEED TO ESTABLISH A COMMON OPERATIONAL PICTURE

M. Chiara Leva[1], Tom Kontogiannis[2], Nora Balfe[1], Emmanuel Plot[3] & M. Demichela[4]

[1]*Centre for Innovation in Human Systems, Trinity College, Dublin, Ireland*
[2]*Dep. of Prod. Eng. & Management, Technical University of Crete, Greece*
[3]*INERIS Parc Technologique Alata, Verneuil-en-Halatte, France*
[4]*Department of Chemical Engineering, Polytechnic of Turin, Italy*

The Total Operations Management for Safety Critical Activities (TOSCA) project aims to develop a safety management framework that integrates best practices, tools and methods for functional analysis, risk assessment, interactive emergency scenarios analysis, performance monitoring, design review, training and knowledge management. The TOSCA approach is described by a T-model based around a central 'Common Operational Picture' (COP) that holds information regarding the operational system, and is used to support risk assessment and management. The information held in the COP may be represented in different ways but should be accessible to all stakeholders in order to analyse and communicate risk, and to support training and procedure design.

Total safety management

Over the recent past, the accumulation of major mishaps, crises and accidents has made it clear that organisations must still improve their capabilities to address safety, not as a stand-alone activity that is separate from the main activities and processes of the organisation, but as an integrated part of total performance management. Total Safety Management uses the basis of Total Quality Management to drive safety within an organisation (Herrero et al, 2002), but in contrast to quality management, Total Safety Management influences performance, quality and safety resulting in a much wider beneficial effect (Cooper & Phillips, 1995).

It is essential that we understand how weaknesses in the technical processes combine with flaws in organisational interfaces and give rise to significant losses and major industrial accidents. The traditional fields of practice, such as risk analysis (RA) and probabilistic safety assessment (PSA), have often been rooted in simplified accident models and failed to conduct a functional analysis that takes into account any dependencies between technical, human and organisational processes. In addition, traditional RAs and PSAs have not provided robust solutions because they have not been embedded within a 'total operations' or 'performance management' framework to deliver solutions that are both innovative and safe. It is not sufficient that production systems are reliable (i.e., their failure probability is acceptably low); they must also be resilient and capable of recovering from irregular variations, disruptions and degradation of working conditions. It is often the case that system vulnerability and resilience arise from the same interactions between socio-technical dimensions. Thus, the mechanisms that create vulnerabilities and resilience cut across traditional disciplinary borders, e.g., engineering, sociology, psychology and political science.

In the last decade, there has been an increasing interest in occupational health and safety in Small and Medium Enterprises (SMEs) accompanied by many European projects supporting their viability (see European Agency for Safety and Health at Work Report 2005). The majority of studies in the literature have found that SMEs have an increased risk of accidents compared to large enterprises. However, Sørensen et al. (2007) found that this relationship only holds for SMEs that are independent. In contrast, for SMEs that are part of larger organisations, the work environment does not seem to present more hazards than the large enterprises. Another survey conducted in Italy (84 small-sized and 25 medium-sized enterprises responded to a questionnaire) reported on the importance of SMEs' perception of safety and identified current safety management priorities and methods. Micheli and Cagno (2009) found that, although 80% of SMEs claimed that safety was among their main priorities, they reported problems in planning safety interventions because of limited financial resources, lack of management tools and a burden of compliance with regulations and codes. SMEs focused their investments on issues associated with purely regulatory or legislative aspects, that is, (1) training and information of workers on safety, (2) upgrading installations to comply with safety standards, and (3) introducing safer production technologies and personal protective equipment. A tendency was observed among SMEs to outsource safety management to compensate for the lack of specific competences within the enterprise; this tendency was greater in small-sized enterprises. Therefore, SMEs can gain considerable benefits in safety and performance by developing a capacity to risk assess actual operations with a practical and resilient methodology as well as a capacity for monitoring operational data in a solution embedded in their everyday data collection process.

The Total Operations Management for Safety Critical Activities (TOSCA) project is a European Project within the context of the 7th Framework Program aimed at developing an innovative approach to integrate and enhance safety, quality and productivity, especially for SMEs in the process industry. The scope of TOSCA is to establish an economically suitable framework in which innovative tools and techniques (e.g. advanced 3D software, virtual reality, innovative theoretical models, updated information exchange protocols, etc.) are used together with established approaches in order to take advantage of possible synergies in processing human factors requirements, fulfilling regulations, improving safety, and enhancing productivity.

To achieve this, the project is developing a theoretical framework for Total Safety Management in the process industry, particularly focused on SME applications.

The aim is to better define and highlight the needs of the industry regarding the development of an integrated methodology for assessing safety, quality and operations management.

Requirements for safety management

Safety management has traditionally focused on correcting safety concerns, problems, or hazards and taking the necessary steps to bring the system back to normal operation. Existing safety approaches seem to rely on what is commonly known as closed-loop feedback control. We need to develop a new approach that not only solves discrepancies between safety goals and current states but also helps us understand the current state of operations and risks involved in a plant. If stakeholders are able to develop a common picture of operations and risks then they will be in a better position to anticipate the effects of corrective actions and risk mitigations. In order to develop this anticipatory or proactive capability, our safety approach should rely on an internal model of the process that predicts the future state of the process and compares alternative actions in terms of effectiveness and cost. This type of model-driven feedback control enables safety analysts to cope with an overload of information and direct attention to critical events in a timely fashion. Safety practitioners should be able to monitor what could become a threat in the near-term and what could impair their abilities to respond (internal performance). This monitoring capability is supported by a common picture of how the technical process works, how people organise their jobs, and how the environment can affect the process and the people.

Several activities that are critical from a safety or productivity perspective may require a strong coordination between many agents (e.g., safety managers, supervisors, operators and external contractors) as well as communication of information regarding possible side-effects, threats and escalation of events. This 'knowledge transformation' process requires that data and information are systematically managed and integrated with people's knowledge of the functioning of the system. Building a common picture of opportunities and threats will allow different agents to understand the systemic causes of safety issues and provide a basis for suggesting practical interventions.

The requirements for safety management in existing and upcoming standards and regulations (example ISO 31000 and its related upcoming revisions, Seveso II directive etc.) call for a proactive strategic approach, demonstrating a capacity to anticipate risks and keep safety at the centre of changes driven by commercial competition as well as ensuring that evidence collected through risk analysis becomes an effective driver of innovation and change. This is particularly important for Major Hazardous Activities where prevention and mitigation are necessary for survival and where the complexity of organisations requires a 'system of systems' perspective. However, there is currently a gap between the principles stated in the available standards and regulations and the actual roadmaps to their implementation. Organisations, especially the safety critical ones, find it difficult to integrate their different functional units in a common programme of operations management.

Common Operational Picture in safety management: COMPRIS

A Common Operational Picture is a single source, usually a display, of relevant operational information. The term originates from the military domain where it is used to describe the complete graphical picture of the battlefield used by commanders to make effective command decisions (Looney, 2001). The aim of a common operational picture is to share situation awareness among distributed stakeholders and the concept has also been applied in emergency management and humanitarian crisis management (McNeese et al, 2006). TOSCA applies this concept to safety management, and is developing a *Common Operations Management and Risk Information System* (*COMPRIS*). COMPRIS provides a representation of information and knowledge about the operational system that can be used to support risk assessment and safety management. The information may be represented in different ways, but should be accessible to all stakeholders involved in a project in order to analyse and communicate risk, and to support training and job design.

In terms of control theory, COMPRIS is a 'mental model of how the system works' that guides the application of a safety management system (SMS) in everyday practice. In this respect, COMPRIS is a 'mental model' of how a specific SMS works, what risks are significant at a particular area, what methods should be used to assess risks, what uncertainties exist in a risk evaluation, and what risk mitigation measures can be chosen to reduce risk to acceptable levels.

Establishing COMPRIS, or a 'risk picture', involves a useful synthesis of the risk assessment, with the intention to provide understandable information to the relevant decision-makers and users about the risk and the risk assessment performed. The 'risk picture' shall be understandable by all relevant personnel, decision makers as well as engineering and/or operating personnel. This may be achieved with the use of tailored documentation and presentations to different groups of internal and external stakeholders. The presentation and documentation of the risk picture shall be a comprehensive, balanced, many-facetted and holistic picture of the risk associated with facilities and operations.

COMPRIS is an internal model that addresses how safety is measured and what 'performance indicators' will be monitored to measure not only the 'final outcome' but also 'antecedents' so that changes are made before undesired outcomes are produced. Performance indicators provide a good basis for integrating measures of safety with productivity and quality control. Figure 1 shows a model-driven safety management system that comprises four functions from resilience engineering (i.e., RESPOND – MONITOR – ANTICIPATE and LEARN):

1. Given a specified safety goal, the safety practitioner has to RESPOND; take control action that changes the technical process in order to produce the desired output. In turn, this is measured by suitable indexes and by feedback (MONITOR).
2. COMPRIS (i.e., the internal model of the process) also needs to enable safety practitioners to ANTICIPATE disturbances so that actions are taken before they

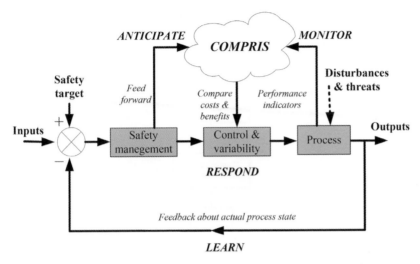

Figure 1. The role of COMPRIS within a safety management system.

occur. The advantage here is that control actions can prevent adverse events from taking place or intervene before their consequences have had time to spread to other parts.

3. Safety practitioners must be also able to control threats from internal variability due to fatigue of personnel, changes in team composition, unavailability of tools and so forth.

4. Safety practitioners must be able to LEARN from experience, which includes many changes in the internal model of the process.

Individual companies have different approaches to building their 'understanding of risks and risk control measures' depending on the resources invested in safety management. SMEs may have greater problems than large companies in getting an accurate 'COMPRIS' (or 'understanding') of their risks and the possible measures to control risks. Among other reasons, SMEs are more diversified than large companies and this makes it difficult to draw on what risk information already exists in the specific industry domain they operate. There is a need therefore to examine the scope of COMPRIS in both SMEs and large companies.

Application of COMPRIS in TOSCA

There is limited information on 'what constitutes a common operational picture' in a SMS. There has been a tendency to consider the 'risk registry' and the 'risk acceptance criteria' as a common operating picture so that all stakeholders and operators are aware of the whole spectrum of risks in a company. However, it is interesting to ask whether the common picture should also address the risk analysis

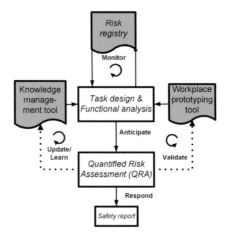

Figure 2.   TOSCA methods and tools to establish COMPRIS.

tools used to identify risks, their limitations, the kinds of people who participated in the risk analysis, and the influence of the conditions under which the analysis took place. Other issues concern the level of detail that should be presented about risk items such as the description of risks, the possible risk control measures, the responsible person, etc. Individuals may have their own boundaries of responsibility and it becomes time consuming for them to consider any 'risks' in other areas beyond their responsibilities. We can view the common operational picture as a database of risk information that should be accessible to key players but 'who should see what information' remains a challenging issue to be resolved. In TOSCA a set of methods and tools are used to establish the COMPRIS and produce a robust risk assessment (Figure 2). These include:

- Task design and high-level functional hazard analysis using a participatory approach. This can be achieved with a modified Business Process Modelling (BPM; White et al, 2008) approach;
- Operational risk screening in order to feed in the results of functional analysis into bow-tie diagrams to identify important technical and human barriers that prevent or control industrial hazards;
- Quantified risk assessment for complex scenarios using tools, such as traditional fault trees or computational fluid dynamics, for consequence and likelihood assessment.

The results of risk analysis can be tested in workplace prototyping tools that create virtual workplaces for operators to interact with planned designs or changes in a 3D virtual representation. 3D maps of the site can also be used to display risk information in an accessible format, including critical risk areas and key safety barriers. Many SMEs complain that risk assessment is a complicated process that can be done only once during the design stage. Updating the risk assessment process every time there is a change management program is very cumbersome. For this reason, TOSCA builds a tool for managing and visualising safety knowledge.

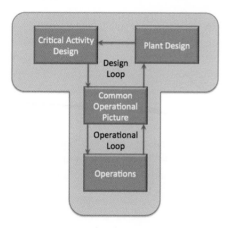

Figure 3.   TOSCA Total Safety Management Framework©.

Finally, TOSCA builds a risk registry for collecting data, monitoring risks and communicating safety knowledge. The risk registry is based on the concept of building a 'business case for safety' where the values of safety are seen beyond the traditional reduction of risks to establish connections with quality and productivity.

Current applications and future developments

The TOSCA T-model© describes the Common Operational Picture at the heart of TOSCA safety management, and links this picture to a design loop and an operational loop (Figure 3). The design loop applies the TOSCA tools to the design of new plant sections, the management of technical and organisational changes, and the risk analysis of critical activities. The operational loop applies the methods and tools to the management of safety barriers, the design of workflows, coordination of teams, and training. Modular products to deliver safety management are under development for each area of the framework.

To follow up the development and the testing of the TOSCA methodology in the three main sections and the deployment of the tools to support it within the project we have currently chosen to develop five test beds to test the effectiveness of the proposed approaches in the areas identified. The test beds include the use of process mapping and task analysis to improve risk assessments for a food processing organisation, the establishment of a risk register and a set of KPIs to support hazard identification and risk monitoring in a energy generation company, the development of a 3D risk map to be used as a knowledge and risk management system for a company producing fertilisers, the use of rapid prototyping to optimise a rare testing procedure for a LPG storage organisation, and the use of VR to review and train operators and contractors for loading and unloading of cryogenic liquids. Three of the test beds will be conducted in SMEs, allowing the TOSCA approach to be tailored and tested for small enterprises as well as large ones. The results of the first phase of testing will be used to refine the individual tools and techniques deployed, as well as the overall approach and the concept of COMPRIS.

Acknowledgments

This research has received funding from the European Commission's Seventh Framework Programme FP7/2007-2013 under grant agreement FP7-NMP-2012-SMALL-6-310201 "TOSCA" and it is © copyright of the TOSCA consortium).

References

Cooper, M. D. and Phillips, R. A. 1995, Killing two birds with one stone: Achieving quality via total safety management. *Leadership and Organisational Development Journal* 16(8): 3–9.

European Agency for Safety and Health at Work, 2005, Report 2004: ISBN 92-9191-141-0 Promoting health and safety in European Small and Medium-sized Enterprises (SMEs)://osha.europa.eu/en/publications/reports/ag05001.

Herrero, S. G., Saldaña, M. A. M., del Campo, M. A. M. and Ritzel, D. O. 2002, From the traditional concept of safety management to safety integrated with quality. *Journal of Safety Research* 33: 1–20.

ISO 31000, 2009, *Risk Management – Principles and guidelines*.

Looney, C. G. 2001, Exploring fusion architecture for a common operational picture. *Information Fusion* 2 (4): 251–260.

McNeese, M. D., Pfaff, M. S., Connors, E. S., Obieta, J. F., Terrell, I. S. and Friedenbery, M. A. 2006, Multiple vantage points of the common operational picture: Supporting international teamwork. In *Proceedings of the Human Factors and Ergonomics Society 50th Annual Meeting*, 467–471.

Micheli, G. and Cagno, E. 2009, Perception of safety issues and investments in safety management in small and medium-sized enterprises: a survey in the Lecco area. *Prevention Today* 4(10): 7–18.

Seveso II. 1996, *Council Directive 96/82/EC on the control of major-accident hazards*.

Sørensen, O. H., Hasle, P. and Bach E. 2007, Working in small enterprises – is there a special risk? *Safety Science* 45: 1044–1059.

White, S. A. and Miers, D. 2008, *BPMN Modeling and Reference Guide*. Future Strategies Inc.

WORKSHOP: MANUAL HANDLING
WHEN IS A TASK OK OR NOT OK?

Wendy Morris & Michael Marshall

Jaguar Land Rover

What is the background to the workshop?

The Manual Handling Operations Regulations 1992 require an employer "so far as is reasonably practicable, to avoid the need to undertake any manual handling operations at work which involve a risk of an employee being injured" (Regulation 4(1)a). To comply with this regulation the employer is required to undertake an initial level of assessment to determine if an employee is at risk of injury from a task that involves manual handling. There is an implication that some manual handling tasks may not present a risk of injury to the employee and an employer can discriminate between those that do and do not involve the risk of injury.

The HSE has developed tools to assist with a manual handling assessment (Risk Filter In L23 and the MAC and VMAC tools). These are based on psychophysical studies carried out by Snook and Cirello (1991) Outside the UK the NIOSH equation (biomechanical basis) is widely used and is referred to in European and International standards (ISO 11228 and EN 1005). Comparison of the tools and guidance identifies some discrepancies between the guideline loads for various frequencies of lifting. The introduction to L23 also notes that the guidelines "should not be regarded as precise recommendations" and "where there is doubt make a more detailed assessment." An ergonomics approach is set out for the more detailed assessment considering "the load, the task, the environment, the individual and other factors". The risk assessment approach does require a level of objective data to be collected (weight, height, distance, frequency) but complex tasks that have variable weights, heights, distances, frequencies then require a subjective judgement on the risk of injury. As an ergonomist with several years of experience in manufacturing industry most tasks have variation and I still find it a challenge to weigh up all the information and determine the level of risk and what may be considered reasonably practicable steps to manage the risk identified.

The MHO regulations definition of manual handling is the transporting or supporting of a load (including the lifting, putting down, pushing, pulling, carrying or moving thereof) by hand or by bodily force. The MHO regulations apply to the workplace but a level of manual handling is part of everyday activities for example getting shopping from the store to the home, caring for family members, maintaining the home environment where 'risk assessment' is generally not a formal consideration. The development of musculoskeletal disorders has been linked to a wide range of physical and psychosocial risk factors and the contribution of both work and personal activities to a person's experience of a musculoskeletal disorder is difficult to determine.

What is the purpose of the workshop?

The purpose of the workshop would be to present some real life examples of industry based work tasks for the group to consider how they would assess them using the published tools available to an employer in the UK and how they would determine whether the task is ok or not ok. In doing so the evidence base behind the tools and the decision making process would be explored and gaps in research identified to assist in the targeting of future research resource.

Delegates attending the workshop will also have an opportunity to bring examples from their areas of work that have been problematic for review in small groups with feedback to the main forum as part of the discussion.

Who is your intended audience?

- Ergonomists / health & safety professionals who undertake manual handling assessments for employers and need to make judgements as to whether a task is ok or not ok and where risks for handling injuries are identified deciding what actions are reasonably practicable to reduce the risk.
- Researchers who are interested in developing new or improving the current manual handling tools to understand the issues practitioners face.
- Policy makers / health and safety executive representatives to support discussion of the interpretation of the legislation and approved code of practice guidance

What key skills, tools or knowledge do you want your participants to take away?

To have gained a greater understanding of the strengths and weakness of the current published manual handling assessment tools. Also had an opportunity to benchmark their judgement on manual handling risk and gain confidence in providing advice to employers on manual handling risk assessment and reasonably practicable control measures.

What activities will you use to engage your participants and help them learn?

Case study presentations, panel of experts for Q&A session and small group work to benchmark manual handling assessment judgements.

What materials will you use or need?

Laptop and projector required for presentation of case studies, flip charts for small group discussions, copies of published manual handling assessment tools (L23 Risk Filter, MAC and VMAC tools). We can supply laptop and copies of manual handling assessment tools but would request projector, flip charts and pens be provided.

THE PERFORMANCE OF PLANT PERSONNEL
IN SEVERE ACCIDENT SCENARIOS

David Pennie[1], Jon Berman[1] & Trevor Waters[2]

[1] *Greenstreet Berman Ltd*
[2] *Sellafield Ltd*

This paper reports on research (prompted by the Fukushima-Daiichi nuclear accident in 2011), primarily in the form of a literature review that considers the response (behaviours and performance) of individuals (specifically operational or front-line staff) confronted by the challenge, stress and emotional impact of tackling a Severe Accident (SA). Five measures were identified that could improve response, these were: 1) enhance family contact and support; 2) ensure that job roles are clearly defined; 3) provide a diverse and robust means of communication; 4) seek to develop and enhance emergency response training to enable operational personnel to practice key coping strategies and 5) use site exercises to explore the psychosocial aspects of operator response.

Introduction

Background

The key driver for this research was the nuclear accident at Fukushima-Daiichi in 2011 and the recognition that knowledge from that event can help inform how high-hazard organisations, such as Sellafield Ltd (SL) (where failures in safety management and risk control could have a major impact upon workers and the public at large) can improve resilience in the face of such disasters. The Weightman Report (2011), instigated by UK government shortly after the accident made recommendations on a number of issues, including the need to consider the psychosocial factors that can affect an operator's willingness and ability to respond during an SA event.

About Fukushima

On 11 March 2011, Japan's east coast was hit by a magnitude 9 earthquake and then about an hour later by a very large tsunami. At the time of the earthquake three reactors were operating; one was on refuelling outage and two were shut down for maintenance. When the earthquake struck, all three operating reactors shut down automatically and shutdown cooling commenced. When the tsunami hit, all electrical power to the cooling systems for the reactor and reactor fuel ponds was lost, including that supplied from backup diesel generators. Over the next few days workers struggled to restore power. Failure to do this caused the fuel to gradually heat up until the cladding around the fuel reacted with steam releasing hydrogen, which ignited, causing several large explosions from 12th to the 15th of March. Containment was breached, leading to a significant release of radioactivity into the environment. The accident continued for over a week with attempts to cool the reactors and spent fuel ponds by using untried and unplanned means. Gradually electrical supplies were reconnected and a degree of control returned. The accident was classed as a Level 7 event, the highest level classification on the International Nuclear and Radiological Event Scale. The following is one of a number of quotations, taken from TEPCO's own accident investigation (2012) and serves as an illustration of the impact of the event on those on site and tasked with responding: *"At around 11:00 when I was about to contact the head office by telephone and report after sampling data there was an incredible blast. I couldn't see anything through the dust. I was dazed for a second and then confirmed the safety of my crew and reported to that office. They told me that Unit 3 had exploded. I figured that the next shift wouldn't arrive and prepared myself to die from long-term exposure".*

Method

Research aim

The overall aim of the research was to provide a sound understanding of what response might be expected during an SA and under extreme threat – illustrated by the above quote. It considered the factors that would influence response and the arrangements that might help maximise an operator's willingness to act and perform effectively. The research aims were distilled down to the following key themes: 1) understanding more about how response is affected; 2) predicting or modelling response and 3) identifying interventions to enhance response. A number of related research questions were developed around the themes.

Data gathering

The primary data gathering method was a review of literature from the UK and abroad. This included: academic; grey (informally published material less likely to be obtained via academic channels) and other media sources. The literature review followed a formal process within which it: used search terms: was question led; applied inclusion & exclusion criteria and used quality appraisal to rate materials (e.g. 'weight evidence').

This initial literature review identified approximately 160 papers and source material. Following quality appraisal process the number of reviewed publications was refined to 80 papers and a further 10 or so relevant other media sources (e.g. newspaper articles and TV interviews).

In addition to the literature review the research was supported by a visit to Sellafield site to gain contextual insight into SA management, and a small number of interviews with subject matter experts and a range of SA management professionals from related industries. Both the interviews and the site visit helped with interpretation of the findings from the literature review.

Findings

Summary

The factors that influence and degrade behaviour and performance during an SA were categorised as 'stressors'. These can be 'external' – the direct and tangible pressures linked to the changed impact event environment (e.g. increased noise, actual physical danger, failure of lighting; information overload/under load etc.) or 'internal' – an individual's reaction to an event, encompassing both physiological and psychological pressures (e.g. fear, anxiety – worry and concern about self and others). Stressors can affect response (e.g. willingness to act; behaviour and performance) in a number of ways. The main ways that these effects are manifest is through impairment to: attention; decision making and memory, and maladaptive behaviours such as: denial; freezing; stereotypical behaviour; perseveration (repetition of a response even though it might be inappropriate) and inappropriate behaviours (e.g. hyperactivity). Willingness to respond will be adversely affected by: threat to life; the unfamiliarity of the event; difficulty in contacting family or loved ones; poor communication. The complex and unique nature of an SA means it is difficult to model or predict exactly how stressors affect response. For example we cannot say for certain that fear and anxiety will always lead to cognitive impairment. We do know, however, that response will vary. For example, some people will fail to respond adequately; most will respond as expected and others will go beyond expectations of not only the organisation but also themselves – as was the case with Chesley Sullenberger who landed Flight UA1549 on the Hudson River. In addition the findings from the literature review suggest that performance will be affected by the presence (or otherwise) of a number of moderating factors, which have been grouped by the researchers into four key areas:

- Organisational culture & support.
- Organisational preparedness.
- Individual characteristics.
- Team characteristics.

Linking stressors, moderators, and response does not seem possible with any great certainty. For example an internal stressor, such as anxiety, is unlikely to be moderated by only one area (e.g. organisational support) or affect response in only one area (e.g. willingness to act). In reality it is far more likely that anxiety will be moderated by a combination of factors and will affect response in a number of ways. This would suggest that although counterintuitive to Human Factors, it may

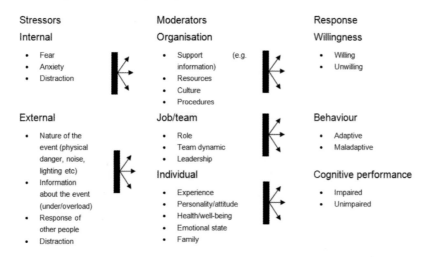

Figure 1.   Model of operator response to a severe accident event.

be better to focus less on the root cause of the problem (e.g. the stressors) and instead seek to enhance moderators where possible and practical to do so. The relationship between stressors, moderators and response is represented in the models below. This model (developed as part of the research and synthesised from existing models) aims to enhance understanding; it does not attempt to provide a causal explanation of operator response.

The key components of response (willingness to act, cognitive performance and behaviour)

Research by Shaw (2001) found consistent evidence that; *"abandonment of disaster-related duties is rare"*. This is supported by a number of studies and, contrary to popular belief, suggests that fear, panic and abandonment of duties is not a normal or common response. One reason postulated by Sime (1983) for why people might stay rather than abandon a situation is because: *"in the face of threat, we are motivated to seek the familiar rather than simply exit; and (ii) the presence of familiar others (i.e. affiliates) has a calming effect, working against a 'fight or flight' reaction"*. There are acceptations, however, notably the case of the captain of Costa Concordia. Fritz and Marks (1954), in a highly regarded study of human behaviour in disasters, state: *"panic flight and other highly uncontrolled forms of behavior appear to occur under quite restricted conditions present only in some disasters and only for some of the persons involved in such disasters."* Other evidence, primarily from first-hand accounts of responders during impact events, indicates that given certain circumstances some people may abandon duties and fail to act when: there is direct threat to safety and life; when the nature of the event and the risk are unfamiliar; when it is difficult to contact family and when poor communication adds to uncertainty about the event. The literature review also indicates that people may be further influenced (positively or negatively) by factors such as: the perceived level of sufficient organisational support; the quality of leadership; the team dynamic and professionalism; belief in their job role and that they can make a difference. Once operators have made a conscious or unconscious decision to act and are in a state of being prepared to do something (e.g. ready to respond) then it is important to consider how cognitive performance and behaviour are affected.

Evidence suggests that cognitive performance is affected in three main areas: 1. Perception and attention – fixated narrowed attention and a failure to attend to alternative information. 2. Cognition and decision making – restricted ability to formulate new schema (e.g. novel or new ways of working). 2. Memory – impaired recall of information (e.g. facts and knowledge).

When faced with a threat the body prepares itself to respond (e.g. 'fight or flight'). There is a physical change in neural and endocrine chemicals (e.g. adrenaline is produced increasing heart rate etc). Initially the effects might be advantageous, such as increasing physical performance but continued production of cortisol can quickly lead to a decrement in performance. For example, although the mechanism is not fully understood there is experimental evidence that correlates continuing high levels of circulating cortisol with poorer cognitive function. The literature also identified five key maladaptive physical behaviours that are exhibited during times of extreme stress, these are:

- Denial – refusing to recognise the situation.
- Freezing – becoming paralysed and unable to decide on an action to follow.
- Stereotypical behaviour – reverting to known or familiar behaviours even though they may not be appropriate to the current situation.
- Perseveration – continuing with a course of action despite contrary and compelling new information that it might be inappropriate;
- Other maladaptive behaviour and action disorders (e.g. jumping from one task to another or simply pacing backwards and forwards).

What individual factors have an impact on operator response

There is a belief, sometimes expressed by lay people and commentators, that certain individuals possess special qualities or characteristics (including personality traits) that make them better able to respond in a crisis situation. With respect to personality traits, the findings from this research indicate that although this is likely to play a role in response, it is difficult to isolate and define what these personality characteristics might be and then to confidently identify them using psychometric tools. The findings do, however, indicate that experience appears to be important in terms of influencing operator response. This is because having previous experience allows a person to select from existing schema (ways of working) rather than to develop new concepts or novel approaches, which are more difficult to form under stressful conditions. It should be noted that having experience does not prevent the extreme response to an event as evidenced by this quote from Chesley Sullenberger during an interview with CBS television. On being asked the direct question about what were his immediate thoughts when he realised he had lost both engines on the Airbus

A320 he was flying due to a bird strike, he said: "*It was the worst sickening pit of your stomach falling through the floor feeling that I have ever felt in my life ... I knew immediately it was very bad ... my initial reaction was one of disbelief. I can't believe this is happening, this doesn't happen to me*" This quote illustrates both the emotional response and the initial reaction of denial. The difference with Chesley Sullenberger as opposed to another pilot is that he was able to recover very quickly from this initial reaction. Notably, Sullenberger's first act was to decide to return to La Guardia but after a very short time he realised he would not be able to fly that far and he was able to change his plan and look at other options, including ditching on the Hudson River, a decision made 2½ minutes into the flight. What was required on water was a very precise landing with the wings directly level and the nose up and at a speed that was survivable. This placed immense pressure on Sullenberger considering what was at stake and the fact that very few pilots had ever successfully landed on water. Sullenberger was a highly experienced pilot with nearly 30 years of experience of flying aircraft, a specialist in accident investigation and he also instructed flight crews on how to respond to crisis situations, as Sullenberger stated in the interview with CBS: "*I think in many ways, as it turned out my entire life up to that moment had been a preparation to handle that particular moment.*"

Selected measures that can enhance response (willingness, behaviour and performance)

According to the literature a key issue that can affect response is concern about family, which can impair cognitive performance (e.g. distracted by anxious thoughts) and increase the risk of role abandonment. If a person is worried about their family they might feel compelled to leave a situation to try and contact them – as was the case with some 9/11 responders (Smith, 2008). Trainor & Barsky (2011) recommend a 3-step strategy to reduce worry and concern about family and to help people to re-focus on emergency response, as follows:

- Facilitate family preparedness.
- Provide effective channels for communication, to and from the family.
- Help organise and plan for responder families' needs.

Job role and professionalism were also considered to be important motivators in willingness to act during an impact event. To some extent undertaking the role expected of you is a product of social norms, and a desire for the familiar and to be connected to others who might be responding as part of a team. Often people will also go beyond their role because it is socially the right thing to do, e.g. save a life, comfort someone, provide timely information. Initial response, as indicated by the accounts provided from emergency responders during 9/11, can simply be because it is clearly part of a designated role. If people are uncertain about role, however, this may make them more hesitant in their response. Ensuring clarity of job role, particularly when operators may have a dual role to perform, is considered to be an important intervention.

A key finding from the literature review was the effect that lack of information and communication can have on response both in terms of performance and willingness to act. At Fukushima, communication between the site and the main control room was limited and hindered the flow of information because communication devices were lost as a result of the tsunami and subsequent failure of power. The INPO report (2012) suggests that multiple and diverse means of communication should be provided to guard against such failure.

Training is considered to enhance performance by helping people to more quickly overcome the initial shock and stress and associated impairment to cognition. Also because this impairment affects the necessary development of novel ways of working, training provides a set of existing schema available for recall and it is easier to make a choice between options already accessible from memory than to develop new ones. Leach (1994) states that: *"training removes the fear which occurs through lack of knowledge while repeated drilling enables the person to function at an automatic level"*.

There is a need to distinguish training and site exercises. Training is concerned with the acquisition of skills and competence. It will comprise a range of different methods, and will provide opportunity for trainees to explore alternative behaviours and approaches. Exercises typically comprise demonstration and practice of acquired competences. In practice the two activities tend to overlap, with exercises providing training opportunities. Evidence from the literature review strongly supports the benefits of simulated exercises for helping people prepare for threat situations. This is because like training, they help to normalise abnormal situations, helping people to quickly recover after the initial shock of an impact event. They also, if designed appropriately, help to mimic the complex multifaceted nature of an event.

A discussion of the value of the research

It should be noted that although there were a number of experimental research papers only a small number (<10) were quality rated as being good. This means the findings are based largely on lower quality rated research and materials such as: accident reporting, behavioural studies and theoretical/conceptual papers and, to a lesser extent, grey material such as: media reporting, books and literature reviews. Without robust empirical evidence there will inevitably be questions over the robustness of the findings. The view of the authors is, however, that despite these reservations a clear theoretical position does emerge, that is: stressors (resulting from an impact event) will be ameliorated by a number of moderating factors which can be influenced and changed through intervention. Therefore, if organisations can do more to enhance and strengthen these moderators then response should be improved.

A further benefit of this research is that the findings provide insight into what these moderators might be, why they influence performance and also how they can be enhanced. Even though it is acknowledged that the review is reliant on less traditional research materials a measure of confidence in the findings is achieved through consensus in the data/information derived. For example, the benefits of training (as a moderator) are strongly indicated from the full range of different research materials with no alternative or no contradictory position expressed. In particular, the 10 minute TV interview with Chesley Sullenberger is highly compelling in terms of illustrating the impact of stressors and moderators (such as appropriate experience and training) in helping a pilot handle such a critical situation. Importantly a more traditional research approach might have discounted information derived from such a media source.

Finally, in addition to the observations concerning this approach to research, the authors consider that high-hazard industries, not limited to the nuclear sector, can be increasingly well-positioned to strengthen their ability to respond to severe accidents in the very unlikely event that they arise.

References

CBS news, 2009, Interview with Chesley Sullenberger.

Fritz, C. and Marks, E., 1954, The NORC Studies of Human Behavior in Disaster *Journal of Social Issues*.

INPO Special Report 2012, *Lessons Learned from the Nuclear Accident at the Fukushima Daiichi Nuclear Power Station*, INPO Special Report 11-005 Addendum August 2012.

Leach, J. 1994, *Survival Psychology*. (Macmillan Press Ltd)

Shaw, R., 2001, Don't panic: behaviour in major incidents. *Disaster Prevention and Management* 10(1): 5–10.

Sime, J. D., 1983, Affiliative Behavior During Escape to Building Exits. *Journal of Environmental Psychology* 3: 21–41.

Smith, E., 2008, Willingness to work during a terrorist attack: A case-study of first responders during the 9/11 World Trade Centre terrorist attacks, *Australasian Journal of Paramedicine* 6(1).

Tokyo Electric Power Company 2012, *Fukushima Nuclear Accident Analysis Report.*

Trainor, J. E. and Barsky, L. E. 2011, *Reporting for duty? A synthesis of research on role conflict, strain and abandonment among emergency responders during disasters and catastrophes.* Disaster Research Center.

Weightman Report, 2011, *Japanese earthquake and tsunami: Implications for the UK nuclear industry.* Final Report HM Chief Inspector of Nuclear Installations.

IS SAFETY CULTURE STILL A THING?

Steven Shorrock

EUROCONTROL, Brétigny-sur-Orge, France

Background

Safety culture has been a theme in human factors research and practice for more than two decades. The concept has resulted in a mass of research and practical interventions, both small and large, in many industrial sectors. Recent thinking has, however, been increasingly critical of the concept and the value of safety culture research and practice. Criticisms of safety culture include the accusations that safety culture is inherently contradictory (Walker, 2010), that the links with systems thinking are not evident (Reiman and Rollenhagen, 2014), that there is insufficient integration with macroergonomics (Murphy et al, 2014), that culture talk can also obscure uncomfortable, yet crucial social phenomena (Szymczaka, 2014). There is also a concern that safety culture, or facets of it such as 'just culture', may be rooted in a traditionalist safety paradigm (Safety-I) and conflict with the emerging paradigms (Safety-II, see Hollnagel, 2014). At the same time, practitioners' experience on the ground is that the concept remains useful and relevant for various practical and political reasons, and should not be discarded. The middle ground may be that our ideas about safety culture need to adapt, both theoretically and practically, in several ways. This should be of interest to many human factors specialists, both for research and practice.

Purpose

The purpose of the workshop is to discuss the value in the safety culture concept in theory and practice, experiences with assessing and working with safety culture in practice, and how research and practice might need to adapt.

References

Murphy L. A., Robertson M. M. and Carayon P. 2013. The next generation of macroergonomics: integrating safety climate. *Accident Analysis and Prevention* 68: 16–24.

Reiman, T. and Rollenhagen, C. 2014, Does the concept of safety culture help or hinder systems thinking in safety? *Accident Analysis and Prevention* 68: 5–15.

Szymczak, J. E. 2014, Seeing risk and allocating responsibility: Talk of culture and its consequences on the work of patient safety. *Social Science & Medicine* 120: 252–259.

Walker, G. W. 2010, A safety counterculture challenge to a "safety climate". *Safety Science* 48(3): 333–341.

AN ANALYSIS OF THE FATIGUE AND SHIFT-WORK ISSUES IN THE BUNCEFIELD EXPLOSION

John Wilkinson & Julie Bell

Principal Human Factors Consultant, The Keil Centre, Edinburgh
Principal Human Factors Specialist, Health and Safety Laboratory, Buxton

Although there have been multiple reports of the 2005 Buncefield explosion and fires, much of the detailed underlying analysis has not yet been fully published, particularly for the human contribution. In this paper the original Health and Safety Executive (HSE) and Health and Safety Laboratories (HSL) investigation analysis of the fatigue and shift-work contribution to the incident is presented in more detail and set in the context of the full incident. The key lessons for the major hazard industries are identified. The authors' experience of being involved, as human factor professionals, in such a lengthy and demanding incident investigation is also addressed, and the key lessons learnt are identified for future investigators.

Background to the incident

This incident was Britain's largest peacetime explosion to date (HSE 2011, p. 11). Although no one died, there were substantial economic and community impacts, e.g. the two major fuel pipelines to Heathrow and Gatwick airports were also controlled from an adjoining site. The investigation was lengthy and complex. Despite this, the incident was really very simple. At around 05.40 on Saturday 11th December 2005 one very large petrol storage tank at the Buncefield oil storage depot in Hemel Hempstead overfilled. The leak was not noticed in the control room or detected outside before a large petrol vapour cloud had formed. This ignited and exploded just after 06.00. All of the safety barriers failed on the night, including the tank level gauge which stuck so that the associated high-level alarm did not sound in the control room. The final independent high level switch (IHLS) on the tank had not been correctly set, and so also failed to work.

The depot operation was essentially simple. Three large pipelines delivered a range of fuels to the site. The smaller one was dedicated to the site and mainly site-controlled and operated; the other two were larger and faster, and mainly controlled from elsewhere. These two delivered fuels to other sites as well and so could be 'on or off' with respect to Buncefield.

Table 1.   **Shift schedule worked by supervisors at the Buncefield Oil Depot**
(N-Night; D-Day; T-terminal; P-Pipeline; RD-Rest day; S/By-Standby).

Shift	Mon	Tues	Weds	Thurs	Fri	Sat	Sun	Total Hrs
1	N(T)	N(T)	N(T)	N(T)	N(T)	RD	RD	60
2	D(P)	D(P)	D(P)	D(P)	RD	RD	RD	48
3	S/BY	S/BY	S/BY	S/BY	N(P)	N(P)	N(P)	36
4	RD	RD	RD	RD	D(P)	D(P)	D(P)	36
5	N(P)	N(P)	N(P)	N(P)	RD	RD	RD	48
6	D(T)	D(T)	D(T)	D(T)	S/BY	S/BY	S/BY	48
7	RD	RD	RD	RD	D(T)	D(T)	D(T)	36
8	RD	RD	RD	RD	RD	RD	RD	0

The investigation

A human and organisational (HOF) specialist (first author) was involved early on in helping assess documents and other evidence, and in preparing the questions for interviews. The investigating inspectors were HOF-trained and aware and so identified emerging HOF issues quickly, allowing the relevant evidence to be captured and professionally assessed e.g. recovery of the shift schedule from the control room wall suggested that there could be an issue. The full evidence emerged or was confirmed through documents and interviews – these took over 6 months to set up and carry out, a long and painstaking process. The HSE's Health and Safety Laboratory (HSL – second author) provided deep topic expertise, essential scientific support in an area that is traditionally perceived by some as a 'soft' science. This incident happened at night and towards the end of a 12-hour shift – the classic 'small hours' period when accidents, due to fatigue, are more likely to occur (HSE 2006 p6) e.g. the pipeline supervisor was focused on another pipeline task on the depot's 'own' pipeline immediately before the release and right at the end of his 12-hour shift; and the initial confusion for both supervisors arose at the end of shift/ handover (HSE 2011). Fatigue can be acute e.g. within-shift, and 'chronic' (or cumulative) i.e. building up over a longer period – the latter was the main issue at Buncefield (see below).

Analysis of the shift schedule as planned and as worked

The interviews confirmed that the 8-week shift schedule was designed for 10 supervisors but was being worked by 8 because of on-going recruitment and retention difficulties. So supervisors were working, e.g. seven 12-hour shifts in a row (weeks 4–5), five successive 12-hour night shifts and increased overtime. They were left to arrange their own holiday and other relief cover. There were also no structured arrangements for within-shift rest and meal breaks. Supervisors would usually eat or take time out at their desk in the control room.

Table 2.   Average overtime worked by the Buncefield supervisors in the 5 months before the incident (HSL 2007).

Time period (2005)	Minimum	Maximum	Average per person
14 Nov–13 Dec	0	114	34.86
14 Oct–13 Nov	0	127	60.35
14 Sept–13 Oct	0	87	34.18
14 Aug–13 Sept	10	92	53.36
14 Jul–13 Aug	6	121	57.4

The shift schedule was 0700-1900/1900-0700, 26 work shifts over an 8 week cycle. There were 7 standby shifts and 23 rest days (RD). Staff worked on either the 'pipeline' or 'terminal' side of the operation. The total number of hours worked in any seven day period ranged from 0 hours to 60 hours. Over 17 weeks, the average was 40.2 hours. *However, this planned schedule of work does not provide a complete picture: specifically, Shifts 4 and 5 combined to make a 7-day period where 84 hours are worked.*

Overtime was also required over a sustained period for on-going project and (ironically) safety improvement work. While individual levels varied the following table (Table 2) shows the scale. The drop off towards December was largely due to an increasing reluctance by supervisors to continue to work these levels of overtime.

HSL also carried out an assessment of the shift schedule against the HSE shift-work guidance (HSE 2006). The original analysis follows and is based on the key dimensions of concern identified in the guidance for shift work.

Shift start times

Planned shifts started at 0700 and 1900. The start time of 0700 was considered adequate; HSE Guidance recommends avoiding start times before 0700 however consideration should be given to commuting times because time required to travel to work will reduce the rest period and the time available for sleep. Early morning starts can reduce sleep and increase the risk of fatigue. Based on current knowledge, a start time of 1900 is adequate.

Shift duration

HSE Shift Work Guidance recommends that employers should avoid shifts that are longer than 8 hours, where work is demanding, safety critical or monotonous and/or there is exposure to work-related physical or chemical hazards. This is because, while research to compare the effects of 12-hour shifts with those of 8-hour shifts has produced equivocal findings, 12-hour shifts are related to greater subjective fatigue, insufficient sleep and performance decrements (HSG256 Table 5, p. 20). However, a number of advantages are claimed for 12-hour shift work, including the reduction in the number of handovers (but see Wilkinson & Lardner 2012 for a full discussion of the handover issues at Buncefield – same person handover can introduce other error potential). In essence though whichever length is chosen it will need managing both to release the potential benefits and minimize the potential disadvantages. The working of 12-hour shifts is acceptable if it is managed correctly, for example if the following are taken into consideration:

- The employees are committed and motivated to work 12-hour shifts;
- Work is organised to minimise the risk of fatigue while undertaking tasks that are sensitive to the effects of fatigue;
- Breaks are taken at regular intervals throughout the shift;
- The number of consecutive shifts is minimised, particularly night shifts;
- Adequate rest days are taken following a period of 12-hour shifts;
- Methods for mediating the effects of fatigue are utilised.

The available evidence indicated that the working hours by supervisors at the Buncefield site were not managed to minimise the risk of fatigue.

Number of consecutive shifts

When working 12-hour shifts it is important to limit the number of consecutive shifts, especially night shifts. The HSE guidance recommends, "in general, a limit of 5–7 consecutive working days should be set for standard (i.e. 7–8 hour) shifts. Where shifts are longer than this, for night shifts and for shifts with early morning starts it may be better to set a limit of 2–3 consecutive shifts, followed by 2–3 rest days to allow workers to recover". (HSE 2006 Table 7, p. 23). The number of consecutive shifts is closely linked to the subsequent number of rest days. However if just the number of shifts is considered there were a number of issues of concern e.g. one shift comprised a block of five 12-hour night shifts; working this on a regular basis can increase the risk of fatigue. Another block of shifts ran into a second resulting in seven consecutive shifts (three 12-hour day shifts, followed by four 12 hour night shifts). This was not appropriate and increased the risk of fatigue.

Rest periods – Number of days between shifts

The number of rest days after a series of consecutive shifts is important because it determines the amount of time the worker has to recover from work. The number of rest days included in the Buncefield shift schedule was variable but there were a number of occasions where fewer rest days were provided than would be recommended by guidance. For example, there was a planned period of seven, 12-hour shifts followed by only 3 rest days before recommencing work.

Breaks

In order to mitigate the effects of fatigue during a 12-hour shift, regular breaks are required. Interview transcripts show that the workers took breaks when the work allowed and that these were not taken at fixed times or away from the control room.

Cumulative fatigue

Cumulative fatigue is the term used by some researchers to describe the accumulation of fatigue based on duty periods, the number of consecutive shifts before a rest period, and the duration of that rest period. There are uncertainties associated with the term but generally it relates to the increase in fatigue associated with working over a number of days or weeks. Supervisors at the Buncefield Oil Depot worked 12-hour shifts without regular planned breaks. At various points in the schedule the number of consecutive shifts was higher than recommended and the number of rest periods following a period of consecutive shifts was inadequate to allow sufficient recovery from work. Based on the evidence available and current knowledge and guidance, the basic, planned shift schedule worked by staff at the Buncefield Oil Depot was likely to have resulted in an accumulation of fatigue. This was of concern because fatigue can increase the risk of errors and accidents.

Workload

It is relevant to note that much of the work undertaken by supervisors at the Buncefield Oil Depot involved the monitoring of systems in the control room with periodic decision-making, for example relating to tank filling. Performance on such tasks is susceptible to the effects of fatigue and errors are more likely. Another significant factor was the absence of a formal policy or procedure for managing fatigue and shift-work. As with looking for fatigue and shift-work in an investigation, if there is no safety management system focus then effective arrangements are unlikely to be in place and there is no baseline set.

Lessons for the major hazard industries

The investigation concluded that cumulative fatigue contributed significantly to the Buncefield event. However many investigations do not adequately address fatigue issues. While fatigue is rarely a 'smoking gun' (especially in organisations with multiple layers of protection), it is pervasive and significant and requires good management like any other risk to safety. And whatever industry sector is involved, the same human beings work there. So they are affected in the same way by fatigue, whether from poorly designed or poorly controlled shift schedules, or from excessive working hours and workload.

The key message for the major hazard industries, on- and off-shore, is 'Keep it simple': design and manage shift schedules both to minimise fatigue and its effects, *and* to optimise manageability i.e. the ease with which any shift system can be run and worked, and its chosen shift schedule populated and managed in practice. A smooth-running shift system is good evidence that the shift schedule is reasonable. This means that for example, there are good arrangements to cover sickness/ absence, and also training, optimisation and other demands on shift staff's time. It also means that safety critical communications are well thought through and established e.g. for shift handover, day/shift, operations/maintenance (and see Wilkinson and Lardner 2012 for a fuller discussion).

But people get it right most of the time . . .

The fatigue and shift-work issues were important contributory causes at Buncefield but were not a 'smoking gun' – they rarely are. But organisational factors like these are often seen as just the 'wallpaper' by employees and others – the way things *are* around here in an organisation (Wilkinson & Rycraft 2014). They contributed to e.g. poor judgement, decision-making, behaviours, effectiveness, vigilance and much more over a sustained period. Many other HOF issues contributed too including shift handover, competency and training, procedures, the control room layout and the human-computer interface, (see HSE 2011). So what went wrong on the night? There was one key error at the frontline: the supervisors' 'situational awareness' (their grasp of the bigger picture) was faulty as evidenced by their shared confusion about the pipeline-tank alignments immediately after the incident. What they said in their initial police statements later changed, though they still expressed surprise and disbelief at the outcome. Typically, where error is involved those concerned can usually describe what they have done and seen but cannot explain, or link it to, the result (HSE 1999).

Due to the poor shift handover arrangements, and a series of poor handovers leading up to the incident (the same two supervisors were involved from 9 December), they had confused which pipeline was filling which tank and this mistaken view had persisted through several handovers. Why? Put simply, *they were not expecting to fill a petrol tank on that night shift* (and in the incorrect configuration they shared, neither tank would have filled that night) and so have to switch tanks, and *therefore the absence of an alarm was no cause for alarm*. You don't look for something if you aren't expecting it and their focus was anyway elsewhere for key parts of the night. Consequently, with the serial failure of all the other barriers, the human – and last – barrier failed too.

What can we do to get shift working right?

The published literature shows that over the last 10 years, there has been a move toward the use of fatigue risk management systems (FRMS), especially for more complex organisations where fatigue is likely to be one of a number of factors that combine rather than the single cause of an accident. The increasing pressure on staffing levels, lack of experienced staff and reliance on overtime, increases fatigue risk. While limits on working hours have a role to play it is equally important to fol-low good practice guidance, using fatigue models, collecting and analysing safety data, subjective reports, and education and training. A risk assessment approach should be taken and there is simple and accessible guidance (HSE 2006). Avoid over-complicating things though. It's vital to gather feedback from the real operating experience through auditing and monitoring, and from those involved in managing, supervising and working the shift system. It is essential to be able to record and analyse the *actual individual* hours and shifts worked, not just the averages or the planned shifts. A policy and procedure are necessary but keep them short and accept they will develop through experience. Equally a shift system needs managing – it's no good leaving those working in it (or an overloaded supervisor) to deal with planned and unplanned absences on an ad-hoc basis. While inevitably there will be exceptions (but they should be just that, exceptions, not the norm), it is how these are managed afterwards that really matters to avoid tired people returning to work without effective mitigation.

There is absolutely no point in tinkering with an existing shift schedule if it isn't being worked as planned. The underlying organisational factors are usually simple e.g. not enough staff to work the rota, or not enough people in some roles, or insufficient flexibility in the schedule to cope with real work demands. Workloads and work planning may also need fixing. So these factors need to be addressed first, *then* check out the shift schedule and review it once there is experience of it working as planned.

Finding a reasonable and workable shift schedule also requires early, active and informed user involvement i.e. those who work the shifts, supervise and manage them. But there is no magic or perfect shift system out there. They are all compromises because humans are day animals – we have a biological need for sleep during the night and so are not 'wired' for shift working. And the range of shift schedule choices is in practice small because of simple arithmetical facts about dividing the available time and days into workable and fair arrangements over a calendar year, e.g. maximising and fairly sharing out the weekends off within a schedule to make the most of the diminished social opportunities created by shift-working. The practical options are set out very clearly in (Miller 2006). Training and awareness play a part, particularly in focusing manager and supervisor attention on what matters in operating the shift system, and in helping users and their significant others to understand the likely impacts on health, safety and their social and domestic lives, and how best to mitigate this.

Finally, considering the social and domestic side of working shifts is a vital (HSE 2006; Miller 2006). A *predictable* shift schedule (Miller 2006) allows those working it, and their families, to make plans to mitigate the inevitable reduction in normal social and domestic contact. Schedules that result in ad-hoc working to fill gaps, cover overtime and so on, are inherently unpredictable.

A good shift system and shift schedule will be one that is characterised by (reasonably) smooth running, i.e. it will work away quietly and not create undue management, supervisory or user difficulties. However, as with all things, change is inevitable, so fatigue management will need effective monitoring and auditing, and regular review against operating experience, improving standards and knowledge, and changing business needs. Good luck . . .

Lessons for investigators

Fatigue is recognised as a contributing factor for accidents, injuries and deaths in a wide range of settings. Published literature supports a positive association between long work hours, risk of attentional errors, accidents and/or injury occurrence, sleepiness and fatigue. While fatigue is acknowledged as a contributing factor for accidents, the evidence for a causal link is more complex. Errors which initially may seem to be about e.g. procedural violation, lack of situational awareness can, when explored in more depth have fatigue as an underlying factor. Fatigue is insidious and, while it may rarely be a 'smoking gun' after an incident, it can set the scene for a making a wide range of failures more likely, but unpredictably. Investigators tend to find what they look for though, and even basic recording of timings, previous shift history and hours, and shift system performance e.g. overtime levels, actual shifts worked vs planned, are often neglected. A good investigation system will look for fatigue and shift-work issues among a wide range of other performance shaping factors.

References[1]

Di Milia, L., Smolensky, M.H., Costa, G., Howarth, H.D., Ohayon, M.M & Philip, P. (2011) Demographic factors, fatigue and driving accidents: An examination of the literature. *Accident Analysis & Prevention* **43**, 516–532

HSE (1999) *Reducing Error and Influencing Behaviour*, HSE Books, London. http://www.hse.gov.uk/pubns/books/hsg48.htm

HSE (2006) *Managing shift work* HSG256, HSE Books. http://www.hse.gov.uk/pubns/books/hsg256.htm

HSL (2007) *Report on the Buncefield Shift Schedule* Bell, J. (unpublished)

HSE (2011) *Buncefield: Why did it happen?* HSE. http://www.hse.gov.uk/comah/buncefield/buncefield-report.pdf

Miller, J. (2006) *Fundamentals Of Shiftwork Scheduling,* Air Force Research Laboratory & Human Effectiveness Directorate, AFRL-HE-BR-TR-2006-0011, NTIS, Springfield VA. http://tinyurl.com/kc5st6m .

Wilkinson, J., & Rycraft, H. (2014) Improving organisational learning: why don't we learn effectively from incidents and other sources? *Proceedings of The Institute of Chemical Engineers' Hazards 24 Conference*, Edinburgh

Wilkinson, J., & Lardner, R. (2012) Pass it on! Revisiting Shift Handover After Buncefield, *Proceedings of The Institute of Chemical Engineers' Hazards 23 Conference*, Edinburgh

Williamson, A., Lombardi, D.A., Folkard, S., Stutts, J., Courtney, A.K. & Connor, J.L. (2011) The link between fatigue and safety *Accident Analysis and Prevention*, **43**, 498–515

---

[1] All web references retrieved 21st November 2014)

HEALTHCARE

HOW DO WE CHALLENGE MYTHS AND MISUNDERSTANDINGS ABOUT HUMAN FACTORS IN HEALTHCARE?

Paul Bowie[1,2], Laura Pickup[3] & Sarah Atkinson[3]

[1] *NHS Education for Scotland*
[2] *Honorary Senior Lecturer, Institute of Health and Wellbeing, University of Glasgow*
[3] *Human Factors Research Group, University of Nottingham*

Background

It is now widely accepted that the scientific discipline of 'human factors' can make a significant contribution to understanding, evaluating and enhancing healthcare safety and performance, but less so around improving value for money. However recent journal publications by human factors experts highlight examples of how the discipline can be unintentionally misrepresented in the healthcare literature [1, 2]. For those of us working to advance human factors understanding in the NHS this provides further confirmation of our professional experiences in the workplace. We often bear witness in conferences, workshop settings, training sessions, websites and policy documents to the unintentional muddled thinking and misunderstandings propagated by 'experts' and 'those with an interest' (mostly clinicians and aviation consultants) that regularly accompanies presentations, discussions and debate devoted to the topic of human factors. The implications are of high importance to improving patient safety, developing and spreading appropriate HF educational content, and the integrity of the wider HF community because of the confusion that arises and the potential of NHS staff to misapply related knowledge and methods. Unfortunately there is very limited HF expertise within the NHS, particularly at senior levels, to counter misunderstandings and also contribute to the design, leadership and implementation of a more informed HF educational strategy for the healthcare workforce. The lack of real HF expertise at strategic and operational levels in the NHS and how to 'close this gap' is clearly also an issue of high relevance to the HF profession.

Purpose

To raise awareness of how the discipline of human factors in healthcare is often unintentionally misrepresented by 'experts', and explore how best to challenge and overcome related myths and misunderstandings so that a more informed representation of the science can be developed and implemented in the NHS by the HF community and other stakeholders.

References

Catchpole K. 2013. Spreading human factors expertise in healthcare: untangling the knots in people and systems. *BMJ Quality & Safety*, 22(10): 793–797.

Russ A.L., Fairbanks R.J., Karsh B.-T. et al. 2013. The science of human factors: separating fact from fiction. *BMJ Quality & Safety*, 22(10): 802–808.

PARTICIPATORY DESIGN OF A PRELIMINARY SAFETY CHECKLIST FOR THE GENERAL PRACTICE WORK SYSTEM

Paul Bowie[1,2] & Sarah Atkinson[3]

[1] *NHS Education for Scotland*
[2] *Institute of Health andWellbeing, University of Glasgow*
[3] *Human Factors Research Group, University of Nottingham*

In UK general practice safe system designs and checking processes can be inadequate. There is limited experience of checklists to help reliable checking of safety issues of importance to health, wellbeing and performance. A prototype checklist was co-developed with care team members and 'safety experts' in Scotland, which contains six safety domains (e.g. medicine management), 22 sub-categories (e.g. controlled drugs) and 78 related items (e.g. stock balancing). Implementation potential is strong, but socio-cultural barriers to adoption need to be considered.

Introduction

In healthcare there is now growing interest in checklists, particularly in acute hospital settings, to standardize checking processes and act as cognitive aids to ensure task completion by clinicians and others, and provide further systemic defences against error and preventable harm to patients (Ko et al., 2011). In UK general medical practice, it is estimated patients may be avoidably harmed in between 1% to 2% of clinical consultations, which is potentially significant given that there are approximately 1 million consultations daily (Health Foundation, 2013). The nature, scale and organisation of patient care in general practice are characterised by an inherent complexity and uncertainty due in large part to the diverse range of (often elderly) patients who are living longer with increasingly complicated co-morbidities and often taking multiple high risk medications; making safe and effective clinical management particularly problematic (Barnett et al., 2012).

Safety incidents clearly have physical and psychological effects on patients, relatives, clinicians and staff, and also impact negatively on organizational performance (Dekker, 2013; McKay et al., 2009). The design quality of systems and related checking processes to support patient safety can be variable, unsafe and ineffective and are cited as important contributory factors in adverse events (Health Foundation, 2013; McKay et al., 2009). Examples include: the management of controlled drugs (e.g. correlation of recorded and actual stock levels); maintaining emergency equipment (e.g. the practice oxygen supply is checked for sufficient levels and functionality as per regulatory requirements); and storage of medications (e.g. non-controlled drugs are stored at correct temperatures and are within expiry dates).

Adopting a consistent, methodical and measurable approach to checking processes may lead to the high reliability and standardisation of task performance and the systems concerned (Hales & Pronovost, 2006). However, there is very limited experience of the use of checklists as a safety intervention in general practice. Against this background, the study aims were as follows:

1. To identify and prioritise workplace hazards that are known to impact on the safety, health and well-being of patients, visitors and GP team members, and organisational performance using a human factors work system model (Carayon et al., 2006).
2. To co-design, develop and content validate a standardised checklist process which reflects system-wide hazards and risks.

Methods

Eighteen experienced general medical practitioners, nurses and managers were recruited from six Scottish NHS Board regions alongside a multi-professional patient safety 'expert' development group (n = 6). The conceptual and participatory design (Hignett et al., 2005), iterative development, validation and testing of the preliminary safety checklist were undertaken using a combination of methods, including: a review of the relevant literature; application of Carayon et al.'s (2006) Safety Engineering Initiative for Patient Safety (SEIPS) work system model; review of safety-related policies, protocols and procedures; consensus generating workshop meetings; and a content validity index (CVI) exercise. Agreement on inclusion of all identified checklist domains, sub-categories and related issues was dependent on a minimum of 15 of 18 participants rating each item ≥ 3 on a 4-point scale.

Results

A range of hazards was identified by participants (e.g. medication management, information systems and health & safety), which reflected the breadth of the general practice work system and informed checklist development. Hazards were identified to the health and wellbeing of patients (e.g. needle stick injury); practice visitors (e.g. accidental injury from loose carpets or tiles); GP team members (e.g. infection due to cross-contamination of bodily-fluids); and the safe performance and productivity of the practice organisation (e.g. out-of-date Business Continuity Plan).

A preliminary safety checklist was developed and validated (Table 1) consisting of six domains (e.g. medicines management), 22 sub-categories (e.g. controlled drugs) and 78 related items (e.g. monthly stock reconciliation undertaken). The contents were judged by participants to be safety issues of priority across the work

Table 1.   Levels of agreement: number of raters' (n = 18) rating each checklist domain and sub-category ≥3 and calculated content validity index (CVI) ratio.

Checklist Domains and sub-categories	Number of Raters ≥3	CVI
Medicines Management	18	1.0
1. Controlled Drugs	18	1.0
2. Emergency Drugs & Equipment	18	0.94
3. Prescriptions & Pads	18	0.94
4. Vaccinations	17	0.83
5. All Other Drugs on Premises	18	0.83
Housekeeping	18	0.83
6. Infection control	18	1.0
7. Stocking of clinical rooms	17	0.94
8. Confidential waste	18	0.94
9. Clinical equipment maintenance	17	0.94
Information Systems	18	0.94
10. The practice Business Continuity Plan is up-to-date?	16	0.89
11. The back-up of all significant IT systems can be verified?	18	1.0
12. Data protection	18	0.94
13. Record keeping	18	0.94
Practice Team	18	0.89
14. Registration checks	18	1.0
15. CPR and Anaphylaxis training	18	0.89
16. Induction processes	15	0.77+
17. All staff have access to ongoing Patient Safety-related training opportunities	18	0.94
Patient Access and Identification	18	0.94
18. Information for Patients on how to access the practice urgently or in an emergency is widely available in different formats	18	0.94
19. Standardised Patient Identification (ID) verification	17	0.89
Health & Safety	18	0.89
20. Building safety and insurance	17	0.89
21. Environmental awareness	18	0.89
22. Staff health and wellbeing	18	0.89

system that required to be routinely checked to minimise the risks of hazards to people and organisational practice performance. Participants unanimously agreed that the checklist should be consistently applied at least three times per calendar year (i.e. once every four months) to ensure necessary checking within acceptable timescales, but that common sense judgements may be applied to items that require checking annual or longer (e.g. Anaphylaxis and CPR training) or may not be applicable to all (e.g. stocking of controlled drugs). The important issue agreed is the implementation of a reliable, consistent but contextualised and flexible checking system. The overall CVI ratio for the preliminary checklist was 0.92.

Discussion

The study aims were achieved in applying participatory methods to identifying and prioritising hazards in the general practice work system that can potentially impact on the safety, health and well-being of patients and GP team members, and negatively affect organisational performance. A robust, user-centred and systems-based approach to the design and development of a necessary safety checklist prototype for the GP work system was then undertaken. However, checklist efficacy in improving safety processes and outcomes is dependent on user commitment, and overcoming socio-cultural barriers to implementation. Further usability testing work is necessary but the initial development should be of interest in the UK and internationally.

Acknowledgements

We offer sincere thanks to all study participants for their ideas, support and enthusiasm for this research work.

References

Barnett K., Mercer S., Norbury M., Watt G., Wyke S., Guthrie B. 2012. The epidemiology of multimorbidity in a large cross-sectional dataset: implications for health care, research and medical education. *The Lancet* 380; 37–43

Carayon P. et al. 2006. Work safety design for patient safety: the SEIPS model. *Qual Saf Health Care* 2006; **15**:i50–i58 doi:10.1136/qshc.2005.015842

Dekker S. 2013. Second victim: error, guilt, trauma and resilience. CRC Press, Taylor & Francis Group: Boca Raton

Hales B.M., Pronovost P.J. 2006. The checklist tool for error management and performance improvement. *J Crit Care*. 21:231–235

Health Foundation (2013). Evidence scan: levels of harm in primary care. Available at: http://www.health.org.uk/publications/levels-of-harm-in-primary-care/ [Accessed 25th September 2014]

Hignett S., Wilson J.R., Morris W. 2005. Finding ergonomic solutions – participatory approaches. *Occupational Medicine*, 55; 200–207

Ko, H.C.H., Turner, T.J., Finnigan, M. 2011. Systematic review of safety checklists for use by medical care team in acute hospital settings – limited evidence of effectiveness. *BMC Health Services Research* 11:211

McKay J., Bradley N., Lough M., Bowie P. 2009. A review of significant events analysed in general medical practice: implications for the quality and safety of patient care. *BMC Family Practice*. 10:61.

HUMAN FACTORS THAT INFLUENCE THE PERFORMANCE OF TELECARE

Peter Buckle

Helen Hamlyn Centre for Design, Royal College of Art, London

Systems approaches and their relevance to the design and to the risk assessment of technology for telecare have been investigated. We have examined risk in five contexts namely: Assessment of telecare need and equipment provision; Installation of equipment; Review of client needs; Client use of telecare; Monitoring and Response. The results provide important areas of systems failures that have impact across the many stakeholders involved in the provision of telecare. The opportunities for improving design and are presented.

Introduction

The Advancing Knowledge of Telecare for Independence and Vitality in later life (AKTIVE) research study (Yeandle, 2014; Yeandle et al, 2014) has afforded a unique opportunity to study how frail older people and those who support or care for them are interacting with today's telecare systems. Those studied include not just the 'end users' of the telecare service, but also the care support group which may be involved with them (e.g. relatives, friends, neighbours), those working in monitoring and response centres, assessors and reviewers of end-user requirements and even those who install and maintain the equipment and services.

Method

The in-depth nature of the study has enabled the social dimensions of the system to be observed and explored providing rich, in-depth data that has not been gathered before. The Everyday Living Analysis (ELA) social methodology has been described elsewhere (Yeandle et al, 2014). This method combined a variety of elements to gain a rich, contextualised account of older peoples' lives and to assess the role of telecare within them, not only from their own perspectives, but also from the viewpoint of those involved in caring for or supporting them (whether as paid home care workers or as family or unpaid carers). This person-centred method enabled the researchers to keep the older person at the centre of their own stories, and provided rich data. These data were supplemented with information collected from working with and observing others within the wider telecare system to obtain a fuller understanding of how and where failures or difficulties might occur.

Within the ELA, almost two-thirds (39) of participants were female and over half (41) were living alone (32 in widowhood); 17 participants were married, eight were divorced and two were single. The participants' health varied from those who were very frail and unable to attend to their own daily needs to those who had few physical impairments but suffered from memory problems. Participants were followed over a period of about a year and six visits/interviews were held with each over this time. Telecare support varied but varied form simple pendant alarms to more complex care packages. These data form the basis for this paper on the human factors that influence the performance of the system, where difficulties or failures have or could occur and how improvements might be made.

Results

Problems that result in potential failures within the current telecare system can occur through procurement, assessment of client needs, installation of equipment, review of equipment, client use, at monitoring centres and even in response settings. While it has been difficult to quantify or even prioritise the scale of these failings, it is nevertheless apparent that significant problems exist. These problems have frequently emerged in discussion with stakeholders throughout the system, and in particular through the ELA.

The data arising from the ELA provide a rich set of insights into the challenges facing those providing telecare services. These are fully described elsewhere (Buckle et al, 2014). The concerns identified from the social research team's work with telecare users (and those who care for or support them) have helped to establish that solutions can only be found through co-operative participative design consultation.

Problems related to client users have demonstrated that a large number of difficulties may be encountered. These have important implications throughout the system. For example, false alarms are a common cause of complaint. They take up a large amount of time at monitoring centres, lead to distrust of equipment by end-users, cause distress to the older people involved and, where applicable, those who form their caring network, and can lead to a complete lack of use of the equipment.

Understanding and correcting the root cause of these faults is essential to drive better design, improve assessment, inform reviews and help provide better information to those operating the equipment. Monitoring centres are particularly well placed to review call data and to classify these issues in a manner that would inform and potentially enhance the performance of their business. Routinely studying the nature and origins of a 'false alarm' and seeking a probable cause, for example, would provide valuable information that could be shared with designers and others to prioritise interventions. The time saved in dealing with false alarms could be better spent providing other client services, such as online support to address social isolation issues, initiate reviews or observe trends in behaviour that might be indicative of a client needing a proactive intervention to prevent a more serious outcome.

During the course of the study it became clear that better training, embracing new technology, is required for many users within the system. The training provided to assessors appears to be fragmented and idiosyncratic and may depend on the locality in which the service is provided. For example, currently, training may be 'bundled' with the equipment provided by manufacturers. Because, as shown in the AKTIVE study, it sometimes fails to consider how one piece of technology links to and interacts with other pieces of technology within the system, such training is often of limited value. When telecare services in a locality have been fragmented, for example through a commissioning process, these problems are exacerbated.

Discussion

Advances in technology have to be seen in the context of how useful and how trusted they are to those that rely on them. Thus an automatic fall pendant that may contain state of the art technology may never achieve its full potential. This can occur, for example, if the user fails to appreciate what the technology is doing and how to use it. Similarly, if there are too many false alarms these generate difficulties and embarrassment for the end-user and their carers and may well lead to reluctance to use telecare at all.

There is also an important debate still needed regarding the relationship between how telecare helps to manage risk and how its reliability may, in turn, influence the perception of risk. What has become evident in this study is that where equipment does fail, the legacy of failure is substantial. In particular, it may lead to a disproportionate reduction in the perception of reliability. While this may not be reflected in the *actual* failure rate, risks perceptions need to be studied more closely to ensure the industry can present data in a way that reassures those who are using, buying or are reliant on such technology.

The study found that design of telecare needed this 'in-depth' analysis of the existing systems to identify risks of failure and opportunities for design improvement. Latent failure types were commonly identified during the study. Latent failures are made by people whose tasks are removed in time and space from operational activities, e.g. designers, decision makers and managers. Examples of latent failures are: poor design of equipment; ineffective training; inadequate supervision; ineffective communications; and uncertainties in roles and responsibilities. Exemplars of these have been identified and design approaches that may overcome these deficiencies have been presented.

A new focus is now required on how future developments in telecare might aid activity, engagement and quality of life for older people and those who care for them. New design initiatives need both to take account of these systematic, human factors risk assessments, and to protect users by building appropriate resilience into the system (Levenson, 2011; Hollnagel et al., 2006.)

References

Buckle, P. (2014) Human factors that influence the performance of the tele-care system. AKTIVE Working paper 7. University of Leeds, available at: http://circle.leeds.ac.uk/.

Hollnagel, E., Woods, D.D. & Leveson, N. (2006) *Resilience Engineering: concepts and precepts,* Farnham: Ashgate Publishing Ltd.

Leveson, N.G. (2011) *Engineering a safer world: Systems thinking applied to safety,* Cambridge, MA: The MIT Press.

Yeandle, S. (2014) *Frail older people and their networks of support: how does tele-care fit in?* AKTIVE Research Report Vol. 2. Working Paper 2, Leeds, CIRCLE, University of Leeds, available at: http://circle.leeds.ac.uk/.

Yeandle, S., Buckle, P., Fry, G., Hamblin, K., Koivunen E.-R. & McGinley, C. (2014) *The AKTIVE project's social, design and prospective haz-ard research: research methods,* AKTIVE Research Report Vol. 3. University of Leeds, available at: http://circle.leeds.ac.uk/.

STRESS IN UK TRAINEE MENTAL HEALTH PROFESSIONALS: A MULTIVARIATE COMPARISON

John Galvin

Centre for Occupational and Health Psychology, Cardiff University, UK

This paper describes a comparison study focusing on stress in two groups of mental health professionals in training: trainee clinical psychologists and psychiatric nursing students. PhD students acted as a control. In order to investigate these groups, a transactional approach was adopted and multiple variables considered. Trainee clinical psychologists reported higher levels of demands but received more support than the other groups. This was associated with higher job satisfaction but more perceived stress and psychological ill health. Psychiatric nursing students reported lower control and more negative coping styles than the other groups, which needs to be addressed. These findings are examined in relation to previous research in this area and future directions are discussed.

Introduction

Occupational stress affects almost every profession. However, it is now widely accepted that mental health professionals are at an elevated risk, with a greater number of potential sources of stress having an impact on their health (Hannigan, Edward & Burnard, 2004; Lim et al., 2010). For these reasons, more attention is now being placed at a stage prior to their incorporation into the workplace: their training period. This paper describes research conducted for the author's ongoing PhD project. In this project, stress is conceptualised as a transaction between a person and their environment (Lazarus & Folkman, 1984). According to this approach stress occurs when an individual appraises their environmental demands as outweighing their abilities to meet those demands. Two groups of trainee mental health professionals are of central concern here; trainee clinical psychologists and psychiatric nursing students. A sample of PhD students acts as an additional control group. These groups have been selected due to the multiple stressors that they face, and the high levels of stress and mental illness reported in the literature (Pakenham, 2012; Pulido-Martos et al., 2014).

Cushway (1992) reported significantly high levels of psychological distress in a sample of trainee clinical psychologists. Specifically, she found that 59% of participants reached the clinical cut-off score on the General Health Questionnaire (GHQ; Goldberg, 1978). Factor analysis of the data indicated six underlying factors: 1) course structure and organisation; 2) workload; 3) poor supervision; 4) disruption of social support; 5) self-doubt; and 6) client difficulties and distress. Similarly, Jones and Johnston (1997) reported high GHQ caseness of between 35 and 44% in a sample of student nurses, with higher distress being reported around the time of clinical placements. Indeed, both trainee clinical psychologists and psychiatric nursing students are required to undertake clinical placements in the National Health Service (NHS) as part of their training, an environment renowned for being stressful. For example, the NHS staff survey for 2013 revealed that 39% of staff reported feeling unwell due to work-related stress, the highest levels reported since the survey started in 2003. Forty-six percent of psychiatric nurses reported work-related stress, compared to 40% of learning disabilities nurses, 39% of adult nurses and 39% of child nurses. For clinical psychologists, 37% reported work-related stress which was higher than many other professions, including the 33% of medical and dental staff who reported work-related stress. Furthermore, only 44% of the staff surveyed said they believe their organisation takes positive action on health and well-being.

However, it should be noted that some of the stressors reported by mental health professionals are often inherent to the job and are therefore unavoidable. For example, the organisational and administrative tasks, frequent patient contact and the nature of caring for the mentally ill (Nolan & Ryan, 2008). For this reason, researchers are often interested in the coping strategies employed by these groups of workers (e.g. Cushway & Tyler, 1994; Tully, 2004). Lazarus and Folkman (1984) described two broad types of coping strategies: emotion-focused coping and problem-focused coping. Emotion-focused coping involves reappraising the relational meaning of the problem. An example of this strategy is avoidance coping, whereby the individual avoids the situation or uses substances, such as alcohol, to deal with the problem (Cohen, Ben-Zur & Rosenfeld, 2008). On the other hand, problem-focused coping involves taking one step at a time, attempting to change the situation at hand, and not allowing emotions and feelings to interfere too much. There is substantial evidence that emotion-focused coping is positively correlated with distress (Cohen, 2002). In contrast, problem-focused coping is positively correlated with positive affect (Ben-Zur, 2002), suggesting that the latter is more favourable than the former.

A multivariate approach

Smith et al. (2004) noted that previous research in the area of occupational stress fails to account for a wide range of potentially influential factors (both positive and negative). For example, previous experiences, personality and health behaviours, in addition to work characteristics and appraisals, are often neglected. It is more common for researchers to concentrate on a smaller number of variables rather than considering a number of factors which may determine stress levels and interact with each other simultaneously. Time, costs and participant attrition are some of the potential reasons why researchers may decide to limit the focus of their research in this way. However, this approach is unlikely to be representative of real-life situations, whereby individuals are exposed to multiple hazards at work (Smith et al., 2004). Due to the increasing popularity of short, practical wellbeing measures (e.g. Williams & Smith, 2012) it is now possible for questionnaires to incorporate a wide range of factors while still maintaining acceptable levels of validity and reliability. Using this approach, work characteristics, coping, personality, health behaviours, childhood experiences, perceived stress, job satisfaction and psychological ill health were assessed in trainee clinical psychologists, psychiatric nursing students and PhD students.

Method

This project was given ethical approval from the School of Psychology Ethics Committee at Cardiff University. Participants were invited to complete an online questionnaire.

Participants

The relevant course directors were contacted and permission sought to approach participants. Five clinical psychology programmes and three nursing programmes agreed to distribute the questionnaire. The PhD students were recruited from Cardiff University. In total, 515 participants completed the questionnaire, including 168 trainee clinical psychologists, 94 psychiatric nursing students and 253 PhD students.

The control group of PhD students were from a number of different disciplines and some worked across disciplines. However, the PhD students invited to take part in the study were only asked to participate if they considered their PhD project to be mainly research-focused. In other words, they were asked to confirm that their work does not involve patient contact or other substantial clinical work.

Measures

The questionnaire used in this study was designed based on data from qualitative interviews collected by the author prior to data collection. This allowed the quantitative measures used in this study to be focused on the most relevant variables for the populations under investigation.

The questionnaire included measures of work characteristics, appraisals, coping and mental health outcomes using global, single-item measures from the Wellbeing-Process Questionnaire (WPQ; Williams & Smith, 2012). Health behaviours, individual differences, childhood experiences and demographics were also measured. For a full list of measures see table 1.

Preliminary analysis

To support the conceptual framework of the study, principle components analyses were conducted. In total, 15 components were extracted with eigenvalues equaling or exceeding the threshold of 1. These were job demands, resources, emotion-based

Table 1.   The variables and associated factors measured in the questionnaire.

Factor	Individual variables	Description of measures
Work characteristics	Demands, control, support, effort, reward, supervisor relationship.	Single-item measures of work characteristics from the WPQ (Williams & Smith, 2012).
Appraisals	Perceived job stress, outside work stress, hassles, uplifts, job satisfaction.	Single-item measures of appraisals from the WPQ.
Coping	Problem-focused, seeks social support, blame-self, wishful thinking, avoidance.	Single-item measures of coping from the WPQ.
Health behaviours	Alcohol, sleep, exercise, breakfast, chocolate, crisps, biscuits, fruit & vegetables.	Single items of health behaviours.
Individual differences	Extraversion, agreeableness, conscientiousness, neuroticism, perfectionism variables (standards, order, discrepancy), imposter feelings, core self-evaluations.	Big 5 Inventory-10 (BFI-10; Rammstedt & John, 2007), Almost-Perfect Revised Scale (APS-R; Slaney et al., 1996), Imposter Phenomenon scale (Clance, 1985), and Core Self Evaluations Scale (Judge et al., 2003).
Childhood experiences	Negative home environment, sexual abuse, physical abuse, own mental health disorder, families' mental health disorder, parent-focused and sibling-focused parentification, perceived benefits of parentification.	Child Abuse and Trauma Scale (CATS; Sanders & Beckers-Lausen, 1995), Experiences of Mental Health (EMH) scale, and Parentification Inventory (PI; Hooper et al., 2011).
Outcomes	Depression, anxiety, burnout, happiness.	Outcomes were measured using single-item measures from the WPQ.
Demographics	Age, gender, ethnicity, year of training, education level	Single item questions

coping, seeks social support, perceived stress, job satisfaction, negative personality traits, conscientious attitude, relational personality style, negative childhood experiences, childhood responsibilities, alcohol consumption, healthy lifestyle, bad diet and psychological ill health. To reduce the variables into manageable units and decrease the possibility of chance effects, component scores were created using the Anderson-Rubin method and these scores were used in later analysis.

Summary of key results

Below is a brief summary of the key findings. For a more detailed description of the results see Galvin and Smith (2014).

Differences between the groups on component scores

Trainee clinical psychologists reported higher levels of demands, resources and perceived stress than the other groups. They also reported higher levels of psychological
ill health than PhD students. Psychiatric nursing students reported significantly
more emotion-based coping and engaging in a less healthy lifestyle than the other
groups. Additionally, psychiatric nurses reported more negative personality traits
than PhD students.

Regression analyses for component scores

A series of regression analyses were carried out to investigate the associations of
the multiple independent variables with three outcomes; psychological ill health,
perceived stress and job satisfaction. These were run for the whole sample and each
group individually. High job demands were associated with high levels of psychological ill health and perceived stress and low job satisfaction in the trainee clinical
psychology group. Individual difference variables such as personality were important in all the regressions for trainee clinical psychologists. For example, relational
personality traits were associated with lower psychological ill health and higher
job satisfaction, and negative personality traits was associated with higher psychological ill health. Negative childhood experiences were also associated with high
levels of psychological ill health for trainee clinical psychologists. For psychiatric
nursing students, high levels of job demands, childhood experiences and alcohol
consumption were associated with poorer outcomes, whereas high resources were
associated with better outcomes.

Univariate analyses

After the overall latent structure was explored, the individual variables were then
considered. Trainee clinical psychologists reported higher levels of demands, but
received more support than the other groups. This was associated with higher job
satisfaction, but also more stress and mental health problems. Psychiatric nursing students reported lower control and more negative coping styles than the other
groups and these items were associated with poorer outcomes. In terms of coping styles, psychiatric nursing students reported engaging in less problem-focused
coping, and more wishful thinking and avoidance coping than the other groups.
Additionally, they reported higher levels of anxiety than PhD students, and scored
higher on the neuroticism scale than both trainee clinical psychologists and PhD
students.

Both trainee clinical psychologists and psychiatric nursing students scored significantly higher on the sexual abuse scale of the CATS than PhD students, and
psychiatric nursing students scored significantly higher than PhD students on the
punishment scale. However, there were no significant differences found for the negative home environment scale. No differences were found on the parent-focused
or sibling-focused parentification scales. However, trainee clinical psychologists

scored significantly higher on the perceived benefits of parentification scale compared to PhD students. Both groups of mental health trainees reported more incidences of psychiatric history in their family than PhD students.

Conclusion

Few studies focusing on stress in trainee clinical psychologists and psychiatric nursing students have considered the combined effects of multiple factors on outcomes. The present research describes a holistic approach to studying stress in trainee mental health professionals, with group comparisons and a transactional approach being preferred. During a time of increased demand and pressure on resources, this area of investigation is particularly important. Trainee clinical psychologists reported higher demands and higher perceived stress than psychiatric nursing students and PhD students. They also reported higher levels of psychological ill health than PhD students. Additionally, psychiatric nursing students reported more emotion-based coping than the other groups, and more negative personality traits and anxiety compared to PhD students. Taken together, these findings support other research that suggests mental health professionals are at a greater risk of mental health problems than other groups (e.g. Cushway, 1992; Hannigan, Edwards & Burnard, 2004). This is likely to be due to the type of individuals going into these professions, and the facets of their job. These findings promote the essentiality for mental health training courses with both academic and clinical components to provide support systems that are safe, well resourced and effective.

It is important for nurse educators to discover the sources of stress and coping strategies used by psychiatric nursing students so that they can be helped to cope effectively in the emotionally demanding environment in which they work. Psychiatric nursing students reported more emotion-based coping strategies than the other groups. Specifically, psychiatric nursing students reported more wishful thinking and avoidance coping, but less problem-focused coping. The potential greater use of academic and theoretical models within postgraduate training could be relevant to this finding. For example, Galbraith and Brown (2011) conducted a systematic review of stress and coping in nursing students and concluded that training providers should only include stress interventions that take into consideration current theories of stress. Furthermore, alcohol consumption was associated with higher perceived stress and psychological ill health, and lower job satisfaction for psychiatric nursing students. However, alcohol did not predict outcomes in any of the other groups' regressions, suggesting that alcohol is specifically influencing the outcomes of psychiatric nursing students. Alcohol consumption could be an important coping strategy employed by this group, and future research should focus on this particular characteristic and possible intervention.

Elliott and Guy (1993) found that qualified mental health professionals reported more problems in their home life compared to other professionals. In our sample of trainees, negative childhood experiences were higher for the trainees compared to PhD students. Of particular note was a greater incidence of sexual abuse and family psychiatric history in these groups. Nonetheless, the groups did not differ on the negative home environment subscale of the CATs or the parent-focused and sibling-focused scales of the PI. Trainee clinical psychologists reported higher levels of perceived benefits of parentification than PhD students suggesting that, in general, trainees may reflect on their childhood experiences as having a positive impact on their development.

This research allows for the examination of associations, but it cannot determine the direction of these effects. To redress this limitation, longitudinal data are currently being collected from these groups. Limitations related to the collection of questionnaire data, such as response bias, could also have an impact on these results. However, with relatively large sample sizes compared to other research in this area, the vast amount of information provided by this research can be useful in the planning and implementation of stress management interventions for these groups of trainee mental health professionals.

References

Ben-Zur, H. 2002. "Coping, affect and aging: the roles of mastery and self-esteem." *Personality and individual differences* **32**, (2): 357–372.

Clance, P. R. 1995. The Imposter Phenomenon: When success makes you feel like a fake. *Bantam Books*.

Cohen, M. 2002. "Coping and emotional distress in primary and recurrent breast cancer patients." *Journal of Clinical Psychology in Medical Settings* **9**, 3, 245–251.

Cohen, M., Ben-Zur, H. and Rosenfeld M. J. 2008. "Sense of coherence, coping strategies, and test anxiety as predictors of test performance among college students." *International Journal of Stress Management* **15**, (3): 289–303.

Cushway, D. 1992. "Stress in clinical psychology trainees." *British Journal of Clinical Psychology* **31**, (2): 169–179.

Cushway, D. and Tyler, P. A. 1994. "Stress and coping in clinical psychologists." *Stress Medicine* **10**, (1): 35–42.

Elliott, D. M. and Guy, J. D. 1993. "Mental health professionals versus non-mental health professionals: childhood trauma and adult functioning." *Professional Psychology Research and Practice* **24**, (1): 83–90.

Galbraith, N. D. and Brown, K. E. 2011. "Assessing intervention effectiveness for reducing stress in student nurses: Quantitative systematic review." *Journal of Advanced Nursing* **67**, 4, 709–721.

Galvin, J. and Smith, A. P. 2014. A transactional study of stress in trainee clinical psychologists and psychiatric nursing students. *Manuscript submitted for publication*.

Goldberg, L. and Bretznitz, S. 1982. Handbook of stress: Theoretical and Clinical Aspects. (New York: Free Press).

Hannigan, B., Edwards, D. and Burnard, P. 2004. "Stress and stress management in clinical psychology: Findings from a systematic review." *Journal of Mental Health* 13(3): 235–245.

Hooper, L. M., Doehler, K., Wallace, S. A. and Hannah, N. J. 2011. "The parentification inventory: Development, validation and cross-validation." *American Journal of Family Therapy* 39(3): 226–241.

Jones, M. C. and Johnston, D. W. 1997. "Distress, stress and coping in first-year student nurses." *Journal of Advanced Nursing* 26(3): 475–482.

Judge, T. A., Erez, A., Bono, J. E. and Thoresen C. J. 2003. "The core self-evaluations scale: Development of a measure." *Personnel Psychology* 56(2): 303–331.

Lazarus, R. S. and Folkman S. 1984. *Stress, appraisal, and coping*, (Springer Publishing Company).

Lim, J., Bogossian, F. and Ahern, K. 2010. "Stress and coping in Australian nurses: a systematic review." *International Nursing Review* 57(1): 22–31.

NHS Staff Survey 2013. Retrieved 27th September 2014, from: http://www.nhsstaffsurveys.com/Page/1006/Latest-Results/2013-Results/

Nolan, G. and Ryan, D. 2008. "Experience of stress in psychiatric nursing students in Ireland." *Nursing Standard* 22(43): 35–43.

Pakenham, K. I. and Stafford-Brown, J. 2012. "Stress in clinical psychology trainees: Current research status and future directions." *Australian Psychologist* 47(3): 147–155.

Pulido-Martos, M., Augusto-Landa, J. M. and Lopez-Zafra, E. 2012. "Sources of stress in nursing students: A systematic review of quantitative studies." *International Nursing Review* 59(1): 15–25.

Rammstedt, B. and John, O. P. 2007. "Measuring personality in one minute or less: A 10-item short version of the Big Five Inventory in English and German." *Journal of Research in Personality* 41: 203–212.

Sanders, B. and Becker-Lausen, E. 1995. "The measurement of psychological maltreatment: Early data on the child abuse and trauma scale." *Child Abuse and Neglect* 19(3): 315–323.

Slaney, R. B., Rice, K. G., Mobley, M., Trippi, J. and Ashby, J. S. 1996. "The revised almost perfect scale." *Measurement and Evaluation in Counselling and Development* 34(3): 130–145.

Smith, A. P., McNamara, R. and Wellens, B. 2004. Combined effects of occupational hazards. *HSE Books*.

Tully, A. 2004. "Stress, sources of stress and ways of coping among psychiatric nursing students." *Journal of Psychiatric and Mental Health Nursing* 11(1): 43–47.

Williams, G. M. and Smith, A. P. 2012. Developing short, practical measures of well-being. In M. Anderson (Ed.), *Contemporary Ergonomics and Human Factors 2012*, (Taylor & Francis, London), 203–210.

WHAT IS THE RELATIONSHIP BETWEEN HUMAN FACTORS & ERGONOMICS AND QUALITY IMPROVEMENT IN HEALTHCARE?

Sue Hignett[1], Duncan Miller[2], Laurie Wolf[3], Emma Jones[4], Peter Buckle[5] & Ken Catchpole[6]

[1]*Loughborough University, UK*
[2]*Sheffield Teaching Hospitals NHS Foundation Trust, UK*
[3]*Barnes Jewish Hospital, St Louis, Missouri, USA*
[4]*Department of Health Sciences, University of Leicester, UK*
[5]*Royal College of Art, London, UK*
[6]*Cedars Sinai Medical Centre, Los Angeles, USA*

A recent initiative in the National Health Service (NHS, UK) has led to an increased interest in Human Factors & Ergonomics (HFE). As part of initial discussions there have been questions about the similarities and differences between HFE and Quality Improvement (QI). We believe that there are considerable advantages from a more structured relationship between HFE and QI in healthcare and have comparatively mapped a range of dimensions (origins, drivers, philosophy, focus, role and methods). Our conclusion is that HFE in healthcare should use four criteria to maximise the benefits from this opportunity, including the use of HFE methods to design systems, environments, products etc. and the direct involvement of qualified (chartered) HFE professionals.

Introduction

There is a renewed interest in applying an HFE approach for patient safety issues in the UK (National Quality Board, 2013). Sixteen healthcare agencies have signed a statement '*that a wider understanding of **Human Factors** principles and practices will contribute significantly to **improving the quality** (effectiveness, experience and safety) of care for patients*'

At two initial meetings it became apparent that there was considerable confusion about whether HFE was a new initiative or was already being achieved through QI projects. This paper explores and maps the relationship between HFE and QI in healthcare.

HFE and QI in healthcare

Human factors & ergonomics

The need for HFE in healthcare has been recognised since the inception of the profession and discipline but development and growth have been slow (Carayon, 2010) with a dysfunctional separation of the human elements in healthcare systems into occupational health (and operational excellence) for staff, and patient safety (Hignett et al, 2013). This misconception and limited application of HFE is apparent within training programmes for non-technical skills for surgeons (NOTSS; Yule et al, 2008). This has, in our opinion, contributed to the tendency to blame *'the failures of people as the underlying cause of adverse events or broken healthcare delivery processes, a stance that is contrary to human factors science and counterproductive for advancing patient safety'* (Russ et al, 2013a) as *'little attempt is made to explore and address the underlying systemic causes that lead to errors'* (Russ et al, 2013b). Catchpole (2013) has commented that *'this behavioural safety approach, while entirely legitimate and increasingly well evidenced, is limited. Yet, it has dominated perceptions of what constitutes HF and shaped the application of HF principles in healthcare. Frequently espoused by well-meaning clinicians and aviators, rather than academically qualified HF professionals, it has led to misunderstandings about the range of approaches, knowledge, science and techniques that can be applied from the field of HF to address patient safety and quality of care problems.'*

Quality improvement

Quality initiatives have been used in the UK healthcare system for over 30 years, with as *'sporadic efforts to implement quality circles and TQM* [Total Quality Management] *… in the 1980s and early 1990s'*(Ferlie and Shortell, 2001). These early initiatives focused more on organisational performance and efficiency than safety. Quality and safety were not explicitly linked in healthcare until the late 1990s as *'the language of error and harm had not entered healthcare discourse'* (Vincent, 2010). Initiatives to link quality and safety in the NHS derived from clinical governance (mid 1990s; Scally & Donaldson, 1998) which gained prominence following the Bristol heart scandal (1984–1995; DH, 2002) and included clinical audit, clinical effectiveness, education and training, research and development, openness, risk management, and information management. The Commission for Health Improvement (CHI, 1999–2004; Day and Klein, 2004) redefined clinical governance as 7 pillars; patient involvement, clinical audit, risk management, staffing and management, education and training, clinical effectiveness, and use of information. This was followed firstly by the Commission for Health Audit and Inspection (CHAI, 2004–2009) also known as the Healthcare Commission (Haslam, 2007), with a slightly different remit, philosophy and expanded role; and then, secondly by the Care Quality Commission (2009–). There have also been advisory and arms-length bodies with a role in the quality agenda including National Quality Board (2009–),

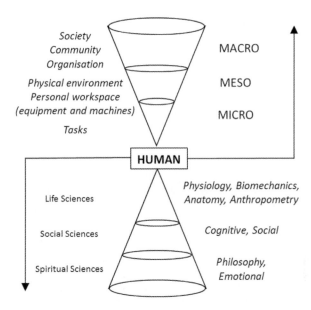

Figure 1.   Human Interactions model (Hignett, 2001).

the National Patient Safety Agency (2001–2012), NHS Institute for Innovation and Improvement (2005–2013; including the Productive series and Quality, Innovation, Productivity & Prevention (QIPP) work streams), and more recently NHS Improving Quality (2013–).

Comparing HFE and QI

A very simple comparison suggested by the authors is that:

- HFE projects look at the humans and the system, and then re-design the tasks, interfaces and the system (Fig. 1), whereas
- QI projects look at the system and then change the humans with the focus on meeting the needs of the customer demand and not necessarily considering the individual worker or optimal efficiency.

This evolution of QI in healthcare has resulted in a range of definitions for QI in healthcare including any change which improves quality (patient experience and/or clinical outcome); a change that uses a generic (e.g. training, setting standards) or specific method for a quality improvement change (e.g. Plan-Do-Study-Act); or a QI approach e.g. re-engineering, Six Sigma, Lean (Øvretveit, 2009).

The term 'Ergonomics' has been used for professional practice in the UK from 1950 (and internationally from 1961) whereas the term 'Human Factors' was used in the USA from 1957. These terms have been harmonised in US, UK, Australia and New Zealand by the inclusion of both Ergonomics and Human Factors in

Society, Institute and Association names. Most non-English speaking countries continue to use the term Ergonomics. In 2000 an international definition was agreed: *'Ergonomics (or Human Factors) is the scientific discipline concerned with the understanding of interactions among humans and other elements of a system, and the profession that applies theory, principles, data and methods to design in order to optimise human well-being and overall system performance'* (IEA, 2000).

The philosophy of the disciplines have diverged from a similar origin of engaging workers in the identification of problems and development of solutions. Both developed in the 20th century; QI in response to production quality control (reduction in errors and cost) and HFE to improve worker safety and performance as an integration of occupational health, engineering design, physical/cognitive behaviours (human and health sciences) and socio-technical systems. QI initiatives often seek to eliminate waste (Lean) and decrease variation (Six Sigma) using systematic data-driven continuous improvements and process redesign. The drivers for QI are mostly linked to performance (commissioning or reimbursement), for example through the Commissioning for Quality and Innovation (CQUIN) framework (UK) which aims to *'support improvements in the quality of services and the creation of new, improved patterns of care. It is intended to complement our approach to the payment system, providing a coherent set of national rules'* (NHS Commissioning Board, 2014). Drivers for HFE are to improve both productivity and human wellbeing (including safety). HFE uses task and systems analyses to understand and map human variance and re-design the interfaces, environments and systems based on principles of individual participation and inclusion.

Eklund (1997) suggested that *'several aspects of TQM ... have potential to violate ergonomics [HFE] principles'*, for example *'standardisation, ... reduction of variability, copying of best practice, reward systems and heavy demands on cost reduction for customer benefit.'* The role of QI and HFE professionals may be similar for the facilitation of change but differ with respect to focus (despite both using the same terminology; micro-meso-macro). In HFE, 'micro' refers to individual, human abilities and limitations, whereas in QI 'micro' refers to a team level (staff-patient). There are some overlaps in the methods used (for example systems analysis with Failure Modes Effect Analysis, FMEA) but the two professions mostly use different methods (Table 1).

Discussion and conclusion

The first challenge in implementing the National Quality Board Concordat (2013) is to spread and embed an understanding of HFE and how it differs from QI across the 16 signatory organisations: Care Quality Commission, Department of Health, Health Education England, NHS Employers, NHS England, NHS Trust Development Authority, Monitor, The Parliamentary & Health Service Ombudsman for England, National Institute for Health and Care Excellence, Public

Table 1.   Comparison of some of the methods used in HFE and QI.

HFE	QI
Systems Concepts Structure and dynamics of systems; sociotechnical systems theory, human-automation interaction (machine/computer), systems analysis and design. Integrated view of human characteristics (physical, psychological, social); Participatory ergonomics.	Understand the process using data and patient experience; understand demand, capacity and flow.
	Business process re-engineering; Plan, Do, Study Act (PDSA); Statistical Process Control.
Design Concepts Task analysis to understand and map human variance (Hierarchical Task Analysis, Link Analysis, Verbal Protocol Analysis, Postural Analysis); Anthropometry; Translation of general design principles, standards, guidelines and regulations into project specific requirements; User-centred design, HFE impact on product design cycle, inclusive (universal) design. Limitations of technology (promises, pitfalls and realities).	*Lean*: Continuous improvement based on short term snapshots. Eliminate waste by continuously striving for value added to the customer; Observe workplace; Value stream mapping; Rapid improvement event; 5S (sort, straighten, shine, standardize, sustain). *Six Sigma*: Decrease variation, by a systematic data driven process – Define-Measure-Analyse-Improve-Control (DMAIC). Gauge Repeatability & Reproducibility SIPOC (Supplier, Input, Process, Output, Customer); Process mapping; Design of Experiment; Voice of Customer; Culture change methods; Stakeholder assessment; Statistics & control limits.

Health England, General Medical Council, HealthWatch England, NHS Litigation Authority, Nursing and Midwifery Council, Social Care Institute for Excellence, NHS Leadership Academy.

There has been a gradual increase in the application of HFE in healthcare (Hignett et al, 2013). As the importance and relevance of HFE in healthcare has grown there are examples of bright and action-oriented healthcare professionals interested in safety and quality rushing off to '*do human factors*' with only a superficial understanding of the fundamental concepts, resulting in '*do it yourself*' HFE (Saleem et al, 2011). This should be of concern to the HFE professional community as '*medical [clinical] education does not necessarily provide a good preparation to understand the ideas and literature of other fields*' (Kneebone, 2002), for example engineering, psychology and design.

We suggest the relationship between HFE and QI could be described as similar to that between the disciplines and professions of Medicine and Dentistry; similar origins and drivers but differences in focus and methods. There are considerable advantages from a more structured relationship between HFE and QI. For example QI practitioners often identify problems and bring expertise in training and facilitating change but may lack an understanding of individual human behaviours (physical and cognitive interactions and capabilities) that are needed to develop solutions and interventions. In contrast HFE practitioners may have less expertise in process redesign, measuring variance and implementing recommendations.

To increase the use of HFE in healthcare we recommend that the following four criteria should be used to benchmark the quality of all HFE training and professional input (derived from Carayon et al, 2014):

1. Use of HFE tools
2. Use of HFE knowledge
3. Application of HFE to the design of equipment, medical devices, products, buildings, vehicles and systems
4. Direct involvement of qualified HFE professionals (registered/chartered member of a federated society of the International Ergonomics Association (http://www.iea.cc/about/council.html)

References

Carayon, P. (2010) Human factors in patient safety as an innovation. *Applied Ergonomics* **41**(5):657–665.

Carayon, P., Xie, A. & Kianfar, S. (2014) Human factors and Ergonomics as a patient safety practice. *BMJ Quality & Safety* **23**:196–205.

Catchpole, K. (2013) Spreading Human Factors expertise in Healthcare: Untangling the knots in People and Systems. *BMJ Quality & Safety* **22**:793–797.

Day, P. & Klein, R. (2004) *The NHS Improvers. A study of the Commission for Health Improvement.* London: The Kings Fund.

DH (2002) *Learning from Bristol. The Department of Health's response to the report of the Public Inquiry into children's heart surgery at the Bristol Royal Infirmary 1984–1995.* London: TSO.

Eklund, J. (1997) Ergonomics, quality and continuous improvement: conceptual and empirical relationships in an industrial context. *Ergonomics* **40**:98–1001.

Ferlie, E.B. & Shortell, S.M. (2001) Improving the quality of healthcare in the United Kingdom and the United States: A framework for change. *The Milbank Quarterly* **79**:281–315.

Haslam, D. (2007) What is the Healthcare Commission trying to achieve? *Journal of the Royal Society of Medicine,* **100**:15–18.

Hignett, S., Carayon, P., Buckle, P. & Catchpole, K. (2013) State of science: human factors and ergonomics in healthcare. *Ergonomics* **56**:1491–1503.

Hignett, S. (2001) *Using Qualitative Methodology in Ergonomics: Theoretical Background and Practical Examples.* Ph.D. Thesis. University of Nottingham.

IEA. (2000) *International Ergonomics Association, Triennial Report.* Santa Monica, CA: IEA Press. 5.

Kneebone, R. (2002) Total internal reflection: an essay on paradigms. *Medical Education,* **36**(6):514–518.

National Quality Board (2013) *Human Factors in Healthcare. A Concordat* http://www.england.nhs.uk/wp-content/uploads/2013/11/nqb-hum-fact-concord.pdf (Accessed 25th August 2014)

NHS Commissioning Board (2014) *Commissioning for quality and innovation (CQUIN): guidance* http://www.england.nhs.uk/wp-content/uploads/2014/02/sc-cquin-guid.pdf (Accessed 10 Oct 2014)

Øvretveit, J. (2009) *Does improving quality save money? A review of evidence of which improvements to quality reduce costs to health service providers.* The Health Foundation. http://www.health.org.uk/publications/does-improving-quality-save-money/ (Accessed 6 Dec 2014)

Russ, A.L., Fairbanks, R.J., Karsh, B-T., Militello, L.G., Saleem, J.J. & Wears, R.L. (2013a) The Science of Human Factors: separating Fact from Fiction. *BMJ Quality & Safety.* **22**:802-808.

Russ, A.L., L.G. Militello, J.J. Saleem, et al. (2013b) Response to separating Fact from Opinion: a Response to 'the science of Human Factors: separating Fact from Fiction. *BMJ Quality & Safety* **22**:964–966.

Saleem, J.J., E.S Patterson, A.L. Russ, & R.L. Wears (2011) The need for a broader view of human factors in the surgical domain. *Arch Surg* **146**(5):631–632.

Scally, G. & Donaldson, L. (1998) Clinical governance and the drive for quality improvement in the new NHS in England. *BMJ* **317**:61

Vincent, C. (2010) *Patient Safety.* 2nd Ed. Chichester, West Sussex: John Wiley & Sons Ltd 35.

Yule, S., Flin, R., Maran, N., Rowley, D., Youngson, G. & Paterson-Brown, S. (2008) "Surgeons' Non-technical Skills in the Operational Room: Reliability Testing of the NOTSS Behaviour Rating System". *World J Surg* **32**:548–556.

NOTTINGHAM UNIVERSITY HOSPITAL GUIDELINES APP – IMPROVING ACCESSIBLITY TO 650 HOSPITAL CLINICAL GUIDELINES

Adrian Kwa[1,2], Mark Carter[3], Duane Page[3], Tony Wilson[3], Michael Brown[4] & Bryn Baxendale[1,2]

[1]*Trent Simulation and Clinical Skills Centre, Nottingham University Hospitals NHS Trust*
[2]*Department of Anaesthesia, Nottingham University Hospitals NHS Trust*
[3]*Application Development Team, Nottingham University Hospitals NHS Trust*
[4]*Human Factors Research Group, Faculty of Engineering, University of Nottingham*

Accessibility to clinical guidelines is a major problem at Nottingham University Hospital NHS Trusts, and a number of critical incidences have been attributed to this. Using a user-centred designed (UCD) approach, we developed a mobile application to improve accessibility to over 650 Trust-approved clinical guidelines. We ran focus groups, performing card-sorting exercise and critical incident techniques. Based on this information we developed prototype apps and evaluated them with 145 staff. The final design achieved a score of 83 on the System Usability Scale (top 5th centile), highlighting the importance of using a UCD approach when designing IT system in healthcare.

Introduction

Clinical guidelines are systematically developed statements designed to help practitioners and patients decide on appropriate healthcare for specific clinical conditions and/or circumstances (Field & Lohr, 1992). These documents are essential to good clinical care, especially when managing less commonly-encountered conditions or emergencies. They are often developed by national agencies such as NICE, or medical colleges, and commonly adapted to fit local needs. At Nottingham University Hospitals NHS Trust (NUH), we have over 650 clinical guidelines covering wide aspects of clinical practice. These are stored within the hospital intranet system, and are accessible only via Trust-approved desktop computers. However this is not practical during medical emergencies, when it would be inappropriate to leave the bedside of very sick patients to search for information. The guidelines menu system is also poorly designed and difficult to navigate. One consequence of this is clinical users working from memory, often missing essential steps in clinical managements (Arriaga et al., 2013). They may also access guidelines and protocols that have not been approved by the Trust, which may not have been verified to be accurate (Hasty et al., 2014), or differ from local guidance. This can potentially lead to patient harm.

Another issue that this initiative seeks to address concerns the relative high turnover of trainee medical staff during their clinical rotations. This presents a problem of how to ensure staff that are new to the organisation are able to access this type of information easily and reliably. This issue is further compounded if short-term agency and locum staff are also considered.

Advancement in mobile technology has turned devices like mobile phones into mini-computers. Most clinical staff carries a mobile device, making it an ideal platform for delivering clinical guidelines at the point of care.

Adopting a User-Centred Design approach, we aimed to develop a mobile application (app) that would allow the users to access clinical information on their mobile devices, in a readily accessible format. This app should improve accessibility to clinical guidelines, and is likely to improve adherence to Trust-approved guidelines, which in turn should support delivery of effective and evidence-based care more reliably.

The design of the app presents a number of specific human factors issues:

1. Improving information retrieval in order to improve patient care, as discussed by Gardner (1997). Users (clinical staff) must be able to quickly search and retrieve specific information from a large collection of documents (over 650) in time-pressured, safety-critical situations.
2. The information architecture of the app must allow documents to be updated by Trust management to ensure they are accurate, while also allowing the documents to be readily available on staff phones when out of internet access (a common issue with web-driven technologies, Balasubramanian & Bashian, 1998).
3. The user-interface design should be take into account small form-factor of most mobile phones, and minimise the use of on-screen keyboard where possible (Buchanan et al., 1991).

Method

We ran focus groups with staff from various clinical background and specialties. We performed critical incident technique, discussing the problems encountered by staff with the current guidelines system. We also ran a series of card sorting exercises, which involve categorising guidelines into "emergencies" or "non-emergencies", and into different clinical disciplines.

Using the information gathered from the focus group meetings, we developed a prototype app. We invited groups of doctors and other healthcare professionals groups (nurses, midwives etc.) to evaluate the prototype. None of the participants had attended the focus group discussions, and none had prior exposure to the prototype. No instruction on how the app functions was given to participants, and no conferring was allowed. A facilitator was present during testing. Participants were given nine clinical scenarios and tasked with finding guidelines for each using the prototype app without the use of search function. Guidelines relevant to these questions may be found in multiple locations, each with a unique 6-digit code. Participants were asked to record these codes on an answer sheet. If they were unable to locate the guidelines on first attempt, they were asked to note down the location where they expected to find the guideline. The prototype was modified after each test until participants were able to locate all guidelines during testing. After the design was finalised, system usability scale (Brooke, 1996) was used to evaluate the app.

Results

We interviewed 30 participants (15 trainee doctors and 15 healthcare professionals) currently employed at NUH in divided groups. Participants told us that the current guidelines system was difficult to use because:

1. Guidelines are collated under directorates grouping. For example, "Cardiology" and "Infectious Diseases" are part of "Cancer and Associated Specialties" directorate. This makes it difficult for users who are unfamiliar to the system to locate the correct guidelines
2. Because of the difficulty described above, staff tend to favour searching for guidelines using the "search" function. However this does not always yield the appropriate result. At NUH, clinical guidelines are managed using Microsoft SharePoint system. Using the search function, the software would search within all documents in its system, selecting ones that contain the keyword(s), and generate results in order of frequency of access rather than relevance. This has left users feeling frustrated, and many have stopped using the system for these reasons.

We also performed card-sorting exercises using all Trust-approved guidelines during the focus group meetings. Participants in all groups agreed that it would be more logical to group guidelines under clinical specialties. There were very few differences in how guidelines were sorted between all groups. It was also apparent that some clinical guidelines are relevant in more than one specialty group. None of the groups were unable to agree on the most appropriate location for these.

We therefore designed a prototype with the following features:

- Three-clicks, multiple point of entry menu system – All clinical guidelines are accessible within three clicks from the home screen, making the menu system quick and easy to navigate. Guidelines are grouped under clinical specialties headings, based on the result from card-sorting exercises. Allowances were made for guidelines of clinical relevance to more than one specialty, making them accessible from multiple points within the menu system.
- In Case of Emergency function – To minimise the use of on-screen keyboard, the "In Case of Emergency" section was developed. Clinical guidelines that have been identified as clinical emergencies during card-sorting exercise may be found within this section. Guidelines within this section is grouped alphabetically, and

can also be accessed within three clicks from the home screen using a bespoke on-screen A to Z keyboard.

A total of 145 staff (92 doctors and 53 other healthcare professionals) participated evaluating the prototype app. We learned that participants from different clinical background and experience search for guidelines in different ways. More experienced and medical staff are more likely to search for guidelines using medical terminology, whereas less experienced and nursing staff are more likely to use layman's terms. There was also wide variation as to how to access guidelines that were previously identified as relevant to multiple specialties. All participants found the app easy to use, and most were able to complete the task in under six minutes.

After further fine-tuning, the final groups of participants (n = 43) were asked to evaluate the prototype using System Usability Scale. The app yielded a score of 83, putting it in the top 5th centile in terms of usability.

Other features were added to the release version of the app. These includes:

- "Search" function – all guidelines are searchable using keywords relevant to the specific guidelines. Keywords tagged with each guideline include medical (and where applicable, layman's) terms to ensure ease of access. In total, over 7500 keywords were used.
- "My Favourite" function – Users were able to create their own shortlist of clinical guidelines relevant to their practice. This also further reduce the number of click to access guidelines from three to two.
- "Auto-disable"/"Update" feature – The app will attempt to connect to the central server for synchronisation at each start up, and alert users of updates. The app will continue to function if there is no internet connection, or if update is not performed when available. However, users must allow internet access or update the app once every 30 days. If this is not performed, most features within the app will be disabled until synchronisation or update has taken place. This enable the Trust to have full control over the content of the app and allow uploading new guidelines and withdrawal of guidelines before its review date. The content within the app is never more than 30 days older than the central server.
- "Auto-expire" feature – Each guideline is assigned an expiry date, after which it will made unavailable automatically.
- Clinical Scoring Aide-memoir – A collection of frequently used clinical scoring system chart are also easily accessible for use within the app.
- "Phone directories" – Phone numbers to clinical areas are searchable within the app.

Conclusion

The app was launched to strong acclaim. It has been downloaded over 1000 times within 8 weeks of launch and is used by up to 50 unique users daily. This is likely to increase adherence to Trust-approved guidance on clinical management, thus assisting improvement in patient safety and reduction in morbidity and mortality rate at our Trust. Based on analytic data from the first 10 weeks of use, guidelines are accessed using the search function less than 10% of cases. This indicates the success of our 3-clicks, multiple point of entry menu system.

This study highlighted the importance of taking a user-centred design approach when developing products in healthcare, and confirmed mobile devices as ideal platform for delivering clinical information for all healthcare professionals at the point of care.

Acknowledgement

We would like to thank colleagues at NUH who have participated in focus groups and testing of prototype. We would also like to thank colleagues at Trent Simulation and Clinical Skills Centre with their help during the testing period.

References

Arriaga A. F., Bader, A., Wong, J. M., Lipsitz, S. R., Berry, W. R., Ziewacz, J. E., Hepner, D. L., Boorman, D. J., Pozner, C. N., Smink, D. S. and Gawande, A. A. 2013, Simulation-based trial of surgical-crisis checklists. *New England Journal of Medicine* 368: 246–253.

Balasubramanian, V. and Bashian, A. 1998, Document management and web technologies: Alice marries the Mad Hatter. *Communications of the ACM*, 41(7), 107–115.

Brooke, J. 1996, SUS: a "quick and dirty" usability scale. In P. W. Jordan, B. Thomas, B. A. Weerdmeester, & A. L. McClelland. *Usability Evaluation in Industry.* London: Taylor and Francis.

Buchanan, G., Farrant, S., Jones, M., Thimbleby, H., Marsden, G. and Pazzani, M. 2001, Improving mobile internet usability. In *Proceedings of the 10th international conference on World Wide Web* (pp. 673–680). ACM.

Field, M. J. and Lohr, K. N. 1992, *Guidelines for Clinical Practice: From Development to Use.* Washington DC: National Academic Press.

Gardner, M. 1997, Information retrieval for patient care. *BMJ*, 314(7085), 950.

Hasty, R. T. et al. 2014, Wikipedia vs peer-reviewed medical literature for information about the 10 most costly medical conditions. *Journal of American Osteopath Association* 114(5): 368–373.

THE MEASUREMENT OF PATIENT SAFETY CULTURE; PROGRESS, BUT STILL A LONG WAY TO GO

Patrick Waterson

Human Factors and Complex Systems Group,
Loughborough Design School, Loughborough University

The first tools for measuring Patient Safety Culture (PSC) emerged in the early part of the millennium. Since then there has been an explosion of interest in PSC and over 200 studies have been published reporting data across a wide range of healthcare settings. This paper follows on from a plenary presentation given by the author at the 2012 IEHF conference in Blackpool and provides an update on progress which has been made in the last few years in relation to PSC. In particular, I point to a set of 10 recurring themes and outstanding challenges (some old, some new), covering theoretical, methodological and practical aspects of PSC.

Ten recurrent themes and future challenges for PSC

1. The controversy surrounding safety culture

The term safety culture tends to elicit very strong opinions – everything from full acceptance of its existence and its validity, through to outright rejection and in some cases, contempt. Many of these debates might be characterized as shedding 'more heat than 'light', however, they also reflect the tensions amongst both academics and practitioners involved in safety improvement. It still remains difficult to define what we actually mean by 'safety culture', even in domains where the term is well established (e.g., the nuclear industry). Some of the difficulties are reflected in the perennial discussions centred on the so-called 'climate vs. culture' debate (Mearns and Flin, 1999), others reflect more fundamental conceptual, theoretical and practical issues (e.g., the relationship between safety culture and the systems approach – Reiman and Rollenhagen, 2013).

2. The continued challenge of 'measuring' medical error

Part of the problem with attempting to measure error and safety culture relates to the characteristics of healthcare environments. Clinical work tasks do not always follow a clear 'linear' path – rather healthcare is a complex sociotechnical system, many of the parts of healthcare delivery are messy and non-linear (Karsh et al., 2010). Even processes which appear at first glance to be simple (e.g., hip replacement surgery), involve a large range of clinical roles, technologies (e.g., electronic record systems) and locations (Eason et al., 2012). PSC will vary a great deal according to characteristics of the work tasks, locations, people involved etc. There is huge variability in the way healthcare systems operate. In our previous work on PSC data from the UK we found that culture on one hospital ward may be radically different to another only a floor above or below it (Waterson et al., 2010). Different types of roles and specialisms may have different attitudes towards safety which makes the measurement of PSC difficult. Hammer and Manser (2014) for example, found that nurses are more likely to rate PSC as high when staffing and workload levels are adequate, whereas physicians are more likely to be influenced by the quality of teamworking in their work environment.

3. Conceptual challenges – widening perspectives

As a number of researchers have pointed out (e.g., see Hammer and Manser, 2014; McDonald and Waring, 2014) there are various different ways in which safety culture have been conceptualized. These include functionalist approaches which assume that the prime function of safety culture is to support management strategies and systems. The measurement of safety culture as seen from this point of view can be reduced to a relatively simple model of prediction and control (Glendon and Stanton, 2000). By contrast, interpretive approaches emphasize the importance of regarding safety culture as an emergent complex phenomenon where issues such as identity, beliefs and behaviour need to be taken into account. Silbey (2009) similarly discusses the influence of three different perspectives on safety culture which have dominated research since the 1990's: (1) 'culture as causal attitude' – here culture is seen as composed on individual attitudes and organisational behaviour which can be decomposed into values and attitudes; (2) 'culture as engineered organisation' – here the emphasis is on how culture leads to outcomes such as reliability and efficiency (e.g., as reflected in research on 'high-reliability organisations' and models of accidents which draw on control theory – Eisenhardt, 1993; Weick et al., 1999; Leveson, 2012); and, (3) 'culture as emergent and indeterminate' – culture from this point of view is seen as an "indissoluble dialectic of system and practice' which "cannot be engineered and only probabilistically predicted with high variation from certainty (Silbey, 2009, p. 356).

What is perhaps most noticeable about these approaches is that within the field of PSC is that the majority of studies so far adopt a functionalist, 'culture as causal attitude' approach towards theory and measurement. Most studies involve the use of quantitative surveys, very few studies in the mainstream medical literature are based on observation or the use of ethnographic studies. Theory still remains poorly defined in most studies (Halligan and Zecevic, 2011), whilst other areas (e.g., contemporary work on 'risk' – e.g., Aven et al., 2011; Ben-Ari and Or-Chen, 2009) are much more developed.

4. Whose safety culture are we assessing?

Related to theme 2, many of the studies which report findings based on PSC surveys focus on a limited set of participants, roles and healthcare specialists. A closer look at the characteristics of respondents reveals that often the largest percentage of these are nurses or healthcare assistants, with a much smaller response rate from physicians and other medical practitioners (Waterson et al., in preparation). Managers and senior administrators are also under-represented. Sample characteristics such as these raise a number of issues, not least, how far we can generalize the findings from these types of studies, based as they are on a limited sample of healthcare staff.

A related issue is the inclusion of groups of people who are often ignored or excluded from studies of patient safety, namely patients and their carers. Very little work of this kind has been so far reported. One exception is the study by Cox et al. (2013) which used an adapted version of the AHRQ HSPSC survey to assess the perceptions of parents of safety on the hospital units where their children were being cared for. Front-line healthcare staff (in particular nursing staff) appear to make up the bulk of respondents to PSC surveys. Accordingly, there needs to be a degree of caution in interpreting the findings from these studies. Both researchers and healthcare practitioners are encouraged to ask questions about the study sample characteristics and respondent profiles. Part of this might be improved involve the development of guidelines or standards covering the reporting of PSC studies (theme 8 – e.g., making sample characteristics more explicit).

5. Methodological gaps and opportunities

Currently, the field of PSC is dominated by the use of a very limited set of methods and tools. The European Network for Patient Safety Project (EUNetPas, 2010) for example, identified 19 different PSC instruments in use throughout the EU member states. The most frequently used of the 19 identified PSC instruments were the Hospital Survey on Patient Safety Culture (used in 12 EU states), the Manchester Patient Safety Framework (used in 3 EU member states) and the Safety Attitudes Questionnaire (used in 4 member states). In contrast to other industries and safety 'a small subset of the types of methods which could be used to assess PSC. LeMaster and Wears (2012) for example, argue that direct observation and ethnography should be used more to assess actual or simulated behaviour (e.g., adverse events, near misses). They argue that there is a need to unpack a wider range of system factors involved in patient safety and to move beyond the limited range covered within culture/climate surveys. Future work should try to adopt a more eclectic approach towards alternative methodologies for PSC. There is a need to move beyond an exclusive focus on surveys and to examine how other tools, methods, interventions and theory (theme 3) might be linked to PSC.

6. 'Culture in the large'

The question of how well PSC instruments developed in one nation translate to another has been raised by a number of authors (e.g., Hammer and Manser, 2014). Many instruments (e.g., HSPSC, SAQ) derive from a US context and analysis of their psychometric properties has shown that in some cases (e.g., measures of staffing and organisational learning) there are difficulties in using them in other national contexts. A variety of reasons might be given why these and other problems occur: the larger health system is very different and its operational functioning may vary a great deal across nations (e.g., private vs. publicly funded); staff may have different attitudes towards safety and this may vary greatly across roles and national contexts (theme 4).

These sorts of considerations underline the need to be cautious of attempts to simply translate and the use PSC surveys without paying attention to national and wider health system characteristics. Surveys need to be tailored and modified in line with national and other cultural dimensions which have been shown to be important (e.g., power distance, individualism – Hofstede, 1993). Pre-testing in the form of the use of focus groups and cognitive interviews is highly recommended.

7. The relationship between PSC and outcomes for staff and patients

Research establishing the link between PSC and a range of clinical (e.g., incident and near miss reports) and organisational outcomes (citizenship and organisational commitment; job and work design variables – Talati and Griffin (2014); Phipps and Ashcroft (2014) is still very much underdeveloped. This criticism is not new (Flin, 2007; Flin et al., 2006), however, it represents a large gap in our understanding about the relationship between PSC and other phenomena and healthcare organisational dynamics. Similar, criticisms can be made of the lack of studies relating PSC to measures derived from staff and patients (e.g., workload, satisfaction with care, hospital re-admission rates). The link between PSC and outcomes for staff and patients is still as yet, relatively unexplored. By contrast, other areas outside of PSC have been more successful in demonstrating relationships between human resource management practices and healthcare productivity (e.g., Patterson et al., 2010). Much could be learned from such examples.

8. Improving the 'science' underpinning patient safety culture

The reporting of psychometric data has improved a good deal since Flin et al.'s (2006) review; most studies now report data covering the reliabilities of survey dimensions for example. However, a number of problems still remain. Many studies appear either not to be aware of established procedures for assessing the accept-ability of survey item validity or reliability. It is common to read a study which concludes that a specific instrument is acceptable to use within their nation whilst at the same time reporting poor levels of reliability for specific dimensions. Like-wise, many studies fail to report the results of exploratory or confirmatory factor analyses. There is also a large degree of variation when comparing the results of studies (e.g., factor structures) which have used a modified or identical instrument which has been developed elsewhere (e.g., the HSPSC – Hammer and Manser, 2014). In combination, these problems indicate that there is still some way to go before we can be confident in assessing the degree to which PSC is being measured by specific instruments and how this compares across multiple levels of analysis including nations, healthcare systems, and locations (Karsh et al., 2014). There are a number of areas for future work which might help to improve the scientific basis of PSC, as well facilitate comparison between studies, these include: (1) the development of a set of procedures covering survey development and evaluation (including psychometric criteria); (2) developing similar procedures or guidelines covering the reporting of PSC data and findings; (3) constructing a comparative database of worldwide studies which have used specific tools and instruments (e.g., HSPSC, SAQ).

9. Learning from other industries

As Kirwan and Shorrock (2014) have shown, there is a huge amount that could be learnt from other safety industries and sectors. Some of the methods which have tried and tested within the nuclear and oil and gas sectors are only just starting to be applied within healthcare (e.g., safety cases and human reliability assessment – Health Foundation, 2012). Other such as the concept of 'safety intelligence' (Fruhen et al., in press) offer the potential for future work, particularly as it relates to senior managers within healthcare. Even a cursory glance at the accounts and recollections of well-known researchers in the field of safety science (e.g., Kletz, 1993; Reason, 2013) is enough to underline the dangers of not learning from other domains. Interdisciplinarity means not only looking at what other scientific research has been done outside 'traditional' studies in healthcare and patient safety. It also means paying closer attention to possible lessons from practitioners working in other industries. The field of PSC needs to pay more attention to past work within the safety-critical industries – some of this may well not translate to healthcare, but in other case there may be potential which is as yet unrealized.

10. Moving beyond 'what went wrong' – from 'safety I' to 'safety II'

A final area which has recently grown in important is studies which have looked at how to improve the resilience of healthcare organisations. The emphasis here is not only on errors or failures to deliver care, but also on identifying successes and other examples where routine practice has resulted in safety and reliability in care delivery. The focus on identifying error and the 'latent conditions and pathways (Reason, 1990) which contribute to mistakes and accidents is sometimes characterized as 'safety I' (Hollnagel et al., 2013; Rowley and Waring, 2013). By contrast, 'safety II' concentrates on everyday performance and how can be seen as a resource necessary for system flexibility and resilience. The safety management principle is continuously to anticipate developments and events (Eurocontrol, 2013). Work on PSC has so far focused on 'safety 1', future work could be carried out on developing tools and methods for identifying and assessing the conditions where performance variability can become difficult or impossible to monitor and control.

Summary and conclusions

It is clear that over the course of its brief history, spanning less than 20 years, the field of PSC has made some real progress. Needless to say there is a long way to go. Measurement issues cut across many of the themes identified in this paper and are an important priority for the future. Recent work by the Health Foundation in the UK following the publication of the Francis report has underlined the importance of accurate diagnosis and assessment of quality and safety in healthcare:

> *"our ability to measure and assess quality of care is improving, [however] . . . there are still many aspects of care, and care services, for which routinely available information on quality is inadequate or non-existent"*. (Health Foundation, 2013, p. 6)

Similar conclusions have been reached by a number of prominent authors in the field of patient safety (e.g., Provonost et al., 2011; Wachter, 2010). Part of 'solution' might be to further develop a core part of current work in PSC and patient safety, namely the application of the systems approach (Carayon et al., 2014; Reiman and Rollenhagen, 2013; Waterson, 2009). The approach remains under-exploited particularly in terms of unpacking the multi-level properties of safety culture (themes 5 and 8), as well as understanding external systemic influences on safety (e.g., the influence of regulators). Perhaps the most important aspect of the systems approach within PSC is that it might facilitate a shift towards a focus on not only senior managers in healthcare organisations, but also other stakeholders including patients and the wider public.

References

Carayon, P., Wetterneck, T.B., Rivera-Rodriguez, A.J., Schoofs Hundt, A., Hoonakker, P., Holden, R. and Gurses, A.P., 2014, Human factors systems approach to healthcare quality and patient safety. *Applied Ergonomics*, 45, 14–25.

Cox, E.D., Carayon, P., Hanson, K.W., Rajamanickam, V.P., Brown, R.L., Rathouz, P.J., DuBenske, L.L., Kelly, M.M. and Buel, L.A. in press, Patient perceptions of children's hospital safety climate. *BMJ: Quality and Safety in Health Care*.

Eason, K., Dent, M., Waterson, P., Tutt, D., Hurd, P., Thornett, A. 2012, Getting the benefit from electronic patient information that crosses organisational boundaries. Final report. *NIHR Service Delivery and Organisation programme*.

Eisenhardt, K. 1993, High reliability organisations meet high velocity environments: common dilemmas in nuclear power plants, aircraft carriers, and microcomputer forms. In K.H. Roberts (Eds.), *New Challenges Understanding Organizations*. New York: Macmillan.

Eurocontrol 2013, From Safety I to Safety II: A White Paper. http://www.eurocontrol .int/sites/default/files/content/documents/nm/safety/safety_whitepaper_sept_ 2013-web.pdf

European Society for Quality in Healthcare 2010. Use of Patient Safety Culture Instruments and recommendations. EUNetPas Project Report, Aarhus, Denmark. Available at: http://90plan.ovh.net/~extranetn/images/EUNetPaS_Publications/eunetpas-report-use-of-psci-and-recommandations-april-8-2010.pdf

Flin R, 2007, Measuring safety culture in healthcare: A case for accurate diagnosis. *Safety Science*, 45, 6, 653–667.

Fruhen, L.S., Mearns, K.J., Flin, R.H. and Kirwan, B. (in press), Safety Intelligence: An exploration of senior managers characteristics'. *Applied Ergonomics*.

Glendon, A.I. and Stanton, N.A. 2000m Perspectives on safety culture. *Safety Science*, 34, 193–214.

Halligan, M. and Zecevic, A. 2011, Safety culture in healthcare: a review of concepts, dimensions, measures and progress. *BMJ: Quality and Safety*, 20, 338–343.

Hammer, A. and Manser, T. (2014), The use of the HSPSC in Europe. In P.E. Waterson (Ed.), *Patient Safety Culture: Theory, Methods and Application*. Ashgate: Farnham.

Health Foundation 2012, Using safety cases in industry and healthcare. Report available at: http://www.health.org.uk/public/cms/75/76/313/3847/Using%20safety%20cases%20in%20industry%20and%20healthcare.pdf?realName=09HlEo.pdf

Hollnagel, E., Braithwaite, J. and Wears, R.L. 2013, Eds., *Resilient Health Care*. Farnham: Ashgate

Hofstede, G.H. 1993, Cultures and organizations: software of the mind. *Administrative Science Quarterly*, 38 (1): 132–134.

Karsh, B.-T., Weinger, M.B., Abbott, P.A. and Wears, R.L., 2010, Health information technology: fallacies and sober realities. *Journal of the American Medical Informatics Association, (JAMIA)* 17, 617–623.

Karsh, B-T., Waterson, P.E. and Holden, R. 2014, Crossing levels in systems ergonomics: a framework to support 'mesoergonomic' inquiry. *Applied Ergonomics*, 45, 45–54.

Kirwan, B. and Shorrock, S. 2014, A view from elsewhere: safety culture in European traffic management. In P.E. Waterson (Ed.), *Patient Safety Culture: Theory, Methods and Application*. Ashgate: Farnham.

Kletz, T. 1993, *Lessons from Disaster: How Organisations Have No Memory and Accidents Recur.* London: Institute of Chemical Engineers.

LeMaster, C.H. and Wears, R.L. 2012, Stepping back: why patient safety is in need of a broader view than the safety climate survey provides. *Annals of Emergency Medicine*, 60, 5, 564–566.

Leveson, N. 2012, *Engineering a Safer World.* Cambridge, Mass.: MIT Press.

McDonald, and Waring, J. (2014), Creating a safety culture: learning from theory and practice. In P.E. Waterson (Ed.), *Patient Safety Culture: Theory, Methods and Application*. Ashgate: Farnham.

Mearns, K. 2010, Safety culture and safety leadership – do they matter? Keynote presentation at *Working on Safety 2010 Conference*. Available at: http://www.wos2010.no/assets/presentations/0-Key-note-Mearns.pdf

Mearns, K.J. and Flin, R. 1999, Assessing the state of organizational safety – culture or climate? *Current Psychology*, Vol. 18, No. 1, 5–17.

Patterson, M., Rick, J., Wood, S., Carroll, C., Balain, S. & Booth, A. 2011. Systematic review of the links between human resource management practices and performance. *Health Technology Assessment*, 14 (51), iv.

Provonost, P.J., Berenholtz, S.M. and Morlock, L.L. 2011. Is quality of care improving in the UK? Yes, but we do not know why. *BMJ*, 342:c6646.

Reason, J. 1990, *Human Error*. Cambridge: Cambridge University Press.

Reason, J. 2013, *A Life in Error*. Farnham: Ashgate.

Reiman, T. and Rollenhagen, C. in press. Does the concept of safety culture help or hinder systems thinking in safety? *Accident Analysis and Prevention*.

Rowley, E. and Waring, J. 2011, Eds., A Socio-Cultural Perspective on Patient Safety. Farnham: Ashgate.

Silbey, S.S. 2009, Taming Prometheus: talking about safety and culture. *Annual Review of Sociology*, 35, 341–69.

Wachter, R.M. 2010, Patient safety at ten: unmistakable progress, troubling gaps. *Health Affairs*, 29, 1, 165–173.

Waterson, P.E. 2009, A critical review of the systems approach within patient safety research. Ergonomics, 52, 10, 1185–1195.

Waterson, P.E., Griffiths, P., Stride, C., Murphy, J. and Hignett, S. 2010, Psychometric properties of the Hospital Survey on Patient Safety: Findings from the UK. *BMJ: Quality and Safety in Health Care*, 19, 1–5.

Waterson, P.E., Burford, E., Jackson, J., Stride, C., Hutchinson, A., Hammer, A. and Manser, T. (in preparation), The Hospital Survey of Patient Safety Culture (HSPSC): a cross-national review of study characteristics and psychometric properties. Unpublished manuscript.

Weick, K.H. (1987), Organizational culture as a source of reliability. *California Management Review*, 29, 112–127.

SMARTPOWERCHAIR: TO BOLDLY GO WHERE A POWERCHAIR HAS NOT GONE BEFORE

Paul Whittington, Huseyin Dogan & Keith Phalp

Faculty of Science & Technology, Bournemouth University, Fern Barrow, Poole, Dorset, UK

A survey was conducted targeting a user community of people in powered wheelchairs (powerchairs) as the requirements elicitation phase of a proposed SmartPowerchair, using online and paper-based methods. Analysis of the survey results using graphs and statistics led to key findings. These showed that opening/closing curtains, windows, doors and operating heating controls were the most difficult tasks to perform from a powerchair and also that an integrated smartphone operated by either touch or head tracking would be the most useful to potential SmartPowerchair users. This research is supported by a usability evaluation case study of a pervasive assistive technology which revealed System Usability Scale (SUS) and NASA Task Load Index (TLX) results.

Introduction

There is an ever-increasing market for assistive technologies (Gallagher et al. 2013), as approximately 500 million people worldwide have disabilities that affect their interaction with society and the environment (Cofré et al. 2012). It is therefore important to improve the lifestyle of people with disabilities.

A SmartPowerchair is proposed, where existing pervasive technologies will be integrated into a standard powered wheelchair (powerchair) with the aim of further improving quality of life for people with disabilities through independent living. To achieve maximum potential, the SmartPowerchair requires optimum usability, defined as "the extent to which a product can be used by specified users to achieve specified goals with effectiveness, efficiency and satisfaction in a specified context of use" (International Organization for Standardization 1998). The usability of a system has greater importance when the users have disabilities (Adebesin et al. 2010). It is therefore important that the requirements for the SmartPowerchair are elicited from the intended user community, i.e. people with disabilities.

State of the art research in assistive technology

A user survey involving people with disabilities was conducted by Ari et al. (2010) to assess the assistive technology needs of students. The aim was to determine how technology enabled equal opportunities of students in higher education, as only a few students were aware of the assistive technologies available. The findings showed that quality of life was increased where all students had access to a computer and the use of the Internet for communication, indicating the relevance of integrating pervasive technology with powerchairs.

An example of an assistive technology requirements elicitation process was conducted by Robinson et al. (2009) as part of the Keeping In Touch Everyday (KITE) project, where the views concerning a proposed armband and electronic notepad were obtained from people with dementia. A needs analysis (i.e. an accurate assessment from the perspective of the user group of how the proposed technologies would facilitate independence) was performed during the scoping stage. The survey results from the SmartPowerchair requirements elicitation process provided similar needs analysis for the tasks that would improve quality of life.

Tanaka et al. (2005), Mihailidis et al. (2007) and Postolache et al. (2009) have previously developed SmartPowerchairs to respectively navigate by either using Electroencephalogram-based control, artificial intelligence or become a form telemedicine to monitor physiological parameters.

Requirements elicitation

It was necessary to approach 32 UK organisations to establish a niche user group for the requirements elicitation survey in order to identify suitable participants between the ages of 12 and 70. The 16 selected participants were a mixture of genders from a variety of backgrounds (including students and the retired) who also had varying disabilities (such as Cerebral Palsy, Arachnoiditis and Hydrocephalus, and Spinal Muscular Atrophy) with either dexterity and/or speech impairment. The participants thereby became a representative sample to accurately elicit the SmartPowerchair requirements.

Method

A survey was provided to the organisations consisting of 37 questions concerning the difficulty of home tasks and possible integrations of pervasive technology. To maximise the number of responses to the survey, the organisations were either approached with an online survey and/or offered semi-structured interviews. The interviews were performed at a special educational needs school where participants were individually interviewed using the same questions as the online survey. The interviews had the advantage of a captive audience compared to the relatively low response rate of the online survey. The SmartPowerchair functionality was developed based on the most difficult tasks identified by the participants.

Results

Tasks: Figure 1 shows that 58% of participants found the most difficult household tasks to be opening/closing curtains and windows. The feedback suggested that

Household Tasks

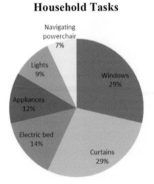

Pervasive Technologies

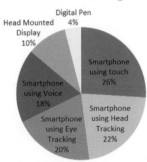

Figure 1.   **Difficult household tasks and potential pervasive technologies.**

causes of this difficulty were due to the curtains/windows either being out of reach, inaccessible due to obstacles such as furniture or requiring a significant level of physical activity to be exerted. Navigating the powerchair around the home was the next most difficult task due to narrow internal doors.

Doors: 27% of participants identified front, back and patio doors to be the most difficult doors in the home to open and close. Garage doors were the second most difficult for 20% of participants. A comment was that opening/closing doors requires concentration to simultaneously drive the powerchair and open/close the door. Participants with dexterity impairments found the door handle position, the weight of the door and locks to be issues. Some participants commented that they could only manage doors if they were left unlocked, but this presents a security risk.

Appliances: Cookers and heating controls were identified as the most difficult household appliances to operate by 38% of participants due to the heat produced by cookers and the small dials on heating controls. Microwaves and kettles were the next most difficult with 25% of participants.

Pervasive Technologies: Figure 1 also shows that 48% of participants stated a smartphone operated by either touch or head tracking had the greatest potential. A smartphone controlled by voice was only popular with individuals who did not have speech impairment. Head mounted displays were the least popular technology at 10% due to being obtrusive and difficult to wear for people with disabilities.

A case study

The case study of SmartATRS (Bournemouth University 2013) is an example of a task that can be supported by pervasive assistive technology.

The rationale behind creating SmartATRS was to improve the usability of the Automated Transport and Retrieval System (ATRS) keyfobs, which are similar to those used to operate automated gates. The objective of ATRS was to create a reliable,

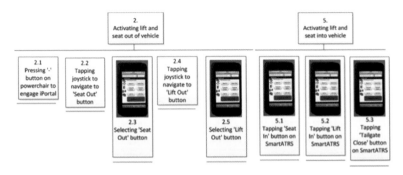

Figure 2.   ATRS operating zones.

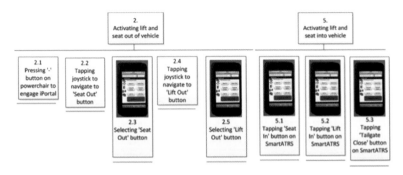

Figure 3.   Extracts from the SmartATRS Hierarchical Task Analysis.

robust means for a wheelchair user to autonomously dock a powerchair onto a platform lift without the need of an assistant (Gao et al. 2008). The system uses robotics and Light Detection and Ranging (LiDAR) technology to autonomously dock a powerchair onto a platform lift fitted in the rear of a standard Multi-Purpose Vehicle while a disabled driver is seated in the driver's seat. Using a joystick attached to the driver's seat, the user manoeuvres the powerchair to the rear of the vehicle until the LiDAR unit is able to see two highly reflective fiducials fitted to the lift. From then on, the docking of the powerchair is completed autonomously. ATRS requires the vehicle to be installed with the three components shown in Figure 2.

User feedback and safety features were incorporated into SmartATRS, which were not present in the keyfobs. Seven command buttons on a Graphical User Interface (GUI) activate each ATRS function and the smartphone can be securely mounted onto the arm of the powerchair, making the system easier to use. Joystick control was developed as an alternative to touch by using iPortal (Dynamic Controls 2014) that communicated with a smartphone via Bluetooth.

Method

A Hierarchical Task Analysis (HTA) was derived to illustrate the tasks supported by SmartATRS, two extracts are shown in Figure 3.

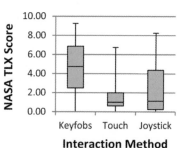

Figure 4.   Comparing mental and physical demand of interaction methods.

The following tasks were performed by the participants: driving the seat out of the vehicle, opening the tailgate, driving the lift out of the vehicle, performing an emergency stop whilst the seat and lift were driving into the vehicle simultaneously and closing the tailgate. A Controlled Usability Evaluation was conducted on ATRS and SmartATRS to assess the usability of the interaction methods: keyfobs, touch and joystick. The evaluation was simulated by forming a user group of 12 participants in powerchairs who could drive a car. The evaluation provided a means to verify the GUI design ensuring that it was "fit for purpose" for users of ATRS.

Results

The Adjective Rating Scale (Bangor et al. 2009) was used to interpret the System Usability Scale (SUS) scores, where keyfobs achieved a score of 51.7 ('OK Usability'), touch achieved 90.4 ('Excellent Usability', bordering on 'Best Imaginable') and joystick achieved 73.3 ('Good Usability'). This clearly highlighted that 'touch' was the most usable; however, the joystick could be seen as a significant improvement to the keyfobs. A second important result identified the safety of the emergency stop function, revealing a standard deviation of 6.8 seconds for the keyfobs, compared to 1.2 seconds for SmartATRS. The box plots in Figure 4 illustrate the NASA Task Load Index (TLX) differences in workload experienced when using keyfobs, touch and joystick. It is evident that 'touch' shows lower mental and physical demand, thus indicating that keyfobs are more demanding and less efficient to use than 'touch'. All TLX workload types were analysed and it was conclusive that 'touch' was the most efficient and least demanding interaction method.

Discussion

The assistive technology research described a requirements elicitation process for a SmartPowerchair conducted through surveys and interviews. It also examined the results of a case study where pervasive technology was applied to increase the usability of an existing assistive technology.

The survey provided efficient needs analysis to elicit the requirements for the Smart-Powerchair. The results identified the difficult household tasks to be supported by the SmartPowerchair: opening/closing doors, windows and curtains, and operating cookers, microwaves, kettles and heating controls.

The SmartATRS case study provided an example of a successful integration of pervasive technology into an assistive technology that replaced difficult to use keyfobs with a smartphone. The HTA was useful for identifying the tasks to be performed in the experimentation. By decomposing the overall goal of SmartATRS into individual tasks and levels, a greater understanding of the processes within SmartATRS was determined. The addition of screenshots of SmartATRS to the HTA highlighted the tasks currently supported by smartphone interaction. Following future developments of SmartATRS, support will be extended to further tasks.

Using the HTA, an appropriate usability evaluation was conducted to accurately assess the usability of the interaction methods: keyfobs, touch and joystick. The NASA TLX results clearly showed significant reduction in mental and physical workload experienced when using SmartATRS compared to keyfobs. It is antici-pated that a SmartPowerchair would provide similar reductions in workload when performing tasks in the home.

The pervasive technology to be integrated into the SmartPowerchair will be a smart-phone operated by both touch and head tracking, as this has the most potential for the user domain. Head tracking will be implemented using Tracking Learning Detection (TLD) (TLD Vision 2014), a real time object tracking algorithm (TLD tracks unknown objects in unconstrained streams such as movies or webcams). The technology automatically tracks the face and learns the appearance from dif-ferent angles. Therefore, TLD is robust as it is not confused by different faces. TLD outputs the positions of the tracked face in real time. It is proposed that TLD will be applied to the SmartPowerchair. By converting the TLD output into XY coordinates; the smartphone cursor will be controlled using a facial feature. The non-obtrusive advantage of TLD will be that the user is not required to wear any accessories. This is an important benefit as the user group for the SmartPowerchair will have disabilities.

Conclusions

The requirements elicitation phase for a SmartPowerchair was performed through surveys and interviews of people with disabilities. The results identified the house-hold tasks that currently present difficulties and pervasive technologies that have the greatest potential to improve quality of life. The most difficult household tasks consisted of opening/closing curtains, windows and doors and operating appliances that generate heat or have small dials. These difficulties will be eliminated by the integration of a smartphone to a standard powerchair that can either be controlled by touch or head tracking. TLD will be utilised to provide non-invasive head tracking whereby the smartphone cursor can be controlled by movements of facial features.

The case study evaluated the usability of ATRS and SmartATRS and showed that the innovative and novel use of pervasive technology significantly improved the usability compared to the small keyfobs. SmartATRS thereby met a functionality metric defined by Metsis et al. (2008), stating that "an assistive technology must perform correctly in order to serve its purpose". Through the knowledge obtained from the requirements elicitation phase and the case study, the proposed SmartPowerchair is anticipated to evolve beyond a standard powerchair creating significant improvements to the quality of life for people with disabilities.

References

Adebesin, F., Kotzé, P. & Gelderblom, H. (2010) *The Complementary Role of Two Evaluation Methods in the Usability and Accessibility Evaluation of a Non-Standard System.* 2010 Annual Research Conference of the South African Institute for Computer Scientists and Information Technologists Bela-Bela, South Africa

Ari, I.A. & Inan, F.A. (2010) "Assistive Technologies for Students with Disabilities: A Survey of Access and Use in Turkish Universities." *The Turkish Online Journal of Educational Technology* 9 (2): 40–45

Bangor, A., Kortum, P.T. & Miller, J.T. (2009) "Determining What Individual SUS Scores Mean: Adding an Adjective Rating Scale." *Journal of Usability Studies* 4 (3): 114–123

Bournemouth University. (2013) "BU student develops mobile app to help wheelchair users get into cars." Retrieved 22nd September 2014, from: http://news.bournemouth.ac.uk/2013/08/07/bu-student-develops-mobile-app-to-help-wheelchair-users-get-into-cars/

Brook, J. (1996) SUS: a "quick and dirty" usability scale. In Jordan, P.W., Thomas, B., Weerdmeester, B.A. & McClelland, I.L. (eds). *Usability Evaluation In Industry.* (Taylor & Francis London), 189–194

Cofré, J.P., Rusu, C., Mercado, I., Inostroza, R. & Jiménez, C. (2013) *Developing a Touchscreen-based Domotic Tool for Users with Motor Disabilities.* 2012 Ninth International Conference on Information Technology: New Generations. Las Vegas, NV

Dynamic Controls. (2014) "iPortalTM Dashboard." Retrieved 22nd September 2014, from: http://www.dynamiccontrols.com/iportal/

Gallagher, B. & Petrie, H. (2013) *Initial Results from a Critical Review of Research for Older and Disabled People.* 15th ACM SIGACCESS International Conference on Computers and Accessibility. Bellevue, WA

Gao, A., Miller, T., Spletzer, J.R., Hoffman, I. & Panzarella, T. (2008) "Autonomous docking of a smart wheelchair for the Automated Transport and Retrieval System (ATRS)." *Journal of Field Robotics* 25 (4–5): 203–222

International Organization for Standardization. (1998) *ISO 9241-11 Ergonomic Requirements for Office Work with Visual Display Terminals (VDTs) – Part 11: Guidance on Usability.* (International Organization for Standardization Geneva, Switzerland)

Metsis, V., Zhengyi, L., Lei, Y. & Makedon, F. (2008) *Towards an Evaluation Framework for Assistive Technology Environments.* 1st International Conference on PErvasive Technologies Related to Assistive Environments. Athens, Greece

Mihailidis, A., Elinas, P., Boger, J. & Hoey, J. (2007) "An Intelligent Powered Wheelchair to Enable Mobility of Cognitively Impaired Older Adults: An Anti-collision System." *IEEE Transactions on Neural Systems and Rehabilitation Engineering* 15 (1): 136–143

National Aeronautics and Space Administration. (1996) *Nasa Task Load Index (TLX) v. 1.0 Manual.* (National Aeronautics and Space Administration Washington, WA)

Postolache, O., Madeira, R.N., Correia, N. & Girão, P.S. (2009) *UbiSmartWheel: a ubiquitous system with unobtrusive services embedded on a wheelchair.* 2nd International Conference on PErvasive Technologies Related to Assistive Environments. Corfu, Greece

Robinson, L., Brittain, K., Lindsay, S., Jackson, D. & Olivier, P. (2009) "Keeping in Touch Everyday (KITE) project: developing assistive technologies with people with dementia and their carers to promote independence." *International Psychogeriatrics* 21 (3): 494–502

Tanaka, K., Matsunaga, K. & Wang, H.O. (2005) "Electroencephalogram-Based Control of an Electric Wheelchair." *IEEE Transactions on Robotics* 21 (4): 762–766

TLD Vision. (2014) "TLD 2.0." Retrieved 22nd September 2014, from: http://www.tldvision.com/product_tld2.php.

JOB DESIGN

THE PSYCHOLOGY OF MOBILE WORKING: PRODUCTIVITY AND WELLBEING IN THE CONTEMPORARY WORKPLACE

Patrick W. Jordan[1] **& Jim Taylour**[2]

[1]*Loughborough University, Loughborough*
[2]*Orangebox Limited, Parc Nantgarw, Cardiff*

Flexible working practices and the changing nature of the office mean that more and more people spend a significant proportion of their working time away from a fixed workstation. A study investigated the pros and cons of this, in terms of productivity and wellbeing. Results suggest that flexible working is popular, increases wellbeing and is more productive. There are drawbacks too, including expectations of longer working hours and lack of visibility within an organisation. Meanwhile offices themselves need to be able to support a variety of types of work and spaces, something that many do not currently do well. Recommendations are made for office and job design that will benefit employers and employees.

Introduction

The office is changing. The traditional model of an employee spending nearly all of their time working from a fixed workplace in the office is becoming less dominant with an increasing number of employees working either outside the office or using the office differently. People are working from home, mobile working (for example from trains and planes or from public spaces such as airports, train stations and coffee shops), working elsewhere in the office such as conference rooms, the new diversity of breakout or third spaces (places in the office that are available for people to use on a flexible basis), sharing a workstation with colleagues, using external co-working spaces or a combination of these (Taylour and Jordan, 2014).

Exactly how many people are working away from a fixed workstation is difficult to quantify but a number of estimates put it at about one-third of the global work-force (e.g. Eddy, 2012; Narayanan, 2012). With these changes come opportunities and threats both in terms of productivity and wellbeing. Different types of work (Table 1) may be better suited to particular environments, meaning that companies can maximise the efficiency with which they use their workspace. Meanwhile having the flexibility to work away from the office can help people achieve a better work-life balance.

Table 1.   Third space taxonomy (7 C's) (Taylour and Jordan 2014).

Type	Third space activity/space
Collaborate	Team working with others formerly/informally
Concentrate	Focussed work to solve a problem individually
Compute	Data input work/head down
Conflabulate	Informal dialogue/fun with colleagues
Consume	Preparing/consuming of food or drink
Create	Creative or project work in a group/individually
Counsel	Customer support (virtual or face to face)

**Table 2.   Percentage of respondents working
in stated locations (n = 131).**

Work Location	%
Dedicated desk	93
On the move	83
Elsewhere in office	74
At Home	73
Shared desk	39

Aims and methodology

The aim of the study was to understand how and why people's work patterns are changing and how this is affecting their performance and wellbeing. The study focussed on professional people whose jobs would traditionally have been mainly, or wholly, carried out in an office or similar environment.

The study included a literature review, three focus groups and an online survey. Two of the focus groups were conducted in London and one in Newcastle. Participants discussed their working locations and hours and the effects of these on productivity and wellbeing. The online survey garnered quantitative data on the same issues. The survey was completed by 131 professional people working in jobs that would traditionally be office based, 58% of whom were men. They were distributed across a wide variety of ages, with 75% being between 25 and 54. The focus groups included a similar profile of worker as well as people with specialist knowledge of working patterns and associated issues. In total, 47 people participated in the focus groups.

Results

Work locations and drivers

Nearly everyone in our survey worked at a dedicated desk in their workplace some of the time but also at other locations as well as summarised in Table 2. Note that the total percentages add up to more than 100 because many people work in more than one location.

Focus group participants cited a variety of reasons for working away from a dedicated desk. These included:

- Difficulty concentrating in an open plan office (which was where most participants' dedicated desks were).
- Disturbing others (for example if their work required talking on the phone or to workplace colleagues).
- Being free from the interruptions of colleagues (who were likely to come over for a chat or to ask them to do something).
- Logistics (it was impractical to do some aspects of their job from the office, for example if they had to do some work on a client's site).
- Collaboration (needing a suitable space for team working).
- Efficiency (optimising space usage using hot-desking and by allowing employees to work outside of the office some of the time).

For many, working away from the office, particularly at home, was seen as an important part of an enlightened company culture and helped with work-life balance and enabled an escape from the distractions of the office. Research by Cisco Systems (Heathfield, 2012) indicates that both businesses and employees prefer flexible working and that it is enabled due to the increasing pervasiveness of digital information and content, the development of good quality and reliable broadband devices and services and mobile technology that supports interpersonal communication. The trend is also being driven shift from a manufacturing to a service economy and the increasing number of working women (who tend to have more demands outside of the workplace than men) according to the Economist (2009).

Mobile working and productivity

The few controlled studies that have been done on flexible working suggest that it boosts productivity (London School of Economics and Political Science 2013). For example, a study by Stanford University Graduate School of Business (2012) found that home workers were 13% more productive, had less sick days and a 50% lower attrition rate than their office-based counterparts.

Participants in our focus groups suggested that one of the reasons why flexible working could lead to higher productivity is because it enables people to work at the times of the day when they feel most productive. However, not all business leaders are in favour of flexible working. Most notably Yahoo! decided to put an end to flexible working practices in favour of the collective insights that come from hallway and cafeteria discussions, meeting new people, and impromptu team meetings (Marks, 2013). Hewlett-Packard have also implemented a policy of recalling remote-working staff to the office.

Table 3.   Causes of stress for those who experience it (n = 73).

Cause of Stress	%
Work-life balance	73
Software problems	68
Noise	60
Connectivity	47
Physical environment (e.g. temperature)	48
Being or feeling on call outside normal hours	43
Feeling 'out of the loop'	40
Comfort	28
Finding somewhere to work	24
Finding power points to charge devices	24
Loneliness	20

While the majority of our focus group participants were in favour of flexible working, there were some who had reservations about it, especially among those in a managerial role. "*One of the things we found is with the people working late. In the morning when you open your email, you find a lot of emails have dropped into your inbox and you feel that if they'd been working the same as me, we could have resolved this rather than the whole email trail being started. So it does have an impact if somebody chooses to work late and clear the decks. They might be clear and ready to go, but actually all they've done is shunted it onto everybody else first thing in the morning.*" This employee raised a similar issue and also talked of the pressure that employees can feel to respond to demands out of hours: "*There are people in the organisation that prefer to work more over the weekend and take some time out during the week. If you switch on your iPad you feel obliged to respond especially if they're senior.*"

Mobile working and wellbeing

When asked, 67% of respondents said that they felt stressed at least occasionally. Mobile working can contribute to work-life balance but also create pressure to work out of hours. These were the first and sixth most mentioned issues.

The second most cited problem was software. The stress associated with unreliable or difficult to use software tended to be particularly acute when people were away from the office with no technical support to hand.

The 'flip side' to feeling always on call is that some employees feel out of the loop when they work away from the office. They miss out on informal conversations and often feel that they are the last to know about many issues.

Some of our focus group participants also expressed the fear that when they were out of the office their contribution was less visible. Studies have shown that the greater the proportion of people's working week is spent outside of the office, the less likely they are to get promoted (Rowe, 2013).

Noise as a stressor was something that tended to affect people more commonly when they were in the workplace than when they were away from it. Focus group participants cited people taking phone calls and conversations between colleagues as stressors when they were trying to work. Research indicates that when doing creative work a background noise level of around 70 dB(A) is optimal in terms of productivity, whereas 55 dB(A) or less is optimal for detailed work (Ravi et al, 2012). 70 dB(A) is typical of the background hum in most coffee shops, which can make them ideal places to go when trying to think creatively. When noise levels drop below this level people tend to think in a more focused but less broad way. Ironically, many organisations provide breakout spaces for doing creative tasks that are quieter than the open plan areas where people are expected to do focused work.

When working on the move, practical issues such as connectivity and being able to charge devices can be a problem. According to the iPass Mobile Workforce Report (iPass 2013), 81% of mobile workers reported having had a bad experience with hotel WiFi in the past 12 months. Similar issues were raised by a number of people in our focus groups. These included losing phone signals when on trains or in cars, and not being able to charge devices or find free, reliable Wi-Fi connections when on the move. This issue overlapped with the difficulty that some people had with finding a place to work.

The physical environment could also be a source of stress, for example unpleasant temperature or lack of fresh air. Again these could apply both outside and within an office. A study by Gensler (2013) suggests that collaborative spaces have become noisy, lacking in privacy and as a result, stressful and unproductive.

Twenty per cent of those who were stressed cited loneliness when working outside of the office. According to a study carried out by London School of Economics for Acas (Beauregard et al, 2013), those who are most extrovert tend to be less happy when they were away from the office.

Conclusion

Nearly all participants in our study work away from a fixed workstation at least some of the time and the vast majority spend a significant proportion of their working week outside of the office. The trend towards flexible working is driven by companies desire to use space more efficiently, the popularity of home working among employees and the belief that some types of work are better done outside of the office. The little evidence that there is available suggests that flexible working does improve productivity.

Flexible working also has drawbacks. Some business leaders believe it hinders creativity and have recalled workers, and some managers feel it is inconvenient.

Some workers feel that they are always on call and that they miss out on informal office chat and some are lonely. There is also evidence that the longer people spend working away from the office the less their chances of promotion.

Offices themselves are often suboptimal in terms of supporting people's work. Noise in open plan offices is potentially set to rise with wearable technology and more voice recognition devices causing disruption and stress if left unchecked.

Not all stress is bad for us and yet high levels are considered to be the biggest hazard in both private and public sectors (TUC, 2012). A criticism of current company health policies from our focus groups was the emphasis on simply reducing absenteeism, injury and disability to get employees to a neutral point of not showing any obvious symptoms of ill health. Stress related issues were thought to be less easy to detect when deploying conventional DSE self-assessment workstation audits in the workplace. In order to be more proactive, organisations need to take a positive approach to achieving high levels of wellness through awareness, education and personal growth (Travis, 2004). They also need to recognise that wellbeing is a joint responsibility between employers and employees (BITC, 2013) and embrace emerging guidelines (Stewart and Berns, 2013).

Recommendations

Employer and employee tips for Wireless Well Working (our term for optimum psychological levels when away from the fixed desk) have been identified.

For employers (Taylour and Jordan, 2014):

- Classify items of work according to type using the 7Cs (see Table 1).
- Create a third space taxonomy with identified work areas in the office that reflect the 7C's and train users on how to get the best out of them.
- For focussed work, ensure that distractions are kept to a minimum and devise 'acoustic etiquette' principles.
- Ensure all work areas in the office away from the fixed desk have a high ergonomic 'footprint' with connectivity, an adequate number of power sockets and optimum ergonomic features to encourage wellbeing.
- Don't adopt a 'one size fits all' approach to flexible working. Tailor it to individuals including personality and generation types.
- Provide adequate technical support for flexible workers including providing support for any of their own devices that they use for work.
- Measure performance according to objective criteria, not by how much a person is 'visible'. Incorporate personal fitness and wellbeing objectives that go beyond neutral to combat stress (Taylour, 2014).
- Provide remote workers with regular opportunities to socialise with their colleagues.

For employees:

Personal wellbeing principles as developed by Jordan (2014) underline the importance of personal responsibility. The model has five principles that employees can apply to work.

- Take responsibility: The quick and ongoing changes in work locations and devices means that employers' policies are likely to be out of date and lack relevance to the current situation. We need to take personal responsibility for our wellbeing.
- Set goals: Aim to achieve wellbeing in all areas of your professional and personal life. Understand what makes for a truly life-enhancing work experience and identify the areas where this experience can be improved.
- Be Positive: Work should make life better and increase our level of wellbeing. Focus not only on how to minimise the negative effects of work, but also on the positive benefits that it can bring in terms of health, wellbeing and personal and professional development.
- Persevere intelligently: Find the information that is needed to help improve productivity and wellbeing and put it into practice. Stay up to date with the latest findings and best practices.
- Connect with others: Identify and seek help from those who can help improve your wellbeing and give the best advice. This includes people within the organisation who can facilitate organisational changes.

References

Beauregard, A., Basile, K. and Caononico, E., 2013, Home is where the work is. Retrieved 8 March 2014 from: http://www.acas.org.uk/media/pdf/f/2/Home-is-where-the-work-is-a-new-study-of-homeworking-in-Acas_and-beyond.pdf

Business In The Community 2013, Employee Engagement and Wellbeing, *BITC Public Reporting Guidelines, BITC, London*

Economist, 2009, Flexible working. Retrieved 19 December 2013 from: http://www.economist.com/node/13685735

Eddy, N. 2012, Mobile Worker Population to Reach 1.3 Billion by 2015: IDC – Retrieved 21 December 2013 from: http://www.eweek.com/c/a/Mobile-and-Wireless/Mobile-Worker-Population-to-Reach-13-Billion-by-2015-IDC-238980/#sthash.S0TYyRC5.dpuf

Gensler, 2013, Workplace Survey. Retrieved 6 May 2014 from: http://www.gensler.com/uploads/documents/2013_US_Workplace_Survey_07_15_2013.pdf

Heathfield, S. M., 2012, Flexible Working, American Society of Resource Management, Human Resources, About.com

iPass, 2013, The iPass Global Mobile Workforce Report. Retrieved 7th March 2014 from: http://www.ipass.com/wp-content/uploads/2013/09/ipass-mobile-workforce-report-q3-2013.pdf

London School of Economics and Political Science 2013, Home workers "happier and more productive". Retrieved 7th January 2014 from: http://www.lse.ac.uk/newsandmedia/news/archives/2013/10/homeworkers

Marks, G. 2013, Why Marissa's Right. Retrieved 7th January 2014 from: http://www.forbes.com/sites/quickerbettertech/2013/03/04/why-marissas-right/

Narayanan, N. 2012, The New Workplace Reality: Out of the Office. Retrieved 20 December 2014 from: http://www.wired.com/2013/06/the-new-workplace-reality-out-of-the-office/

Ravi, M., Zhu, R. J., and Cheema, A. 2012, Is noise always bad? exploring the effects of ambient noise on creative cognition, *Journal of Consumer Research*, 39 (4), 784–799.

Rowe, E. 2013, Flex-Time: Without It, MBAs Say They're Less Ambitious. Retrieved 4th March 2014 from: http://www.businessweek.com/articles/2013-07-09/flex-time-without-it-mbas-say-theyre-less-ambitious

Stanford Graduate School of Business 2012, "Striking the Balance". Retrieved 18 December 2013 from: http://www.gsb.stanford.edu/news/research/striking-the-balance.html

Stewart, T. and Berns, T. 2013, The Human Centred Organisation. Retrived 9 January 2014 from: http://www.ergonomics.org.uk/slider/the-human-centred-organisation/

Taylour, J. A. and Jordan, P.W. 2014, *Mobile Generations*, Orangebox, London

Taylour, J. A. 2014, *Pocket guide to Wireless Well Working*, Orangebox, London

Travis, J. 2004, Illness Wellness Continuum. Retrieved 3 June 2014 from: http://www.thewellspring.com/wellspring/introduction-to-wellness/357/key-concept-1-the-illnesswellness-continuum.cfm

TUC 2012, *Focus on Health and Safety*, Trade Union Trends Survey 04/03, *Organisation & Service Department.*

ERGO WORK: EUROPEAN PERCEPTIONS OF WORKPLACE INCLUSION AND APPLICATION OF ERGONOMICS

Louise Moody & Janet Saunders

Department of Industrial Design, Coventry University

The picture across Europe is variable in terms of the implementation of ergonomics to support workplace inclusion. The ERGO WORK project aims to encourage cooperation between universities, businesses and other organizations to improve learning, teaching and knowledge transfer in respect to ergonomic workplace design. This paper reports on a survey undertaken to explore inclusion and user needs in the 6 European countries involved in the project. The results support a variable picture across Europe in terms of the adaptations made to the workplace and suggest that provision could be improved. These findings will be taken forward to inform the design of teaching materials which will be tested in Poland and Slovenia.

Introduction

The ERGO WORK project is a collaboration between academic and industrial partners in six European countries, Belgium, Italy, Poland, Slovenia, Spain, UK, focused on understanding barriers to inclusion of disabled people in the workplace, and tackling some of these barriers through ergonomics education.

The EC Employment Equality Framework Directive 2000/78/EC established a general framework for equal treatment in employment and occupation. With regard to disability, this Directive places a focus on inclusion and recognizes that failure to provide 'reasonable accommodation' in the workplace can constitute discrimination. In practical terms 'reasonable accommodation' includes measures to adapt the workplace to individual workers with a disability, for example by adapting premises and equipment, as well as patterns of working time etc. in order to facilitate their access to employment.

The directives themselves do not contain definitions of disability. For the purposes of this project, we have used the UK Equality Act 2010 definitions of disability, defined as physical and sensory impairment, learning difficulties, mental health needs and autism. (Equality Act 2010). The 2000 Directive built on earlier directives, but its impact across EC member countries has been both slow and variable (Greve, 2009). It seems likely that later joining members, such as Poland and Slovenia (joined in 2004) would have greater adjustments to make in both policy and practice.

In the UK the employment rate of working-age disabled people is 47.8% compared with 75.9% of non-disabled people (Papworth Trust 2012). In contrast in Poland for working age people with disabilities it is 14.8%, compared to 50.7% non-disabled (GUS). In many EU countries disabled people have been traditionally employed within sheltered employment schemes, and despite a move towards more integrated employment policies, this remains the case in many countries (Kołodziejska 2012, Greve 2009). For example, in Slovenia in 2008 the major employers of disabled people were the 'disability companies' (Greve 2009). In Poland in 2012 of the 237,000 disabled employees registered; 70% worked in special companies of sheltered work, and 30% in the open labour market (Kołodziejska 2012). The shift towards 'supported employment' is a recent development, in which various agencies assist people with significant disabilities to access real employment opportunities in an integrated mainstream employment market (Wistow & Schneider 2007). This move however, requires a better understanding of how workplace and job design should take into account individual needs and ergonomic principles.

The ERGO WORK project aims to encourage cooperation between universities, businesses and other organizations to improve learning, teaching and knowledge transfer in respect to ergonomic workplace design. The project seeks to understand barriers to workplace inclusion, and tackle these through education, knowledge transfer and collaboration in the field of Ergonomics. The EU (LLP–Erasmus) funded project involves collaboration between partners from Slovenia, Poland, UK, Italy, Spain, and Belgium. The initial phase of the project described here seeks to explore the perceptions and needs of stakeholders in terms of workplace inclusion and ergonomics education.

Method

An anonymous online survey using Bristol Online Surveys software was developed to ascertain the needs and views of a range of stakeholders. This enabled efficient collection and analysis of data from a dispersed sample across Europe. The study was approved by the Coventry University Ethics Committee.

Following a brief demographics section (12 questions), the survey consisted of a further 61 questions. The questions were grouped, based on whether or not the participant had a disability and their job role, so respondents completed the relevant group of questions. The majority of the questions were multiple-choice, with 13 open questions asking for more detailed responses. Completion took approximately 20 minutes. The survey was developed and piloted in the UK, then translated by native language speakers into each partner language. Minor adaptations were made after local piloting to ensure culturally appropriate language, but the question set and structure was consistent throughout to allow comparability of results and easy collation of responses.

The survey was distributed by the project partners in Slovenia, Poland, UK, Italy, Spain, and Belgium, to stakeholders including individuals with and without

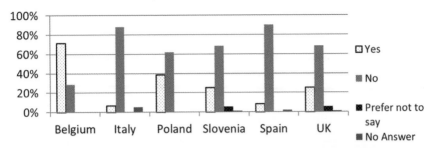

Figure 1. **Percentage of participants with and without a disability.**

disabilities, workers and management within companies and higher educational institutions. It was completed by 520 participants across the countries. The results were collated in the UK. As this was an exploratory survey descriptive statistics rather than tests of statistical inference were undertaken.

Results

The survey participants with and without a disability varied per country.

Of the Belgian respondents, 71% acknowledged a disability whereas in the other countries the proportion was lower, from 7% in Italy to 38% in Poland (See Figure 1 above). All participants were asked how well they thought their workplace was adapted for people with disabilities. Across countries, just under half (49%) of *all* participants believed that their workplace had been well adapted, and a further 28% felt their workplace was about average. A comparison by country showed that in the UK 69%, Poland 53%, Slovenia 51%, Belgium 39%, Spain 38%, and Italy 37% thought their workplace accommodated disabilities 'fairly well' or 'very well' (see Figure 2 below).

However, participants with a disability did not feel well-provided for in the workplace. When asked if they agreed with the statement 'disabled people are not well accommodated in terms of workplace design', the highest proportion of answers agreed with the statement. UK respondents were the least negative with 36% agreeing with the statement.

The results to further questions about workplace contentment indicated that across countries those with a disability rated their workplace lower, felt less included and were less happy at work than participants without a disability.

The graph in Figure 3 shows that participants with a disability felt that the design of the workplace was a barrier to employment opportunities. Participants indicated the greatest barriers to asking for improvements were fears about job security and about being stigmatized or isolated. These reasons were selected by over half the

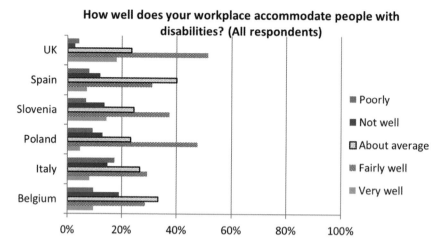

Figure 2.   Perceptions of workplace accommodation.

Table 1.   Disabled people are *not* well-provided for in workplace design.

	Agree or Agree Strongly	Neutral	Disagree or Disagree Strongly
Belgium	60%	20%	20%
Italy	67%	33%	0%
Poland	39%	42%	18%
Slovenia	55%	27%	18%
Spain	50%	39%	12%
UK	36%	33%	30%

respondents in all countries. Other barriers included: lack of knowledge about what adaptations were possible, difficulty finding the right person to ask, fears about cost to the employer and worry about promotion. Relatively few participants thought there were 'no barriers' to asking.

The most common workplace adaptation made in every country was a physical adaptation to the buildings. A general view of the data suggests that most physical adaptation is perceived to have happened in the UK. Considering individual knowledge about potential adaptations for different kinds of impairment, there was most knowledge about physical impairment, and low knowledge among all participants about hearing or visual impairment, with even less knowledge regarding mental health needs and intellectual disabilities.

Across all countries, adaptations had also been made to jobs and the way they were carried out. The UK participants identified changes to job tasks, job role, pace and working hours, much less frequently than stakeholders in the other countries. Interestingly, Slovenians were the most likely to identify changes to the job role and

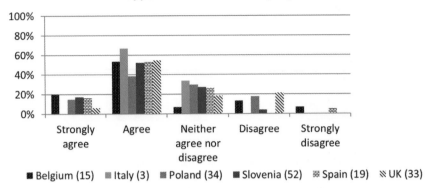

Figure 3.   Workplace design is a barrier to employment.

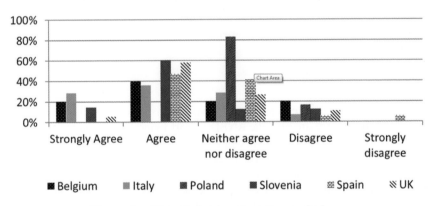

Figure 4.   Knowledge to adapt the workplace.

hours to suit individual needs, whilst the Polish participants particularly recognized adaptations to the pace of work.

Participants with a disability felt that employers need better knowledge about their obligations and possible adaptations in the workplace. As shown in Figure 4 employers and managers had a tendency to believe their knowledge was adequate, or were unsure.

Across most of the partner countries, employer participants considered that they understood the legislation about employing people with additional needs. On the subject of how well they understood legislation about *adapting workplaces*, there was a reasonable amount of confidence: half or over answering 'very well' and 'fairly well' in UK (66%), Slovenia (53%), and Poland (50%).

Would you be interested in training for your organisation in creating more ergonomic workplaces?

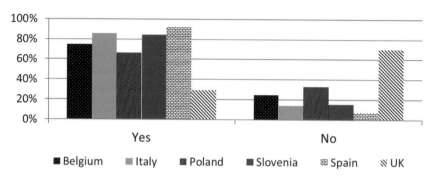

Figure 5.   Interest in ergonomics training.

Employer participants were asked if they were interested in training for themselves, or their organizations in creating ergonomic workplaces. In most countries the majority were very interested (see Figure 5), except in the UK, where 71% selected 'No' to training for their organization. This response from UK perhaps reflects the training and awareness activities that have already taken place although this still leaves around 30% who would be interested in more.

Discussion

Previous research illustrates a paucity of reliable statistics on disability in both the workplace and society. In 2012 the ANED Report found huge variations between member states in the equality gap in employment. These variations were not correlated to levels of economic activity, education or poverty-risk for disabled people in those member states. So a relatively strong performance on the EU2020 measures overall is not associated consistently with a strong performance on equality for disabled people (Priestly, 2012). Within this context, the main purpose of our research was to take a snapshot of stakeholder views in our partner countries to inform the academia to business knowledge transfer activity planned in the project, and to present a broad picture of stakeholder needs.

The data presented here has certainly shown a variable picture across the 6 European countries in terms of the perceptions of the adaptations made to workplaces. Employees with disabilities generally did not feel their workplace was well adapted for their needs. It seems reasonable to suggest there is still some way to go in making adequate provision and adjustments. Participants who reported having a disability were much less confident than employers as to the adequacy of knowledge about provision and adaptation in the workplace, and felt improved knowledge about their obligations and the possibilities of adaptation in the workplace was needed.

There may be cultural differences in response to employee needs; Poland and Slovenia especially appear to favour changes in role or work pacing, and further study would indicate whether this was a positive or negative impact in the workplace. Overall there was most awareness of the issues faced by those with physical impairment, and less about sensory impairments, mental health needs or intellectual impairments. Improved knowledge is therefore needed about adaptations that go beyond purely physical adaptations to buildings, and cater for sensory and cognitive impairments.

From the employee perspective, the design of the workplace was perceived as a barrier to employment opportunities and there continue to barriers which prevent them from asking for adjustments such as fears about job insecurity and stigmatization despite EU law. There is still some uncertainty about workplace legislation; even in the UK a sizeable minority of employer participants was still uncertain about their understanding of the legislation. There is clearly room for awareness-raising and training activities.

The awareness and knowledge of Ergonomics was found to be relatively low. This is something that the ERGO WORK project seeks to address. Given the different pictures between countries, European collaboration and transfer of knowledge and practice in this area should be beneficial. An audit of Ergonomics Education in the partner countries is also being undertaken as part of the project.

There are limitations to the data as it stands. The sample is small and variable across countries and employee groups. The evidence is therefore being used cautiously to point to broad conclusions and suggest actions that might be taken to improve Ergonomics education and to raise general awareness of workplace inclusion in some countries. Further exploration of the data and the broader context is needed to understand these differences and see where there might be practices that can be transferred between countries.

Improved workplace inclusion is a European priority. The ERGO WORK project is focused on understanding barriers to inclusion, and tackling these through education and collaboration between academia and industry. The next stage of project will develop ergonomics curriculum material that can be used to update existing teaching contents in Slovenia and Poland. It will then test this material through 4 pilot projects involving multidisciplinary working groups of students, academics, company staff and disabled users to address workplace inclusion issues.

Acknowledgements

We would like to thank those who participated in the survey, as well the ERGO WORK project network that translated and distributed the survey.

References

Equality Act (2010) Retrieved 3rd December 2014 from: www.gov.uk/definition-of-disability-under-equality-act-2010)

EC (2000) Employment Equality Framework Directive 2000/78/EC

Greve, B. (2009) *The labour market situation of disabled people in European countries and implementation of employment policies: a summary of evidence from country reports and research studies.* Report prepared for the Academic Network of European Disability experts (ANED), (University of Leeds, Centre for Disability Studies, UK)

GUS-Central (2011) "Statistical Office – data for the end of 2011" Retrieved 3rd December 2014 from http://www.stat.gov.pl/gus/index_ENG_HTML.htm

Kołodziejska, A. (2012) *"Niepełnosprawni a polski rynek pracy"*, Retrieved 3rd December 2014 from http://rynekpracy.org/wiadomosc/763104.html

Papworth Trust (2012) "Disability in the United Kingdom 2012", Retrieved 3rd December 2014 from www.papworth.org.uk/downloads/disabilityinthe unitedkingdom2012_120910112857.pdf

Priestly, M. (2012) *Targeting and mainstreaming disability in the context of EU2020 and the 2012 Annual Growth Survey,* Report prepared for the Academic Network of European Disability experts (ANED), (University of Leeds, Centre for Disability Studies, UK)

Wistow, R. & Schneider, J. (2007) "Employment Support Agencies in the UK: current operation and future development needs", *Health & Social Care in the Community*, **5**(2): 128–135.

SUBJECTIVE PRODUCTIVITY IN DIFFERENT STATES OF THERMAL COMFORT

Etianne Oliveira, Antonio Xavier & Ariel Michaloski

Department of Higher Education, Federal Institute,
Maranhão, Brazil
Programs in Masters degrees of Production Engineering,
Federal University of Technology, Paraná, Brazil

This paper presents the results of research that evaluated the subjective productivity of office workers subjected to different levels of thermal comfort. TLX NASA-Workload tool measured subjective productivity and the Confortímetro Sensu equipment was used to measure thermal variables within the work environment, and the measures were compared (subjective thermal comfort). After statistical analysis of the scores, the results showed that there is no statistically significant difference between the mean, although states of mild discomfort scores show greater variability. In the sample, the data suggest that the thermal conditions do not interfere significantly in the scores of subjective productivity of office workers.

Introduction

The aim of this article is to present the results of the analysis of statistical parameters from the scores of workers' subjective productivity in actual work situations, subjected to different states of thermal comfort. Experimental research in the lab, which simulates the office environment, has provided evidence stating that the temperature affects workers' performance and is measurable within a range of thermally acceptable settings (Tham, 2010; Lan, 2009, 2010).

In lab studies, Lan et al. (2009, 2010, 2011), Araújo (2012) and Batiz (2009) used analytical and subjective evaluations of performance to measure the productivity of workers. These evaluations allow us to gauge how much an individual is being asked to accomplish a task, called the 'human cost'. It can be physical, psychological, cognitive and emotional. One of the ways to evaluate the subjective productivity is by measuring the mental workload, which involves physical and cognitive aspects, not only from labor but also from personal environmental and social-cultural factors. According to the authors, there is no standard procedure among ergonomists to measure the workers' productivity. The NASA-TLX Workload tool is widely used to measure the subjective productivity (Cardoso and Gontijo, 2012; Araújo, 2012; Lan et al., 2010; Lan et al., 2009).

On the other hand, Gaoua (2010) questions the effects of thermal conditions over cognitive performance. It states that these effects are still vague due to the methodological variances to evaluate whether the exposure to heat itself has an adverse effect over the cognitive function and in which environmental conditions these effects occur, highlighting the need for more research in this field. Brazilian researchers have stated that there is little research to investigate the characteristics of mental workload in real work situations from ergonomic studies (Cardoso and Gontijo, 2010).

Owing to the sensitivity in human responses and variability of intervening variables, Chen (2012) defends the need for search criteria, especially related to the workplace and the selection of the sample group. Cultural traits can skew the results of the survey, for example subjective aspects in relation to the satisfactory level of comfort.

This research used NASA TLX Workload tool, measure the productivity through measurement of subjective workload. This has been applied worldwide for over 25 years (Hart, 2006). The monitoring of environmental thermal conditions of spaces where workers perform their tasks was done with the Confortímetro Sensu equipment. The insulation of the workers' clothing of was measured via the questionnaire of subjective evaluation of thermal comfort and the metabolic rate was a search criteria previously established. Analytical comfort was calculated and grouped into 'comfortable', 'slight discomfort by heat' and 'slight discomfort by cold'. The groups were statistically analysed using parametric and non-parametric analysis as follows: ANOVA and Kruskal-Wallis.

The article begins with a description of the site. The criteria and tools used in the research are presented thereafter. Later, the results are presented with sample characterisation, sampling groups, analysis of normality of the sample groups, and testing statistics of ANOVA and Kruskal-Wallis. The results are discussed with similar searches; it ends up presenting the findings obtained, work limitations and suggestions for future work.

Methodology

Site

The research was done in five different companies that make use of naturally ventilated buildings. The companies are located in a region with a thermal regime that allows human thermal comfort (Figure 1–5). The selected region has annual average temperature between 17° and 18°C, with average minimum of 13° and 14°C and maximum 24° and 25°C. These thermal characteristics foment the usage of buildings with natural ventilation systems and favourable to human comfort during sedentary activity (Lan et al., 2010). It reaches 49 evaluations of workers.

Figure 1. Company 1.   Figure 2. Company 2.   Figure 3. Company 3.

Figure 4. Company 4.   **Figure 5. Company 5.**

Search criteria

All the companies have sedentary activity as the main task performed by its workers. This activity has metabolic rate of $70\,\text{W/m}^2$, typical for office work. The mental workload is greater than the physical load. All the companies develop their activities in naturally ventilated environment. The employees work under the same thermal environmental conditions. The companies are located in the same geographic region, which ensures a similar thermal acclimation for workers. Workers had performed their duties for at least six months ensuring familiarity with the task performed. All protocols of Ethics Committee on Research have been met according to Resolução 466/12 of National Health Council–BR.

Tools

The subjective productivity assessment tool used was NASA TLX – Workload, this consists of six questions that identify load mental, physical demand, forecast, performance, effort, frustration (NASA AMES, 1986; Lan et al., 2010; Cardoso and Gontijo, 2012).

The tool for measuring thermal comfort was Confortímetro Sensu. This was installed in the workplace and measured four thermal environmental variables from the location from which the workers were performing their tasks. The data were collected during the work day. Data collection procedures followed ISO 7730 (2005) thermal comfort standards.

Along with NASA TLX – Workload, a seven-point subjective thermal comfort evaluation scale was also applied. With this evaluation it was possible to identify the clothing used by the worker. This is important because he insulation level of the garment is one of the variables of the thermal comfort analysis equation (ISO 7730, 2005).

Outcomes

Sample characteristics

Four offices from the first company were evaluated, with each office accomodating a variable number of people. In total, ten workers from the first company participated in the research. The second, third and fourth companies had open plan offices; seven, eight and four workers in each company office respectively. In the fifth company, four workers were evaluated in four different rooms. In total, there were 33 workers, generating 49 evaluations.

The sample comprised workers of both genders with an average age of 31 years $(+/-11.79)$, most with higher education, and working in the same function for an average of five years.

Sample groups

Analytical evaluation of comfort of workers ranged from 1.43 to -1.39; values according to ISO 7730 (2005) represent, analytically, slightly warm and slightly cool respectively. The average score was 0.05 $(+/-0.66)$ indicating a thermally neutral state.

The sample groups were divided into three groups: Group 1 – scores of productivity of workers in a neutral thermal state; Group 2 – scores of workers' productivity in a slightly warm thermal state; Group 3 – scores of workers' productivity in a slightly cool thermal state, with the limit of neutral thermal state considered as -0.5 to $+0.5$, ISO 7730 (2005).

Normality

Table 1 shows the normality statistics for the three groups of data (no outliers). The normality of both groups was analysed by means of statistical tests of normality: K-S, Shapiro-Wilk and Lilliefors. All showed values below the critical values and reference P-value >0.05, except for second group, which presented low normality.

The results tests of normality of first group were: K-S $= 0.10668$ p-valor >0.05, Lilliefors p-value >0.05 and Shapiro-Wilk- 0.96769 p-value >0.05, confirming the normality of this group. The second group results, with $\alpha = 0.05$ was: K-S $= 0.32453$ p-value <0.02; Lilliefors p-value <0.01, and Shapiro-Wilk- 0.78438 p-value <0.007, rejecting the normality. It is noted that the data do not show normality with $\alpha = 0.05$, but present normality with $\alpha = 0.01$. The conclusion

Table 1.   Statistical parameters of comfort and productivity of the sample.

	Group 01		Group 02		Group 03	
Parameters	Neutral thermal	Workload	Slightly warm	Workload	Slightly cool	Workload
Statistics	$(-0.5 < x < +0.5)$	score	$(x > +0.5)$	score	$(x < +0.5)$	score
Average	−0.01	70.63	0.83	69.45	−0.92	61.64
Deviation	0.31	8.59	0.35	14.02	0.28	14.50
N	26		11		12	

Table 2.   Scores of subjective productivity per group.

	N	Average (μ)	Minimum	Maximum	Stand. Dev. (s)	Variance	Sum
Group 1	26	70.63	56.00	87.67	8.59	73.845	1906.93
Group 2	11	69.45	45.00	82.00	14.02	196.51	763.97
Group 3	12	61.64	35.66	81.00	14.50	210.33	739.64

is that the data is not strict, so normalcy can make use of testing parameters with low requirement of normality or non-parametric tests that do not require normality for data analysis. In third group, the result of the tests of normality was K-S $= 0.19607$ p-value >0.05, Lilliefors p-value >0.05 and Shapiro-Wilk- 0.92073 p-value >0.05, confirming the normality.

Statistical parameters

The statistical parameters of the study groups show that in a state of thermal comfort the workers of first group had a higher workload request. The similar average obtained in a state of slight warm. The requirement of the workload in a state of slight cool was about 12.86% lower. It was observed that in a state of slight warm or cool the scores show variability much higher than in a state of neutral (Table 2).

Hypothesis: The workload scores of an office worker performing his task in three different states of analytical thermal comfort derived from samples with the same averages.

The Parametric test (ANOVA):

H0: $\mu 1 = \mu 2 = \mu 2$ (original statement)   H1: one of the averages is different

The non-parametric test (Kruskal-Wallis):

H0: equal medians (original statement)   H1: different medians.

Table 3. ANOVA – single factor.

Variation	SQ	gl	MQ	Fa	p-value	Fc
Among the groups	693.2036	2	346.6018	2.627967	0.082816	3.195056
Within the groups	6198.815	47	131.8897			
Total	6892.019	49				

Interpretation:

ANOVA single factor statistics presented *p-value* $= 0.083 > 0.05$ reference value and Fa $<$ Fc, which leads to support the affirmative that the original data comes from sample with equal average (Table 3).

The non-parametric test Kruskal-Wallis supported the results presented by ANOVA, once it presented *p-value* $0.214 > 0.05$.

It is not possible to state, statistically in this sample, that workers in these states have greater workload, because the difference between the averages of the three groups is not significant.

Discussion

The results of this research show different workload averages, but not statistically different enough to support the state of comfort changes productivity or subjective workload (Table 3); an explanation for the difference of the results can be in the type of research.

Different results have been found in similar experiments performed in the laboratory, where all variables were controlled and the respondents were under previously defined and idealised working conditions (Lan, 2009, 2010).

This research data collection occurred in actual work environments where external factors and several variables can interfere with workers. Lan (2009, 2010) notes that motivated people can maintain high performance for a short time under adverse environmental conditions (warm or cool). Human sensitivity can lead to different results according to Chen (2012).

The data from this research presents different results from experimental research, which shows that more research in real environments is needed. It confirms Gaoua's (2010) position, that the effects of thermal conditions on cognitive performance are still vague; due to methodological discrepancies in evaluating whether the exposure to heat, by itself, has adverse effect on cognitive function.

The activation of the nervous system raises the state of mental alert. This is a preferred state of mind in carrying out tasks that require attention, time and demand performance. How long this ideal state of alert lasts, and the effect of thermal comfort, are issues that laboratory and applied research seeks to address (Tham, 2010).

Conclusion

Based on the results of this survey the state of thermal comfort has not been able to generate significant differences in average subjective scores of workers' productivity performing the same tasks under different thermal comfort conditions. This showed the need for more research in real environments.

It was observed that in a state of slight warm or cool the scores show variability much higher than in state of neutral, but no significant effects were found between the state of thermal comfort and workers' productivity performing in a real work environment.

It is believed that intervening variables affect significantly the results, which does not happen in a simulated environment in the laboratory.

The results show the need to identify situations where workers are in more thermal discomfort by warm and cool. In order to assess whether these thermal sensations are able to change the scores of productivity, since in this survey the states of discomfort obtained analytically were slightly warm and slightly cool, the states of analytical comfort found are too close to the state of comfort preferred by many workers. More research about thermal comfort and workload is necessary in real work environments.

References

Araújo, M.E.M. 2012, *Desempenho Cognitivo em Ambiente Moderado*/Maria Elisa Machado Araújo. Portugal: UMinho, 2012. Tese (mestrado em Engenharia Humana) Universidade do Minho. Escola de Engenharia.

Batiz, Eduardo Concepción., Goedert, Jean., Morsch, Junir Junior., Junior, Pedro Kasmirski., Venske Rafael. 2009, Avaliação do conforto térmico no aprendizado: estudo de caso sobre influência na atenção e memória. *Produção*. v. 19, n. 3, set./dez.

Cardoso, Mariane de Souza, Gontijo, Leila Amaral. 2012, Avaliação da carga mental de trabalho e do desempenho de medidas de mensuração: NASA TLX e SWAT. *Gest. Prod.,* São Carlos, v. 19, n. 4.

Chen, Ailu, Chang, Victor W.-C. 2012, Human health and thermal comfort of office workers in Singapore. *Building and Environment* 58.

Gaoua, N. 2010, Cognitive function in hot environments: a question of methodology. *Scand J Med Sci Sports.* v. 20 (Suppl. 3): 60–70.

Hart, Sandra G. Nasa-task load index (NASA-TLX); 20 years later. 2006, *Proceedings of the human factors and ergonomics society 50th annual meeting.*

International Organization for Standard. 2005, *ISO 7730 – Ergonomics of the thermal environment* — Analytical determination and interpretation of thermal comfort using calculation of the PMV and PPD indices and local thermal comfort criteria. Geneva.

Lan, Li, Lian, Zhiwei, Pan, Li, Ye, Qian. 2009, Neurobehavioral approach for evaluation of office workers' productivity: The effects of room temperature. *Building and Environment*, 44.

Lan, Li, Lian, Zhiwei, Pan, Li. 2010, The effects of air temperature on office workers' well-being, workload and productivity-evaluated with subjective ratings. *Applied Ergonomics,* 42.

Lan, Li, WargockiI, Pawel, Lia, N. Zhiwei. 2011, Quantitative measurement of productivity loss due to thermal discomfort. *Energy and Buildings,* 43.

Tham, Kwok Wai, Willem, Henry Cahyadi. 2010, Room air temperature affects occupants' physiology, perceptions and mental alertness. *Building and Environment,* 45.

THE GOOD JOB SCORE: ASSOCIATIONS WITH POSITIVE AND NEGATIVE OUTCOMES

Andrew P. Smith & Emma J.K. Wadsworth

*Centre for Occupational and Health Psychology,
School of Psychology, Cardiff University,
Cardiff, UK*

There has been considerable research on the relationship between job characteristics, stress and negative affect. Many factors need to be considered and these job characteristics often have additive effects with the sum of negative features of the job being a good indicator of negative outcomes. Less is known about positive job characteristics and positive outcomes. A survey of 1563 local government employees was conducted to address this topic. A "good job" score was derived by calculating the difference between positive work characteristics/appraisals and negative work characteristics/appraisals. This score was found to be a good predictor of physical health, positive mental health and the absence of negative affect.

Introduction

Long established job characteristic models of stress have shown clear links between adverse working conditions and poorer employee health and well-being. Research suggests that job characteristics often have additive effects and that stress and negative outcomes can be predicted by the combined effects of these factors (Smith et al, 2004). Transactional models of stress (e.g. the Demands-Resources-Individual Effects, DRIVE, model, Mark & Smith, 2008 suggest that it is important to examine the combined effects of occupational and individual variables that can influence both positive and negative outcomes.

The emphasis of early research was to identify occupational hazards. More recently, however, there has been a growing awareness that work, compared to worklessness, is generally good for health and well-being (Waddell & Burton, 2006). The important provisos are that account must be taken of the nature and quality of the work. A literature review on wellbeing at work (Wadsworth et al., 2010a) showed that research to date:

- Has no clear conceptual background
- Often fails to distinguish between the absence of negative effects and the presence of positive ones

- Identifies the multi-factorial nature of concepts of well-being
- Largely involves cross-sectional studies
- Rarely controls for other influential factors (such as job and individual characteristics)
- Has not established a "gold standard" for positive measures of job characteristics or well-being.

Secondary analyses were conducted on existing survey data (see Wadsworth et al., 2010b) and the results showed strong, independent associations between both positive and negative measures of job characteristics and well-being. They also identified two distinct patterns of associations showing that in some relationships the presence of positive effects and absence of negative effects had similar effect sizes, while in others either the presence of positive effects or the absence of negative effects had a significantly greater effect size than its opposite. Analyses of effects of individual differences in coping and attributional styles (Mark & Smith, 2011, 2012) showed that these had independent effects on well-being and health. This suggests that they can be added to any score based on job characteristics or appraisal of job characteristics. The present study, therefore, was designed to include a range of relatively new measures of positive job characteristics, appraisals and outcomes, together with more traditional negative measures and factors measuring potentially confounding factors.

Method

The research was carried out with approval from the ethics committee, School of Psychology, Cardiff University, and with the informed consent of the participants (recruited from local government organisations). Surveys were carried out electronically, though organisations were able to take part using some (or all) traditional paper questionnaires if they preferred. Electronic participation involved organisations e-mailing employees with the covering letter which included a link to the questionnaire. The questionnaire could then be completed online and submitted directly, though provision was also made for downloading a paper version on an individual basis which could be returned using a Freepost address. Participating organisations offered respondents entry into a prize draw as an incentive to participate.

Participants

In total 1563 people completed and returned a questionnaire. Most of the respondents were female (1074, 70%) and their mean age was 43.33 years (SD = 11.66, minimum 17, maximum 72). Most were married or living with a partner (1140, 74%) and just over half were educated to degree level or higher (810, 52%). The majority of respondents worked full-time (1184, 76%) and had permanent jobs (1307 (85%). On average they worked 34.63 hours per week (SD = 11.28).

Measures in the survey

These are described in detail in Smith et al (2011). The relevant measures can be summarized as follows:

Job characteristics:

- HSE management standards (Mackay et al., 2004) – Seven scales measuring: demand, management support, peer support, control, role, relationships and change.
- Effort-Reward Imbalance (Siegrist, 1996)
- Job-Demands-Control-Support (Karasek et al., 1998)
- Uplifts (Mayberry et al., 2006)

Appraisals:

- Work involvement (QWL – Warr, Cook & Wall, 1979)
- Job Descriptive Index (JDI – Smith et al., 1969)
- Job satisfaction scale (JSS – Spector, 1985)
- Work stress (Smith et al., 2000)
- Perceived stress scale (Cohen, Kamarck & Mermelstein, 1983)

Outcomes:

- General health (Smith et al., 2000)
- Symptom checklist (Smith et al., 2000)
- Fatigue (Smith et al. 2000)
- Hospital Anxiety and Depression scale (Zigmond & Snaith, 1983)
- Positive and Negative Affect Scale (Watson, Clark & Tellegen, 1988)

Statistical analyses

Overall measures were calculated by summing all the scales for: HSE management standards, uplifts frequency and intensity, QWL, Warr and JDI (a total JSS score is already part of the JSS scoring system). Factor analysis was then used to consider whether the positive and negative measures in the three groups (job characteristics, appraisals and outcomes) were measuring different concepts or opposite ends of the same concept. Three analyses were run each including all the positive and negative measures of either job characteristics, appraisals and outcomes. The analyses grouped similar measures together into broader factors, so measures assessing similar concepts, either positively or negatively, were categorised together. If the positive and negative measures of job characteristics, appraisals or outcomes were measuring different concepts, then very few of the factors produced by the analyses would contain both positive and negative measures.

Table 1.   Job characteristics, appraisals and outcomes factors.

Factor	Comprising
Job characteristics	
1 **Manager support**	Manager support; Intrinsic reward; Uplifts (from supervisor)
2 **Demand**	Demand; Extrinsic effort; Intrinsic effort; Unsociable and unpredictable hours
3 **Control**	Control over work organisation
4 **Role**	Role; Night work
5 **Peer support**	Peer support; Uplifts (from co-workers)
6 **Skills**	Skill discretion; Uplifts (from skills)
Appraisals of job	
1 **Management**	Satisfaction with management/supervisor, quality of working life in terms of management
2 **Work**	Intrinsic job motivation and higher order need strength, enjoyment, satisfaction with work and the nature of work, and quality of working life in terms of autonomy
3 **Stress**	Stress, satisfaction with conditions, quality of working life in terms of balance
4 **Reward**	Satisfaction with pay, promotion and benefits, and quality of working life in terms of pay
5 **Peers**	Satisfaction with co-workers
Outcomes	
1 **Negative mental well-being**	Negative mood, anxiety, depression, cognitive failures at work
2 **Physical health**	Physical health and symptoms, pain killers, fatigue, minor accidents at work
4 **Positive mental well-being**	Positive mood, quality of work in terms of autonomy and happiness

Combined effects modelling

The aim of these analyses was, first, to create a total score which reflected the combined effects of job characteristics and appraisals. This was expressed as a "good job score" with high scores reflecting more positive job characteristics and appraisals. Backwards stepwise logistic regression was used to assess the association between the good job score and the outcomes. The regression model also included the demographic (age, gender, marital status education), work (full-time, contract, pattern) and general stress factors to control for any potential influence of these individual characteristics.

Results

Factor analyses

First, three factor analyses were run representing job characteristics, appraisals and outcomes. These initial factor analyses showed that the job characteristics items together explained 76% of the variance, the appraisal items 77% and the outcomes items 72%. The factors and the items comprising those factors are shown in Table 1.

Table 2.   Associations between the Good Job Score (JC&A score) and Negative mental well-being.

		OR (CI)	p
Total JC&A score	Low	1.00	<0.0001
	Second Quartile	2.48 (1.35–4.55)	
	Third Quartile	8.01 (4.34–14.76)	
	High	23.70 (12.45–45.12)	
Age		1.02 (1.00–1.04)	0.02
General stress	High	1.00	<0.0001
	Low	9.48 (6.22–14.42)	

Table 3.   Associations between the Good Job Score (JC&A score) and Physical health.

		OR (CI)	p
Total JC&A score	Low	1.00	<0.0001
	Second Quartile	2.79 (1.67–4.68)	
	Third Quartile	3.80 (2.27–6.35)	
	High	6.87 (4.04–11.69)	
Age		1.02 (1.01–1.04)	0.01
Gender	Male	1.00	0.02
	Female	0.63 (0.42–0.93)	
Full-time	Full-time	1.00	0.003
	Part-time	2.05 (1.28–3.30)	
Pattern	Fixed hours	1.00	0.05
	Other*	1.45 (1.00–2.08)	
General stress	High	1.00	<0.0001
	Low	4.42 (3.10–6.31)	

*Flexi-time or shift work

Combines effects

Negative mental well-being was strongly associated with total job characteristics and appraisals. A dose-response type relationship was suggested, with those in the highest (most positive) quartile for the total job characteristics and appraisals score being over 23 times as likely to have low (positive) levels of negative mental well-being compared to those in the lowest (least positive) quartile (Table 2). This pattern of association was apparent for both physical health and positive mental well-being as well (Tables 3 and 4). There were also significant associations with both individual characteristics (such as age and general stress [our measure of negative affectivity]) and elements of type of work, emphasising the importance of controlling for other factors.

Components of the Good Job Score
The next set of analyses attempted to bench-mark the components of the "Good Job Score" with the outcomes. Total job characteristics scores were associated with

Table 4.   Associations between the Good Job Score (JC&A score) and Positive mental well-being.

		OR (CI)	p
Total JC&A score	Low	1.00	<0.0001
	Second Quartile	2.89 (1.24–6.75)	
	Third Quartile	5.24 (2.23–12.33)	
	High	22.83 (7.73–67.42)	
Age		1.03 (1.00–1.07)	0.05

all the outcomes in a dose response manner. The strength of these associations was not as great as seen in the analysis of the "Good Job Score". A similar conclusion applied to the analyses using the total appraisal score. The job characteristics and appraisal scores were then sub-divided into positive and negative components. Both job characteristic scores and appraisal scores were included in the same analyses. The strongest predictor of negative mental health was negative appraisals of the job. Positive appraisals had a smaller independent contribution. The same applied to physical health outcomes, although the effects of negative and positive appraisals were more similar. Finally, positive mental health was predicted largely by both positive and negative appraisals but also by negative and positive job characteristics.

The next set of analyses compared "Total Positive Scores" with "Total Negative Scores". These two scores represented the combination of the presence of positive job characteristics and appraisals compared to the presence of negative job characteristics and appraisals. The results showed an independent effect of the positive and negative scores, with the impact of negative job characteristics and appraisals being greater for negative mental health whereas positive job characteristics and appraisals had a bigger effect on positive mental health (and positive and negative scores had similar effects on physical health).

Finally, the influence of the individual factor scores was examined. Not surprisingly, many had significant effects but these were generally small (OR < 2) compared to the effect sizes seen with the combined scores.

Conclusions

The results from this new data collection have shown that the best predictor of health outcomes is the combined measure of total job characteristics (both presence of positive features and absence of negative) and the appraisals of these. We have, therefore, called this the "Total Good Job Score". This study has been important in that it is one of the few that have examined both the presence of positive as well as the absence of negative characteristics. One weakness of the study has been that it is a cross-sectional study which makes it difficult to draw conclusions about causal mechanisms. There was also a lack of measurement of individual differences (e.g. personality) and these may influence perceptions of work characteristics, appraisals and subjective reports of health and wellbeing.

References

Cohen, S., Kamarck, T. and Mermelstein, R. (1983) "A Global Measure of Perceived Stress," in *Journal of Health and Social Behavior* **24**: 385–396.

Karasek, R., Brisson, C., Kawakami, N., Houtman, I., Bongers, P. and Amick, B. (1998) The Job Content Questionnaire (JCQ): an instrument for internationally comparative assessments of psychosocial job characteristics. *Journal of Occupational Health Psychology* **3**(4): 322–55.

Kristensen, T. S., Hannerz, H., Høgh, A. and Borg, V. (2005) The Copenhagen Psychosocial Questionnaire – a tool for the assessment and improvement of the psychosocial work environment. *Scandinavian Journal of Work Environment & Health* **31**(6): 438–449.

Mackay, C., Cousins, R., Kelly, P., Lee, S., and McCaig, R. (2004) Management Standards and work-related stress in the UK: Policy background and science. *Work and Stress* **18**(2): 91–112.

Mark, G. M. and Smith, A. P. (2008). Stress models: A review and suggested new direction. Vol. 3. EA-OHP series. Houdmont, J. and Leka, S. (Eds.). Nottingham University Press. 111–144.

Mark, G. and Smith, A. P. (2011) Effects of occupational stress, job characteristics, coping and attributional style on the mental health and job satisfaction of university employees. *Anxiety, Stress and Coping* **25**: 63–78. *doi: 10.1080/10615806.2010.548088*

Mark, G. and Smith, A. P. (2012) Occupational stress, job characteristics, coping and mental health of nurses. *British Journal of Health Psychology* **17**: 505–521. *doi: 10.1111/j.2044-8287.2011.02051.x*

Maybery, D. J., Jones-Ellis, J., Neale, J. and Arentz, A. (2006) The Positive Event Scale: Measuring Uplift Frequency and Intensity in an Adult Sample. *Social Indicators Research* **78**(1): p61–83.

Siegrist, J. (1996) Adverse health effects of high-effort/low-reward conditions. *Journal of Occupational Health Psychology* **1**: 27–41.

Smith, A., Johal, S. S., Wadsworth, E., Davey Smith, G. and Peters, T. (2000) *The scale of occupational stress: The Bristol stress and health at work study*. Sudbury: HSE Research Report 265.

Smith, A., McNamara, R. and Wellens, B. (2004) *Combined effects of Occupational Health Hazards*. HSE Contract Research Report 287. HSE Books. ISBN 0-7176-2923-6

Smith, A. P., Wadsworth, E. J. K, Chaplin, K., Allen, P. H. and Mark, G. (2011) *The relationship between work/well-being and improved health and well-being*. Report 11.1 IOSH. Leicester.

Smith, P. C., Kendall, L. M. and Hulin, C. L. (1969) *The measurement of satisfaction in work and retirement*. Chicago: Rand McNally

Spector, P. E. (1985) Measurement of human service staff satisfaction: Development of the Job Satisfaction Survey. *American Journal of Community Psychology* **13**: 693–713.

Waddell, G. and Burton, A. K. (2006) *Is work good for your health and well-being?* The Stationery Office: Norwich

Wadsworth, E. J. K., Chaplin, K., Allen, P. H. and Smith, A. P. (2010a) What is a Good Job? Current Perspectives on Work and Improved Health and Well-being. *The Open Health & Safety Journal* **2**: 9–15.

Wadsworth, E. J. K., Chaplin, K. and Smith, A. P. (2010b) The work environment, stress and well-being. *Occupational Medicine* **60**: 635–639. *doi: 10.1093/occmed/kqq139*

Warr, P., Cook, J. and Wall, T. (1979) Scales for the measurement of some work attitudes and aspects of psychological well-being. *Journal of Occupational Psychology* **52**: 129–148.

Watson, D., Clark, L. A. and Tellegen, A. (1988) Development and validation of brief measures of positive and negative affect: the PANAS scale. *Journal of Personality and Social Psychology* Jun; **54**(6): 1063–70.

Zigmond, A. S. and Snaith, R.P. (1983) The Hospital Anxiety and Depression Scale. *Acta Psychiatrica Scandinavia* **67**: 361–370.

STANDARDS FRAMEWORK TO SUPPORT JOB SYNTHESIS ASSOCIATED WITH HCI

Mike Tainsh

Lockheed Martin, Ampthill, UK

The ISO 9241 set of documents provide one of the most coherent bodies of design and assessment techniques for Human Computer Interaction (HCI) that is currently available. They tend to be prescriptive in the way that HCI design should be carried out, and tend to be associated with specific design topics and processes. This paper summarises some of these results and presents them in a framework to support job synthesis i.e. job, task and activity design.

Introduction

Background: User architectural considerations

Human-Computer Interaction (HCI) and its associated User and equipment characteristics, is only ever part of a larger system, but has its own lifecycle. This is reflected, for example, in the UK MoD's technical guidance which is contained mainly within DEFSTAN 00-250. It starts with a consideration of UK MoD's Architectural Framework i.e. MODAF. This sets a context, in terms of its lifecycle, for job synthesis i.e. the design of HCI, and the jobs, tasks and activities associated with it. The ISO standards form a complementary body of detailed technical information.

The ISO documents relevant to HCI are listed in the contents section of ISO 9241: 100. This is not a complete set of relevant documents as others lie outside this area. However, it is a coherent set of documents, if only because of the way that ISO standards are written. Hence, it makes a useful starting point for the generation of a framework for assessment standards supporting job and task design.

General

Design techniques and their assessment are closely linked. This paper reviews HCI design standards to support the construction of a framework for assessing job, task and activity characteristics. The scope of the review is defined by the standards listed in ISO 9241-100, and those associated with them. The results are supported by systems engineering experience within vehicles programmes.

Hierarchy of descriptions

In line with MODAF and previous work (Tainsh, 2013 and 2014), the review was conducted at three system design levels:

- Description of Requirements and Principles;
- The elements of the system and their organization;
- Consideration of system design at the level of job and role characteristics, task characteristics, and activities with the specification of the equipment aspects. These three levels are considered below.

The principles of HCI assessment

ISO 15288, System life cycle

The System Life Cycle standard provides a comprehensive process-based description of all the activities that are required for system realisation, utilisation, evolution and disposal. They are Agreement, Project, Technical and Organisational Project Enabling. The standard recognises that the lifecycle processes may be tailored to meet individual project needs. The full set of Technical Processes includes: Requirements Definition, Requirements Analysis, Architectural Design, Implementation, Integration, Verification, Transition, Validation, Operation, Maintenance and Disposal.

The standard does not make reference to specific disciplines. The project will define which need to be included.

ISO 9241–210, Human-centred design for interactive systems

This standard addresses issues of principle and stands alongside ISO 15288. This combination of standards enables further technical issues to be flowed down into the design process. It considers that the following should be true in any implementation:

- The design is based upon an explicit understanding of users, tasks and environments;
- Users are involved throughout design and development;
- The design is driven and refined by user-centred evaluation;
- The process is iterative;
- The design addresses the whole user experience;
- The design team includes multidisciplinary skills and perspectives.

This is the human sciences contribution throughout the system life cycle. It provides the top down contribution to complement the functional input associated with ISO 15288.

Organising the ergonomics variables

ISO 26800: A general model for organising ergonomics variables

ISO 26800:2011 outlines a general organisational approach to handling ergonomics variables. It is expressed in the form of a model that can be employed by ergonomists working in a systems context. The principles, approach and concepts are all described at a high level. They refer to the broader aspects of working rather than the details of solving particular problems. This provides the statements of relationships between items within the model:

The organisational techniques include:

- Fitting the User's characteristics to tasks and environment, including equipment design characteristics;
- The user, task, equipment and environment are seen as elements of a work system.

This can be the basis of the organisational approach and, if used may be critical to the validity of the design output.

ISO 9241–210: User centredness in assessment

ISO 9241–210, besides stating the principles of User Centred Design, provides a high level technical overview of methods. Three main types are described:

- Users can be presented with models, scenarios or sketches of the design concepts and asked to evaluate them in relation to a real context.
- Inspection based evaluation which is ideally performed by usability experts who base their judgment on prior experience of problems encountered by the Users, and their own judgments of ergonomic guidelines and standards.
- Long term monitoring of the use of the product, system or service.

These are general categories of assessment technique.

ISO 18152–2003: The process and programme

The organisational approach needs to be tailored, taking account of the needs of a systems engineering programme. This standard provides advice on the processes required of a programme and the appropriate assessment techniques:

To assess if the practices have been achieved in a particular case, the assessor is advised to apply the following contextualisation procedure. For each process the assessor and person being assessed:

- Review the process, its outcomes and practices;
- Review the project/organization activities that fulfill the purpose and outcomes for a particular system or business objective;
- Consider the outcomes, practices and notes in that context.

Hence the organisational approach such as described in ISO 26800 should be reflected in the programme characteristics.

Ergonomics design and assessment of user and job characteristics

ISO 22411–2011: Fit to the users' characteristics

As part of the consideration of the system, it is necessary to define the characteristics of the Target Audience – the Users who will be working within the system. Currently, this is achieved with the concept of accessibility (Ref 10) which is described in an entirely general way. Therefore, we always have to ask what account do we take of the majority but also those "on the tails" of the distribution – the extreme 5% to 10%.

This standard requires that the system characteristics should be:

* Equitable – useful to persons with diverse abilities;
* Flexible in methods of use;
* Simple and intuitive, to accommodate a wide range of literacy and language skills;
* Enable perceptible information. It should communicate necessary information.

The need for tailoring the HCI should also be taken into account (ISO 9241, Part 129).

ISO 9241–2: Assessment of the user's jobs when working with the computer based equipment

ISO 9241–2, 1993 emphasises the need for consideration of the tools and techniques for carrying out the assessment and mentions three important characteristics, without details of techniques to be applied, to cover:

* Length and distribution of working periods;
* Autonomy/discretion in determining work requirements;
* Dependence of the work on the tools provided.

ISO 10075–2000: User's mental work load

The concept of mental workload is developed here but the assessment techniques are not described in detail. It states: "Design principles can be related to different levels of the design process and the design solution in the order to influence the intensity of the workload: at the task and or job level, at the level of technical equipment, at the environmental level, at the organisational level, the duration of the exposure to the workload, at the level of the temporal organisation of work." It provides guidance on job and task design although the specific characteristics of these variables remain imprecise.

ISO 9241–410: Assessment of devices

Jobs are composed of tasks and hence it is necessary to assess the components of the jobs directly dependent upon the devices.

This standard describes the assessment issues associated with input devices. There are four generic sets of characteristics to be assessed:

- Appropriateness – this addresses anthropometric issues and the ability of the device to perform without the need for additional tools;
- Operability – obvious, predictable, consistent, feedback;
- Controllability – responsiveness and reliability;
- Biomechanical load – forces required.

These assessments are based on the match between the product and the standard. They include some specific physical characteristics of the devices.

ISO 9241–411: Evaluation methods for devices

This standard provides a large number of tests devised to show that devices that have been introduced in ISO 9241–410 have been designed correctly. It is a compendium of design criteria based on the physical properties of devices.

ISO 9241–143: Assessment of form dialogues

This standard states: "Form based dialogue evaluations generally fall into the following categories:

- When users and user tasks are known, evaluators evaluate the product or observe representative user of the product in the context of accomplishing typical or critical tasks in a typical usage environment;
- When specific users and user tasks are not known, evaluators evaluate all forms used in the product being evaluated;
- Determination of whether a product meets a given requirement or recommendation can be based on the set of forms encountered during the evaluation."

The assessment process addresses and emphasises the match between the product and requirement.

ISO 9241–420: Selection between input and output devices

This standard states: "Occasions when user tests can be warranted include the following:

- Utilizing a given device with others that have not been designed for concurrent use (e.g. in the case of mutual exclusion);
- Selecting others or settings for a given device under a new context of use (subdividing a tablet into smaller zones to reduce postural problems for smaller persons, selecting mouse settings for unstable support surfaces etc.);

- Testing alternative arrangements for work equipment;
- Selecting alternatives of the same type of device;
- Testing for a critical task primitive to select a dedicated device."

This assessment process progresses through a consideration of alternative designs for the equipment.

Assessment of HCI activities

Tasks may be considered to be composed from sets of subtasks (ISO 9241, Part 303), and in turn activities. There are no standards which support assessment at the level of activities. This is the lowest level of task description i.e activities associated with unitary HCI artefacts. One approach is to define an activity in terms of:

- An object (visual object, auditory signal etc) which has semantic properties which are understood or potentially understood by the User. An example would be a graphics object such as specified in ISO 7000.
- A function which is associated with the object and is reflected in the design of the computer equipment, and links one object to another in permitted sequences. From the User perspective this set of functions will be reflected in the semantic and syntax properties of the objects.

The definition of these activities is vital as they will need to be reflected in ergonomics assessments i.e. measures of performance or cognition or other, in line with the organising principle There are other possible approaches such as the consideration of "mental elements" (what the User might consider a unit of perception).

The framework

The ISO documents enable the construction of a framework as shown in Table 1. The key elements of the framework are:

- The hierarchy as defined by MODAF and similar architectural frameworks;
- The ISO standards with their potential for extension to HCI activities.It provides the correct spacing after the list;
- The lifecycle and the use of the standard within it.

The framework includes an indication of the risks associated with failing to comply with the standards on job synthesis i.e. job, task and activity design.

Conclusion

It would appear that at the level of principles, organising approaches, jobs and tasks, there is a set of standards that we can organize into a framework along with the

Table 1.   Framework to show architectural/hierarchy position, with standards, design process and risk to outcome.

Hierarchy Position	Standards	Role in Design Process	Risk to Design Outcome if omitted or not complete
Principles	ISO 15288, ISO 9241–210	Considered as precondition	High risk of ineffective design.
Processes and Programme	ISO 26800, ISO 18152, ISO 9241–210	Well understood processes associated with qualified specialists	High risk of ineffective design.
Design and Assessment – TAD	ISO 22411, ISO 9241–129	Well documented TAD which is valid.	High risk of ineffective design, especially if combined with a lack of principles and processes.
Design and Assessment – Jobs	ISO 9241–2, ISO 10075	Legacy jobs that are well understood	There is a high risk of organisational failure.
Design and Assessment – Tasks/ Activities	ISO 9241–143, ISO 9241–303, ISO 9241–410, ISO 9241–411, ISO 9241–420	Predecessor tasks well understood	Even if the tasks are well defined, additional effort will be required at the level of activities.

risks for failing to apply them. Their application is relatively well understood along with the risks.

However, more work needs to be carried out at the level of activities. It is possible that more work is required on the standards associated with the concept of "job" as the current concepts of jobs appear somewhat simplistic when supporting a comprehensive approach to job synthesis.

Acknowledgement

I acknowledge the support of Lockheed Martin at Ampthill.

References

DEFSTAN (00-250) (Parts 0, 1, 2, 3 and 4). 2008. *Human Factors for Designers of Systems*. MoD, UK.

MODAF 2005. MoD *Architectural Framework-Technical Handbook*.

ISO 9241-Part 100, 2010, *Ergonomics of human system interaction: Introduction to standards related to software ergonomics*.

Tainsh, M 2013. *Hierarchical System Description (HSD), using MODAF and ISO 26800*. Contemporary Ergonomics and Human Factors, Taylor and Francis.

Tainsh, M 2014. *User Systems Architecture, Function, Job and Task Synthesis*. Contemporary Ergonomics and Human Factors, Taylor and Francis.

ISO 15288: 2010, *Systems and software engineering — System life cycle processes*

ISO 9241-Part 210, 2010, *Ergonomics of human-system interaction: Human-centred design for interactive systems.*

ISO 26800, 2011. *Ergonomics — General Approach, Principles and Concepts.*

ISO 18152, 2010. *Ergonomics of human-system interaction — Specification for the process assessment of human-system issues.*

ISO 22411, 2011. *Ergonomics data and guidelines for the application of ISO/IEC Guide 71 to products and services to address the needs of older persons and persons with disabilities.*

ISO 9241-Part 129, 2010. *Ergonomics of human-system interaction. Guidance on software individualisation.*

ISO 9241-2, 1993. *Ergonomic Requirements for office work with visual display terminals (VDTs) – Guidance on task requirements.*

ISO 10075-Part 2, 2000. *Ergonomic principles related to mental workload: Design principles.*

ISO 9241-Part 410, 2008. *Ergonomics of human system interaction: Design criteria for physical input devices.*

ISO 9241-Part 411, 2012. *Ergonomics of human-system interaction: Evaluation methods for the design of physical input devices.*

ISO 9241-Part 143, 2010. *Ergonomics of human-system interaction: Form-based dialogues.*

ISO 9241-Part 420: 2011. *Ergonomics of human-system interaction: Selection of physical input devices.*

ISO 9241-Part 303, 2011. *Ergonomics of human-system Interaction: Requirements for electronic visual displays.*

ISO 7000, 2014. Graphical *symbols for use on equipment – Index and synopsis.*

PHYSICAL DISCOMFORT AND MOBILE WORKING: WELLBEING APPROACH TO HEALTHY BEST PRACTICE

Jim Taylour[1] & Patrick W. Jordan[2]

[1] *Orangebox Limited, Parc Nantgarw, Cardiff*
[2] *Loughborough University, Loughborough*

Rapid changes are occurring in people's work environments and the devices that they use such as tablets, smartphones, laptops and emerging wearable technology. This raises issues for how to mitigate work-related discomfort including new strains of potential ill health brought on by new ways of working. The current Display Screen Equipment (DSE) regulations have not been updated since 2002 and have little or nothing to say about either these devices or mobile working practices. This study identifies the issues and challenges faced by employers and employees and makes recommendations in the absence of up to date regulations, set against a climate of changing organisational attitudes towards proactive health and wellbeing.

Introduction

In the last few years a number of significant changes have affected the workplace. While previously most office workers spent the vast majority of their time at a fixed workstation and worked on a desktop computer, people are now working in a wider variety of locations and using a wider variety of devices. This includes working from home, on the move, at co working spaces or elsewhere in the office (Taylour and Jordan, 2014). Meanwhile desktop computers have become supplemented with or superseded by laptops, smartphones and tablets as common work tools.

This diversity of work locations and work devices brings advantages in terms of flexibility and productivity that can benefit employee and company alike (Heathfield, 2012). However, these changes also raise some potential issues with excessive use of some devices and the challenges of sub-optimal ergonomic environments. The health challenges are potentially compounded with young recruits joining the workplace now and in the near future especially Generation Z (born from 2000) who have been immersed in technology from birth with little or no guidance on usage or best practise at home or at school (Thrasher, 2014).

This study looked at the implications of these new working practices and environments for employees' physical wellbeing. The primary focus was on physical discomfort but psychological health issues were also investigated.

The Health and Safety Executive mandate that UK businesses must adhere to the Display Screen Equipment (DSE) regulations of 1992 for all workers who regularly use display screen equipment as part of their job. They were revised in 2002 but

Table 1.   Percentage of respondents working in stated locations (n = 80).

Work Location	%
Dedicated desk	93
On the move	83
Elsewhere in office	74
At Home	73
Shared desk	39

this was limited and the regulations say nothing about smartphones or tablets. A replacement Musculo Skeletal Disorder Directive has been muted (FSB, 2012) but its popularity and progress is unclear (EEF, 2013).

Aims

The aim of the study was to understand the effects of new technologies and working practices on perceived physical discomfort and to make recommendations about how to address these. This included answering the following research questions: 1. What environments are professionals whose work was traditionally carried out at a fixed workstation currently working in? 2. What devices are they using for work? 3. What physical symptoms are workers displaying? 4. How can these be mitigated in the current working climate?

Methodology

A combination of three methods was used:

- A literature review of the relevant issues.
- Three regional focus groups involving a total of 49 participants. These included professionals in traditionally office-based jobs as well as experts who had knowledge relevant to the research questions, for example physiotherapists, Health and Safety and other occupational health practitioners, office designers, architects and human resource managers.
- An online questionnaire. This was completed by 80 professionals in a cross section of knowledge and process office work roles in full time employment.

Results

Work locations

The results of the online survey showed that the vast majority of people spent at least some of their working week at a dedicated desk. However, nearly everyone also spent some time working in other locations both inside and outside the office as summarised in Table 1.

Table 2.   Percentage of respondents using various devices during their working week (n = 80).

Devices Used	%
Smartphone (work)	78
Laptop	75
Desktop	69
Tablet	66
Smartphone (personal)	65
Standard mobile phone (personal)	49
Standard mobile phone (work)	45
Other	9

Table 3.   Percentage of respondents affected physical symptoms associated with device use (n = 80).

Symptoms Experienced	%
Upper back or shoulder pain	70
Headaches	69
Lower back pain	63
Neck pain	59
Visual or hearing impairment	30
Finger pain	29

Table 4.   Issues associated with symptoms (n = 80).

Effects of Symptoms	%
Negative effect on productivity in last 12 months	27
Received treatment	44
Days off in last 12 months due to symptoms	10

Devices used

The survey showed that smartphones and laptops have overtaken desktops as the most used work devices (see Table 2).

Physical symptoms

Most survey participants had experienced some physical symptoms associated with device use (see Table 3).

Effects of symptoms

Respondents were asked if these symptoms had affected their productivity in the last year, whether they had had treatment for them and whether they had had time off in the last 12 months due to them (see Table 4).

Discussion

Prevalence of physical discomfort

These results underline the potentially widespread nature of physical discomfort. A recent UK survey found that in the previous 12 months 51% of the UK workforce had suffered from back pain and that one in five has taken time off work because of it (Pruyne et al, 2012). Of the 131 million working days that were lost in the UK to sickness in 2011, the greatest number were contributed to MSDs such as back pain, neck pain, and upper body problems (ONS, 2014).

Presenteeism

'Presenteeism' refers to employees who are unwell, injured or disengaged and performing significantly below their best, but they come into work anyway. Although only 10% of respondents had taken time off due to the listed symptoms, 27% said that their productivity had been affected and 44% had had treatment for their symptoms. Presenteeism costs £15 billion per year in the UK, twice as much as absenteeism (Sainsbury Centre for Mental Health, 2007).

Working days lost due to illness and injury

Presenteeism may also partially explain the on-going reduction in working days lost due to illness or injury. The Office of National Statistics' records show that only 4.4. days per employee were lost to illness and injury in 2013 compared with 7.2 days per employee in 1993 (ONS, 2014). While the ONS figures might be encouraging, the increased use of laptops, tablets and smartphones and the increasing amount of work being done outside the office are potentially bringing new strains of occupational ill-health.

Laptops

Unless special equipment, such a separate keyboard/input devices and laptop raisers are used, using laptops is likely to result in a head-bent-forward working posture (Eltayeb et al, 2009). From the focus group it emerged that many companies provide employees with equipment which enables them to use laptops and other equipment more safely, but many did not use them.

It is notable that only 28% of those provided with laptop raisers use them. (The 56% of people who said they used a remote keyboard when provided with one may be because people were using it mainly with tablets rather than laptops).

Smartphones

There has been a sevenfold increase in the number of smartphones sold since 2006 and sales for 2014 were expected to be in the region of 1.3 billion globally (Forrester,

Table 5.   Percentage of employees using equipment when provided with it.

Postural aids for mobile technology use	%
Mouse or input device	70
Docking station	61
Remote keyboard	56
Laptop raiser	28
Headset	18

2013). Indeed, smartphones were the most commonly used work devices among those who completed our survey.

Associated with the rise of smartphone use are increasing reports of thumb and index finger injuries – sometimes referred to as 'BlackBerry thumb'. A report in *Economic Times* suggests that in one in ten doctor's appointments, patients present symptoms caused by smartphone or computer use. In India there was an 89% increase in smartphone ownership from 2012 to 2013 with 51 million devices in use (Khosla, 2013). Also, during the past year, younger adults lost 1.5 more days of work from back pain, believed to be caused by increased use of computers, particularly mobile devices. Younger adults spent as many as 8.83 hours per day using computers and other digital devices (Mientka, 2013).

Tablets

At the moment, the biggest growth in device sales is in tablets; 200 million were sold worldwide in 2013, almost double the sales of the previous year (Gartner, 2013). This number is expected to increase to 450 million by 2017 by which time, according to Forrester's projections, there will be over 900 million in use for work and at home (Forrester, 2013). The tablet's versatility may also be one of the greatest problems from a discomfort viewpoint. 'iPad neck' from over flexing to read the screen, nerve ending damage to fingers from excessive typing on glass and increased wrist and hand injuries are being noted due to dimensional mismatch and over use (Young et al, 2012; Stawarz et al, 2013).

Visual issues were also reported in the online survey and reflect the potential risks affecting the young exposed to excessive screen use. Symptoms being increasingly reported include dry eye syndrome from holding tablets 10cm closer and not blinking enough, computer vision syndrome, a form of repetitive strain injury affecting eye muscles constantly refocussing and 'cybersickness', similar to motion sickness (Lueder and Rice, 2007; Wagner, 2014).

Recommendations

Step 1 for employers: identify what the issues are.

The four corner stones of Wireless Well Working (our term for optimum ergonomic levels when away from the fixed desk) have been identified to help break down the

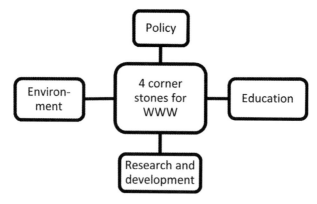

Figure 1.   Wireless Well Working PEER corner stones, Taylour (2014).

categories to tackle ill-health and outlines where responsibilities sit for each area as per Figure 1 below:

Policy includes personal, departmental, company, governmental and in the case of children, parental. *Have you got these in place, how do they shape up, can you glean more from others around you or externally, do you comply?*

Education includes peers, schools, parents and business. *Do employees have access to appropriate training, are they encouraged to attend with CEOs setting the example, do you as an employer need to embrace the bigger wellness picture to gain employee motivation and acceptance?*

Environments include designers, architects, specifiers, health and safety practitioners, human resources and facilities managers. *How does yours fare, are you getting health-conscious advice from the professionals, is it fit for purpose, is there a strong ergonomic 'footprint' away from the fixed desk?*

Research and development includes organisations, consultancies, and academic institutions. *Where there are gaps, can you find out more or sponsor some research on topics such as fresh anthropometric data in the light of increased diversity, acoustic control or long term effects of new strains of ill health?*

Step 2 for employers and employees: how to address the issues.

The Wireless Well Working circle of influence model identifies what you already have in place and what you can draw upon to reach a state of 'wireless well working' as per Figure 2 below.

1) Government policy and International Standards publications and drafts such as the new human centred organization ISO draft (Stewart, 2013) can be followed but a good starting point is the Department for Health's Physical Activity Network Deal (https://responsibilitydeal.dh.gov.uk/)

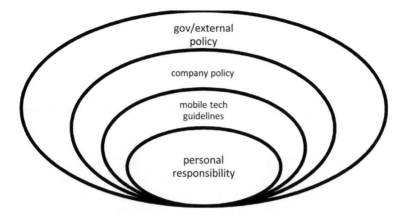

Figure 2.   Wireless well working circles of influence (Taylour & Jordan 2014).

2) Company policy can then be revisited to reflect the Network Deal. Introduce policies based on recent research (Stawarz et al, 2013) and guidance (BITC, 2013). A good starting point for England is Dame Carol Black's wellbeing Charter (http://www.wellbeingcharter.org.uk/index.php)

3) Mobile Technology guidelines can be issued to all personnel using sound ergonomics principles on set up and use in the absence of revised DSE guidance. Practical tips from the findings of this survey are published (Taylour and Jordan 2014) and those for British Telecom are in the public domain. (BT, 2011)

4) Personal responsibility must be encouraged. Add improved wellness to personal or department objectives and introduce incentives that aren't just lip service as outlined by associations such as the Chartered Society of Physiotherapy or the British Heart Foundation (BHF, 2014).

Conclusions

Although an essential starting point, a criticism of the DSE regulations is the focus on simply reducing absenteeism, injury and disability to get employees to a neutral point. With changing work cultures, new technologies, alternative work environments, greater user diversity (weight, ethnicity, ageing, accessibility etc) and generational differences, it is important for organisations to embrace new attitudes and guidelines to achieve high levels of wellness and productivity. This study exposes the need for further research into both the long term physical and psychological effects of mobile technology, changing work practises and the need for innovative ways of affecting change for both individuals and companies interested in improving health and productivity at work.

References

British Heart Foundation, 2014, Health at work: Resources, https://www.bhf.org.uk/health-at-work/resources-2.

British Telecom 2011, Get fit for mobile working http://www.insight.bt.com/en/features/get_fit_for_mobile_working.

EEF, Institute of Mechanical Engineers 2013, Government must do more to ensure Health and Safety works for business http://www.imeche.org/news/blog/government-must-do-more-to-ensure-health-and-safety-works-for-business.

Eltayeb, J. B. Staal, A. Hassan, R. A. de Bie 2009, Work related risk factors for neck, shoulder and arms complaints: a cohort study among Dutch computer office workers. *Journal of Occupational Rehabilitation* 19 (4): 315–322.

Federation of Small Business, June 2012, European Commission proposal for a new Ergonomics Directive, http://bit.ly/1xDMPH9.

Forrester 2013, World tablet adoption forecast 2012 to 2017 (Global), http://www.forrester.com/Forrester+Research+World+Tablet+Adoption

Gartner 2013. Gartner says worldwide tablet sales grew 68 percent in 2013, with android capturing 62 percent of the market, http://www.gartner.com/newsroom/id/2674215.

Heathfield, S.M., 2012 Flexible working, American Society of Resource Management, Human resources, About.com.

Khosla, V. 2013, Smartphone addiction can give you text neck, BlackBerry thumb, http://bit.ly/1woTmkx

Lueder, R. and Rice, V. 2007, *Ergonomics for Children: Designing products and places for toddlers to teens.* (Taylor and Francis, Boca Raton, Fl)

Mientka, M. 2013, iPosture: 84% of young adults getting back pain from using iPhones, iPads, and other mobile devices, http://bit.ly/1wotr7p.
office for national statistics 2014, 131 million days were lost due to sickness absences in the UK in 2013, http://www.ons.gov.uk/ons/rel/lmac/sickness-absence-in-the-labour-market/2014/sty-sickness-absence.html.

Pruyne, E., Powell, M. and Parsons, J., 2012, Developing a strategy for Employee Wellbeing, Ashbridge Business School and Nuffield Health 2012.

Stawarz, K. M. and Benedyk, R. 2013, Bent necks and twisted wrists: Exploring the impact of touch-screen tablets on the posture of office workers. *The 27th Int British Computer Society HCI Conference: The Internet of things.*

Stewart, T. and Berns, T. 2014, Human centred organisations, http://www.ergonomics.org.uk/slider/the-human-centred-organisation/

Taylour, J. A and Jordan, P. W. 2014, *Mobile Generations, Orangebox Office Wars 2015 White Paper publication*, November.

Thrasher, J. 2014, *Children's Use of Tablet Computers, the Health Risks*, unpublished Masters Thesis (HCI with Ergonomics) 2014 , UCL.

Wagner, R. S. 2014, Smartphones, video display terminals and dry eye disease in children, *Journal of Pediatric Ophthalmology and Strabismus* 51 (2):76.

Young, J.G., Trudeau, M., Odell, D., Marinelli, K. and Denneriein, J. T. 2012, Touch-screen tablet user configurations and case-supported tilt affect head and neck flexion angles. *IOS Press* 41: 81–91.

ASSESSING MULTIPLE FACTORS OF WELL-BEING USING SINGLE-ITEM MEASURES

Gary M. Williams

Centre for Occupational and Health Psychology, Cardiff University

This paper presents the results of research to investigate a practical approach to measuring multiple facets of well-being. Measures of work characteristics, coping, and personality were assessed individually and in combination for their relationship with well-being including depression, anxiety, mood, stress, and satisfaction. The analysis demonstrates individual and combined associations between predictor variables and outcomes. The result is a set of single-item measures which can be used to assess well-being and the factors associated with various well-being outcomes.

Introduction

This paper presents the results of research designed to assess a multi-faceted approach to well-being prediction using single-item measures. The purpose was to create a practical well-being tool that can apply the understanding of well-being as a multi-faceted concept while limiting the impact on time and questionnaire length. Previous research had demonstrated the validity and reliability of single-item measures compared to multi-item measures and the research presented here expands on this by demonstrating the significance of the multi-dimensional approach using these items. As a basis for this research, a multi-faceted approach to well-being which defines well-being as the combination of positive, negative, emotional, and cognitive elements was used. This approach is based on previous research which has shown that well-being consists of related but distinct factors and allows for results to be applied to specific aspects of well-being.

The combination of the variables used here represented nationally monitored psychological well-being outcomes and frequently assessed negative well-being outcomes in the workplace (stress, depression, and anxiety) (Mark & Smith, 2012a; Mark & Smith, 2012b) as well as the positive mood, negative mood, and satisfaction factors that make up subjective well-being, which is widely considered to be a significant element of well-being as a whole. Using the single-item measures a six-item well-being outcome assessment was also used (Lucas, Diener, & Suh, 1996).

Previous models of well-being were also used as a basis for selecting the predictor measures. The DRIVE model (Mark & Smith, 2008) was used as the theoretical framework for the research and variables representing demands, resources, and negative coping style were mirrored from previous research using this framework. Work demands were represented by a demands item and an effort item, and resources were represented by reward, control, and support. Negative coping style was represented by 5 items consisting of problem focused, seeking social support, avoidance, self-blame, and wishful thinking. Although the DRIVE model provides a multi-faceted approach to measuring the factors associated with well-being, the use of single-item measures allowed for variables which have been implicated in well-being research but not accounted for in the DRIVE model to be included. As such, role understanding, supervisor relationship and consultation on change were added to the resources group based on the HSE management standards (MacKay, Cousins, Kelly, Lee, & McCaig, 2004), while bullying was added to the demands group based on previous research showing its importance, particularly in nurses (Quine, 1999).

Personality has also been identified as a significant aspect of well-being outcomes, particularly when considering subjective well-being in terms of mood. The most commonly measured factors of personality are the 'big 5' factors of extraversion, neuroticism, agreeableness, conscientiousness, and openness; however other factors including optimism, self-esteem, and self-efficacy have also been considered in well-being research as more narrow concepts of personality, and were also included. The research therefore took a broad approach to the variables considered due to the fact that much of the research looking at these variables in relation to well-being outcomes does so without accounting for other factors already considered to be important in the process. For example, although narrow personality variables have been associated with well-being outcomes, this association may already be accounted for by broader factors such as neuroticism or extraversion and therefore their independent value when assessing well-being in practice should be confirmed in order to reduce redundancy in items and therefore the length of the questionnaire. The ability for single-item measures to distinguish between the effects of these closely related variables also needs to be established to support the use of single-item measures as alternatives to longer measures.

This research provides the opportunity to demonstrate their worth or redundancy alongside each other so that the most effective variables can be selected for the final measure. The research presented here consists of two sets of analyses, the first set of analysis is concerned with the utility of each item, in order to determine the relative contribution of each item to minimise redundancy in the measures and ensure the validity of each item as a predictor of well-being. The second of the two studies is concerned with the predictive validity of the items when combined, to demonstrate the utility of the questionnaire as a whole in predicting and measuring well-being outcomes. The analyses were conducted on data from two studies, one from a sample of university staff and the other a sample of nurses.

Method

Participants

Study 1: One hundred and twenty university staff members aged 20–64 (mean 37) participated in the study as part of a larger survey on well-being measurement.

Participants from all areas of the university were able to participate, including finance, teaching, accommodation, and security, although the role of specific respondents was not recorded.

Study 2: One hundred and sixty female and seventeen male nursing staff aged 19 to 69 completed the questionnaire. The mean age of the sample was 40 years. Participants from all areas of nursing responded to the survey, including practitioners, educators, and managers.

Materials

The questionnaire consisted of single-item measures, developed in-house, and established multi-item scales of the same measures. Items that were structured in the form of an initial statement each had the response scale from "Disagree strongly" to "Agree strongly" while those with an initial question (e.g. "On a scale of one to ten, how depressed would you say you are in general?") each had the response scale from "Not at all" to "Extremely". All items had a response scale from 1-10, chosen for practical and statistical reasons.

Design

A cross sectional design was used for both studies.

Procedure

University staff responded to an internal advertisement on the university online noticeboard. Nursing staff were recruited through UNISON via newsletter. Those interested were sent a link to an online questionnaire which they could complete in their own time. The questionnaire was expected to take approximately one hour to complete. Participants were instructed that they could skip any questions that they were not comfortable answering, although all data were provided anonymously. Ethical approval was obtained from the Cardiff University Psychology Ethics Committee. Informed consent was achieved within the questionnaire where participants could not continue beyond the consent page without agreeing. Following the consent page participants were presented with an instructions sheet and following the questionnaire a debrief sheet was provided.

Results

Univariate relationships

Demographic variables were not correlated with outcomes, with the exception of education and negative emotional well-being which were significantly correlated in the university staff sample ($r = .189$, $p < .05$, $n = 117$). The majority of the predictor variables were significantly correlated with the well-being outcomes in at least one of the samples. The exceptions were: demands and bullying were not correlated with positive emotional well-being, emotional stability was not correlated with negative cognitive well-being, and conscientiousness, agreeableness and openness were not correlated with negative cognitive or negative emotional well-being. These variables were therefore not included in the regression models for their respective non-correlated outcome. Correlations were in general stronger for the nurses sample than the university staff sample.

Multivariate relationships

Multiple regression analysis was performed to assess the significant unique contribution of each predictor variable to well-being outcomes in both the staff and nurses samples. Only those variables with a significant univariate correlation were entered into the regression and a hierarchical approach was used with variables entered in the order: demographics, work characteristics, coping style, personality. The results showed independent contributions of predictor variables which were largely dependent on the specific aspect of well-being involved. For positive cognitive well-being (satisfaction with life and work), control, effort, reward, wishful thinking, emotional stability, and optimism contributed significantly (greater than .05). For negative cognitive well-being (life stress and work stress), demands, control, effort, and wishful thinking contributed significantly. For positive emotional well-being (positive mood), avoidance, self-esteem, self-efficacy, and optimism contributed significantly, while for negative emotional well-being (negative mood, depression, and anxiety), education, bullying, self-blame, avoidance, self-esteem, self-efficacy, and optimism contributed significantly.

Combined effects

The significant unique predictors for each outcome were combined into a predictor measure for that outcome. Participants were scored for their total score on each measure and scores were split into tertiles to examine the difference in mean well-being for those in high, medium, and low risk groups.

Discussion

The results show the significant relationships between single-item measures of a variety of predictor variables and well-being outcomes. Previous research has demonstrated a wide range of variables associated with well-being outcomes, which creates issues for practical application of research findings in the form of impractically lengthy questionnaires and a degree of uncertainty as to the relative contribution of those variables included and the cost of not including others (Williams & Smith, 2012). A number of variables were assessed against well-being outcomes. The univariate analyses showed that the variables had significant

Positive CWB

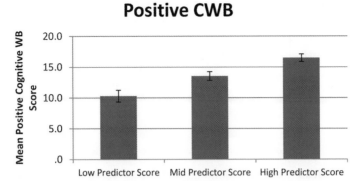

Figure 1. Mean positive cognitive well-being score for those with low, medium, and high predictor variable scores. (Error bars show 95% confidence intervals).

Negative CWB

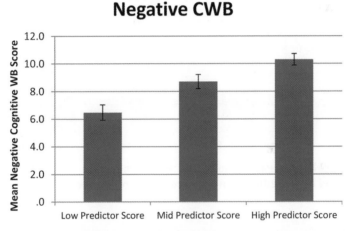

Figure 2. Mean negative cognitive well-being score for those with low, medium, and high predictor variable scores. (Error bars show 95% confidence interval).

relationships with outcomes, as expected based on previous research. The multi-variate analyses, however, showed how the variance in well-being predicted by these outcomes is not always uniquely significant. This was the case across work characteristics, coping style, and personality variables, where supervisor relationship, support, positive coping styles, and broad personality characteristics did not show significant unique associations with outcomes when all variables were included in the analysis together. These results suggest that while a number of variables may be associated with well-being when considered alone, much of the variance is shared amongst them, suggesting limited relative value in the same questionnaire. There are, however, alternative explanations, such as a lack of validity in the single-item measures or an inability for a single-item to measure discrete differences between closely related variables. Results from previous research are contrary to the former

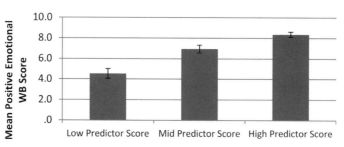

Figure 3. Mean positive emotional well-being score for those with low, medium, and high predictor variable scores. (Error bars show 95% confidence intervals).

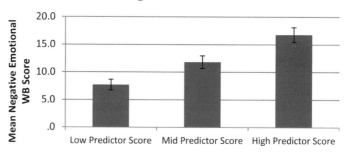

Figure 4. Mean negative emotional well-being score for those with low, medium, and high predictor variable scores. (Error bars show 95% confidence intervals).

however, as for example the supervisor relationship item scored one of the highest for validity and estimated reliability.

While this research does support these broad differences between emotional, cognitive, positive, and negative well-being outcomes and their associated predictor variables, most notably a greater emphasis on personality for emotional outcomes and circumstances for cognitive outcomes, some interesting caveats are found. Bullying, for example, appears to be also associated with emotional well-being outcomes, while control and effort are associated with positive and negative cognitive well-being. This suggests that some circumstances may have implications beyond one's cognitive assessment of stress or satisfaction, while others may have both a presence and absence effect. The difference between the existence of positive factors and absence of negative factors has previously been discussed by Smith et al (Smith, Wadsworth, Chaplin, Allen, & Mark, 2010).

The results therefore support the approach of using briefer measures to assess a broad range of variables in the well-being process, as the results indicate a broad range of variables contribute significantly to outcomes and the ability to identify those contributions is not lost in the brevity of the measures. The second stage of the analysis demonstrated the application of the items in terms of the level of well-being in respondents depending on their combined scores on predictor measures. This analysis shows how, when the range of work, coping, and personality variables are considered, differences in well-being outcomes can be expected between high, medium, and low scoring groups. The results show how the combination of items can not only identify individual contributions of variables but also how these variables can be combined into a complete measure that identifies those at high, medium, or low risk of positive or negative well-being outcomes in a brief, practical way. Beyond the practicality of the measures, the research also demonstrates the importance of the multi-dimensional approach to well-being assessment. As can be seen from the results, the associations between predictor variables and well-being outcomes are not consistent across different aspects of well-being. Although there is the possibility that this can occur from statistical variations, the results are in line with previous research demonstrating such differences across subjective well-being outcomes. For instance, personality variables are more consistently associated with emotional well-being in these data and this is supported by previous research. In contrast, circumstances such as work demands and control are more strongly associated with cognitive well-being outcomes, which is also in line with previous research and the theoretical concept of cognitive well-being which involves a subjective assessment of one's circumstances (Diener, Suh, Lucas, & Smith, 1999).

The implications of this research are that many previous models of well-being in the workplace (such as the HSE management standards approach) consider only certain circumstances, and these results demonstrate that this has limited application to aspects of well-being that involve an emotional aspect such as mood or depression. Further to this, the results show that a more complex approach that includes individual differences and personality can be applied without a significant impact on practicality using single-item measures while being relevant to a range of well-being outcomes. The result of this research is a set of measures, previously validated, that can be used to assess well-being outcomes and associated factors with limited time burden and demonstrated predictive utility.

References

Diener, E., Suh, E. M., Lucas, R. E., & Smith, H. L. (1999) Subjective wellbeing: 3 decades of progress. *Psychological Bulletin* 125(2): 276–302.

Lucas, R. E., Diener, E. & Suh, E. (1996) Discriminant Validity of Well-Being Measures. *Journal of Personality and Social Psychology* 71(3): 616–628.

MacKay, C. J., Cousins, R., Kelly, P. J., Lee, S. & McCaig, R. H. (2004) 'Management Standards' and work-related stress in the UK: Policy background and science. *Work & Stress* 18(2): 91–112.

Mark, G. & Smith, A. P. (2012a) Occupational stress, job characteristics, coping, and the mental health of nurses. *British Journal of Health Psychology* 17(3): 505–521.

Mark, G. & Smith, A. P. (2012b) Effects of occupational stress, job characteristics, coping, and attributional style on the mental health and job satisfaction of university employees. *Anxiety, Stress & Coping* 25(1): 63–78.

Mark, G. M. & Smith, A. P. (2008) Stress models: A review and suggested new direction. In Houdmont, J. and Leka, S. (eds.) *Occupational Health Psychology* Nottingham University Press, 111–144.

Quine, L. (1999) Workplace bullying in NHS community trust: Staff questionnaire survey. *British Medical Journal* 318(7178): 228–232.

Smith, A., Wadsworth, E., Chaplin, K., Allen, P. & Mark, G. (2010) WHAT IS A GOOD JOB? The relationship between work/working and improved health and well-being. *The Open Occupational Health & Safety Journal*, **2**, 9–15.

Williams, G. M. & Smith, A. P. (2012) Developing short, practical measures of well-being. In M. Anderson (Ed.), *Contemporary Ergonomics and Human Factors 2012* (pp. 203–210). London: Taylor & Francis.

THE SOURCES OF RISK TO HEALTH ASSOCIATED WITH NEW TECHNOLOGIES IN THE OFFICE ENVIRONMENT

Thomas A. Winski[1,2], Joanne O. Crawford[2], Richard A. Graveling[2], Terry Lansdown[1] & Guy Walker[1]

[1]*Heriot Watt University*
[2]*Institute of Occupational Medicine*

Using display screen equipment (DSE) arguably has the potential to cause musculoskeletal disorders (MSD) and discomfort. However, the sources of risk to health associated with 'new' DSE technologies are not clearly understood. This paper presents the background research and describes the search strategy adopted for a review of the literature studying this association. After completion of the systematic search and review fourteen papers were identified. Three categories of new technology were identified: hand-held devices, tablets, touchscreen input. MSD's were found as the dominant risk to health relating to new technologies; however, the sources of the risk differ from traditional DSE.

Introduction

This paper is a work-in-progress which presents the background research and describes the search strategy adopted for a systematic review of the literature on the inherent risks to health for people who are working with 'new' display screen equipment (DSE). There is a general understanding that computer work and work with traditional DSE has a relationship with symptoms of musculoskeletal disorders (MSD) and discomfort. However, clear and robust evidence to support this 'knowledge' is quite sparse.

The EU Occupational Safety & Health Directive for DSE (90/270/EEC) was first enforced in 1992 to protect the workforce from risks associated with DSE work. Those risks have been researched throughout the lifespan of the, afore mentioned, Directive. Melrose et al. (2007) aimed to develop a better understanding of the data on DSE work related ill health. The study gathered questionnaire responses from 1,327 DSE users throughout the whole of the UK. They found that for their whole sample population 73% reported one or more MSD symptoms; with a high prevalence in neck, shoulder and back symptoms. Additionally, they found a positive relationship between prevalence of symptoms reported with both more hours of PC use per week and those participants who did not take hourly breaks. The aims of their data analysis were to estimate the prevalence and incidence of the symptoms they identified. As well as examining how symptoms related to current computer work, and resultant time off work. The results identified what appeard to

299

be contradictory information, that while the incidence of symptoms was high the majority of those who reported symptoms did not take time off work.

Melrose et al. (2007) summarised the findings by stating that over half the sample population reported symptoms of head and eye discomfort. Symptoms were reported more frequently by women than men, however, there was little difference in symptoms reported between companies of different sizes and industries. In addition, recorded psychological distress (namely anxiety, depression, and stress) positively correlated with prevalence of symptoms. Therefore, despite the efforts of legislation on DSE risk prevention the level of MSD's amongst computer workers was still cause for concern.

High quality research looking specifically at technology types and their associated risk appears to be scarce. A research report by Heasman, Brooks & Stewart (2000) looked specifically at the health and safety hazards of portable DSE use. This study specifically examined the effects of the technology used. It was a large-scale study, which adopted a mixed-methods approach to investigate those effects; including a review of the literature, a survey, direct observations and statistical analysis. The main aim of the study was to determine just how regularly portable computers were used within organisations, at the time and identify any associated health and safety risks using this technology in comparison with a desktop computer. They also aimed to determine the features of portable DSE which are desirable, as well as undesirable, from the users' perspective. To determine how much portable DSE was used in the workplace they completed a market review and telephone survey. At the time of the study approximately one in five computers purchased was portable, and they were purchased for the business market more than the small office, or home market.

As mentioned previously this paper intends to systematically review the literature on the inherent risks to health for people who are working with new DSE technologies. The intention of the review is to answer the following research question: what are the sources of risk to health associated with new technologies in the office environment?

Method: Database search

There were two large-scale online databases employed: Google Scholar® and Discovery®. A period of scoping research was performed to create the terms and structure of the search string adopted in this review; Google Scholar® was used to do this. Based on the papers discussed during the introduction of this review the most common risk from DSE interaction observed was MSDs, this led to the decision to include MSD/musculoskeletal disorder in the search string. When a test search was performed with these terms and a selection of variations of DSE and new technology the results identified over 100,000 papers. Therefore, the search had to be narrowed down. This was attempted by identifying specific terms for some new technologies, this phase led to the identification of the terms for new technologies used in the final search string. However, with these specific technology terms and

MSDs the number of papers found was still in the 100,000 range. At this stage this author consulted the Centre for Reviews and Dissemination's (CRD) guidance for undertaking reviews in health care to identify how to narrow the search field; in addition advice was sought from a work colleague with experience creating search strings. The conclusion from this was to specify a number of categories of terms. The categories identified in this scoping period were location, population, cause, effect and condition. The final search string used was as follows: (travel OR commute OR home) AND (worker OR employee OR occupation OR staff) AND ("tablet computer" OR smartphone OR touchscreen) AND (msd OR "musculoskeletal disorder") AND (risk OR "ergonomic risk"). From this search 156 papers were found, after duplications were removed. These papers were reviewed using the criteria described below.

Exclusion criteria

Papers which were clearly not relevant to the research questions were excluded; the assessment of relevance was completed on an individual paper by paper basis. Some of the more frequent reasons for exclusion are as follows; medical based research including cancer and multiple sclerosis, research on traditional DSE technology, political documents. Exclusions, also, included case reports, letters to editor, policy papers, and book reviews. Furthermore, papers which considered the relationship between MSDs and conventional DSE were not included as the focus of this review is new technologies. Additionally, papers which considered the reliability or validity of risk assessment methods for DSE risk were excluded.

Inclusion criteria

Only publications written in the English language were included. As the research in the related area is relatively minimal both peer-reviewed and non-peer-reviewed papers were included. Seven papers were included from this database search.

Manual searches

Two stages of manual searches followed the database search described above. These manual searches used the references from those seven papers. The first was a search of the most recent issues of the identified Journals due to the potential delay between the publication of a journal issue and inclusion in the online databases. The second was a 'cited by' search where Google Scholar®was employed to identify any relevant papers which had cited the papers identified by the database search.

Results

From the adopted search strategy described above fourteen papers were identified for this review, these fourteen papers are included in the references at the end of this paper. The findings from this review are as of yet incomplete; however, several interesting outcomes have been identified. These findings presented at this time consider the identified new technologies and the identified areas of risk.

New technologies of interest

There were three categories of technology identified from the review papers; hand-held devices (smartphones), tablets, and touchscreen input. These three technology groups will be treated as the 'new technologies' of concern when attempting to address the research question.

Risks to health

All risks to health discussed in the identified literature fall under the category of MSD's; while the location of the symptoms on the body differed, they were all measured or reported to be in the upper-body. This included the head and neck, the upper extremities with a particular focus on the thumbs, and the general upper body.

Conclusion

As this paper is a work-in-progress any decisive conclusions on the findings would be premature. Therefore, this paper cannot confidently answer the research question; what are the sources of risk to health associated with new technologies in the office environment? However, it does shed some light on how the research question will be answered. The identification of the three 'new technology' categories has highlighted that touchscreen technology is a current focus in the literature. As touchscreen input is a relatively new way of interacting with technology this will be a focal point in this review; to consider whether or not this type of interaction exposes the users to new risks. At the present stage of the review it is becoming apparent that the primary risk to health associated with these new technologies are MSDs, which is not a new risk to health when compared with traditional DSE technology. However it appears that the ways in which a user is exposed to the risks is different, or new. For example static working postures and input interaction with new technology differs physically with those adopted when using traditional DSE technology. Identification of new risks, or new ways in which users are exposed to risk, will result in the need to examine the current controls for DSE risk; namely the DSE Directive (90/270/EEC).

Acknowledgments

This author wishes to thank Heriot Watt University (Dr Terry Lansdown and Dr Guy Walker) and the Institute of Occupational Medicine (Dr Joanne Crawford and Dr Richard Graveling) for their continued support during this project.

References

Centre for reviews and dissemination. 2009. *"CRD's guidance for undertaking reviews in health care."* York Publishing Services Ltd.

Heasman, T., A. Brooks, and T. Stewart. 2000. *Health and safety of portable display screen equipment.* The Health and Safety Executive, England: London.

Melrose, A. S., Graveling, R. A., Cowie, H., Ritchie, P. Hutchison, P. and Mulholland. R. M. 2007. *"Better Display Screen Equipment (DSE) workrelated ill health data."* The Health and Safety Executive, England: Norwich.

References: review papers

Abdelhamed, Abeer Ahmed. 2014. "Upper extremities symptoms among mobile hand-held device users and their relationship to device use." Paper presented at the 5th Health and Environment Conference in the Middle East, Atlantis, March 3–5.

Albin, Thomas, J., and Hugh, E. McLoone. 2007. "The effect of tablet tilt angle on users' preferences, postures, and performance." Paper presented at the 10th Applied Ergonomics Conference, Dallas, March 13–15.

Berolo, Sophia, Richard, P. Wells, and B. Amick. 2011. "Musculoskeletal symptoms among mobile hand-held device users and their relationship to device use: A preliminary study in a Canadian university population." *Applied Ergonomics* **42**: 371–8.

Gustafsson, Ewa, Peter, W. Johnson, and Mats, Hagberg. 2010. "Thumb postures and physical loads during mobile phone use – A comparison of young adults with and without musculoskeletal symptoms." *Journal of Electromyography and Kinesiology* **20** (1) (2): 127–35.

Hammer, Matthew Justin. 2007. "Ergonomic Comparison of Keyboard and Touch Screen Data Entry while Standing and Sitting." MSc Diss., University of Cincinnati.

Kim, Jeong Ho, Lovenoor Aulck, Michael C. Bartha, Christy A. Harper, and Peter W. Johnson. 2012. "Are there differences in force exposures and typing productivity between touchscreen and conventional keyboard?" Paper presented at Proceedings of the Human Factors and Ergonomics Society Annual Meeting.

Kim, Gyu Yong, Chang Sik Ahn, Hye Won Jeon, and Chang Ryeol Lee. 2012. "Effects of the use of smartphones on pain and muscle fatigue in the upper extremity." *Journal of Physical Therapy Science* **24** (12): 1255–8.

Korpinen, Leena, Rauno Päkkönen, and Fabriziomaria Gobba. 2013. "Self-reported neck symptoms and use of personal computers, laptops and cell phones among Finns aged 18–65." *Ergonomics* **56** (7): 1134–46.

Pereira, Anna, Tevis Miller, Yi-Min Huang, Dan Odell, and David Rempel. 2013. "Holding a tablet computer with one hand: Effect of tablet design features on biomechanics and subjective usability among users with small hands." *Ergonomics* **56** (9) (09/01; 2014/07): 1363–75.

Shin, Gwanseob, and Xinhui Zhu. 2011. "User discomfort, work posture and muscle activity while using a touchscreen in a desktop PC setting." *Ergonomics* **54** (8) (08): 733.

Woods, Victoria, Sarah Hastings, Peter Buckle, and Roger Haslam. 2002. "*Ergonomics of using a mouse or other non-keyboard input device*" Health and Safety Executive. England: Norwich.

Xiong, Jinghong, and Satoshi Muraki. 2014. "An ergonomics study of thumb movements on smartphone touch screen." *Ergonomics* **57** (6) (06/03; 2014/07): 943–55.

Young, Justin G., Matthieu B. Trudeau, Dan Odell, Kim Marinelli, and Jack T. Dennerlein. 2013. "Wrist and shoulder posture and muscle activity during touch-screen tablet use: Effects of usage configuration, tablet type, and interacting hand." *Work* **45** (1) (05): 59.

Young, Justin G., Matthieu Trudeau, Dan Odell, Kim Marinelli, and Jack T. Dennerlein. 2012. "Touch-screen tablet user configurations and case-supported tilt affect head and neck flexion angles." *Work: A Journal of Prevention, Assessment and Rehabilitation* **41** (1) (01/01): 81–91.

MANUFACTURING

YOUR NEW COLLEAGUE IS A ROBOT. IS THAT OK?

Rebecca L. Charles, G. Charalambous & Sarah R. Fletcher

Industrial Psychology and Human Factors Group, Cranfield University

Human robot collaboration is a concept under development that will be applied within manufacturing environments in the near future to increase efficiency and quality. While there have been significant advances in technology to enable this progress there is still little known about the wider human factors issues of employing such systems in high value manufacturing environments. This paper sets out our current understanding of key organisational and individual factors that need to be explored.

Introduction: Human robot collaboration – where are we now?

Combined developments in technology and standards (such as the ISO 10218-2; ISO, 2011) have increased the potential for closer interaction between humans and robots in industry. Improved sensor and high speed computer processing capabilities will allow real time monitoring of the environment around automated equipment to remove the need for traditional fixed guarding and move towards true human-robot collaboration (HRC). These developments mean it is almost inevitable that HRC will become particularly attractive as a means of improving work flow in high value manufacturing (HVM) systems. In recent work at Cranfield University a HRC demonstrator system has been developed for the aerospace manufacturing sector with integrated 3D vision monitoring and control safety systems as a safer alternative to traditional physical guarding (Walton, Webb and Poad, 2011). However, although much previous research has shown us that workforce reaction and acceptance of new technology is a significant determinant of success (Venkatesh and Davis, 2000) little is currently known about the human factors challenges of introducing such a radical manufacturing change.

Our concept of HRC differs from other human-robot interaction (HRI) in that it represents working cooperatively at the same time and in the same work space through perception, recognition and intention inference, whereas HRI can be considered as anything requiring communication in close proximity or remotely. HRC needs to reflect human-human collaboration in which people use many different ways to communicate successfully and share adequate feedback which has a major impact on task performance (Horiguchi, Sawaragi, and Akashi, 2000). This will allow robots to be used in cases of process variability where current fixed, non-collaborative systems cannot. Literature from comparable contexts, such as the implementation of advanced manufacturing technologies (AMTs), has shown that failure to address human factors has proved detrimental to successful adoption (Lewis and Boyer, 2002; Castrillón and Cantorna, 2005; McDermott and Stock, 1999; Zammuto and O'Connor, 1992). In absence of an existing guiding structure Charalambous, Fletcher and Webb (2013) developed a theoretical framework of human factors that are likely to affect implementation of industrial HRC at *organisational* and *individual* levels.

Organisational level factors

A broad review of the literature initially revealed that the key organisational factors relevant to industrial automation and technology implementation are: (i) *communication of the change to employees*, (ii) *operator participation in implementation*, (iii) *training and development of workforce*, (iv) *existence of a process champion*, (v) *organisational flexibility through employee empowerment*, (vi) *senior management commitment and support*, (vii) *impact of union involvement*. The importance of these factors to successful adoption of new automation and technology was then supported by the findings of an exploratory case study in an aerospace HVM facility where a traditionally manual work process was being replaced by automation. In addition this study revealed factors not previously identified in the literature, such as: *awareness of the manual process complexity by the system integrator* and *capturing the variability of the manual process prior to introducing the automated system*. It also highlighted that organisational factors should not be considered as a selective 'tick-in-the-box' activity, but rather as a set of inter-related issues. For instance, capturing the variability of the manual process in advance will serve as a vehicle to provide sufficient information to the system integrator to understand the complexity of the process and provide a process capable automated system.

Individual level factors

Additionally, a set of key individual level factors were also identified from the literature: (i) *trust in automation*, (ii) *mental workload*, (iii) *situation awareness*, (iv) *levels of automation*, (v) *automation reliability*, (vi) *attitudes towards automation*, (vii) *perceived attentional control*. All of these issues directly concern the workforce/individuals who will interact with HRC systems and new automation and not only involve human aspects that are more often addressed by engineers, such as physical ergonomics and human computer interaction, but also comprise deeper psychological constructs. This is an important problem as HRC raises interesting questions and challenges regarding user psychology. Early work in the development of the Cranfield HRC demonstrator cell confirmed the potential relevance of key psychological constructs, such as mental workload, situation awareness, and users' subjective comfort (Walton, Fletcher and Webb, 2012). Building on this early work we now focus on operator 'awareness' and 'trust and acceptance' as our work thus far indicates these are fundamental.

Awareness

When a human interacts with a system they accumulate knowledge and build mental models from which they infer the state of the system and their expectations; awareness is therefore vitally important in order for operators to develop and apply working strategies. In the example of a human-robot system, the human would have a level of knowledge about the processes, rules and procedures relating to their job (however they may not be routinely the same) as well as a level of knowledge about the robot and associated safety features. When a human carries out a task and repeats it several times, they develop a rich knowledge of the task, and of peripheral aspects. In HRC systems it should be possible to identify the system state when an action has occurred and when it was given and maintain awareness of state and state changes. However, this may become problematic if the robot does not follow this same learning process as the knowledge becomes unbalanced. It will, therefore, be important for robots to learn how to adapt to the actions and changing states of the human operator in a similar way to how a human learns to respond. Operators will need to be able to teach robots how to perform tasks by demonstration, without engaging in complex robot programming, and maintain fluid interaction thereafter using shared plans and expectations of future states. The technology required for this level of sophistication in a HRC system is available but it is also important to remember that failures become more complex; if failure occurs it is not necessarily the human or robot at fault, but could be a software failure. When the user has built an adequate level of awareness, they can start to use their mental models and existing knowledge to start to predict situations. The type, complexity, familiarity and workload of a task will also influence system efficiency. In the case of collaborative environments, existing models do not describe adequately the state of situation awareness in relation to automation, let alone multi-robot or multi-operator systems. The aforementioned early Cranfield work considered awareness merely in terms of a single human operator's cognisance of robot activity (Walton et al., 2012) and some distributed models of situation awareness start to touch on this issue (e.g. Salmon, Stanton, Walker and Jenkins, 2009). However, there appears to be no current model of situation awareness in true collaborative environments that mutual awareness of human or robot state changes. Thus, further work to explore awareness between humans and robots in real time dyadic and multi-agent HRC systems is needed.

Trust and acceptance

Trust and acceptance are often considered as related constructs (Lee and Moray, 1994), or acceptance a direct result of high trust (Venkatesh and Davis, 2000). Trust between humans and robots has been considered in many domains but very little has been explored in the specific context of HVM applications. Freedy DeVisser, Weltman, and Coeyman (2007) developed a performance model that captured key components of the human-robot system considering team performance and processes at three levels (the individual human; the team human; and the collective human/robot team) and found the most important construct to be trust. Muir and

Moray (1996) propose six components to a good level of operator automation trust (predictability, dependability, faith, competence, responsibility, and reliability) and found that trust is sustained if initial faith is in place but the other components can then be developed through experience. However, low reliability has the potential to undermine developed trust in a system and outweigh any potential benefits that the system could provide by increasing mental workload due to 'workarounds' that can quickly become 'the norm' (Parasuraman and Riley, 1997). By considering an individual's trust in a system and having adequate tools to do so, designers and implementers of HRC can start to predict the level of acceptance of a system at an individual level, and therefore consider the impact at an organisational level. More recent work to develop a psychometric tool to measure human trust in industrial robots has identified a three-factor structure (Charalambous, 2014). Further work will continue to explore how these antecedent factors affect the initiation of trust, and their relevance in the development of operator acceptance and situation awareness in HRC environments.

Conclusions

HRC seems an inevitable step for optimising HVM in the near future but asking people to work so closely with robots raises new questions in the psychology and human factors domain at both individual and organisational levels. This paper highlights some initial investigations and indications for issues that must be addressed to bring this concept closer to reality and ensure the technology can be implemented safely and successfully. In particular, we have emphasised the critical importance of fully investigating the psychological and social prerequisites for effective HRC adoption and use within close-knit teams.

Acknowledgements

This work is funded by the UK Engineering and Physical Sciences Research Council Centre for Innovative Manufacturing in Intelligent Automation.

References

Castrillón, I. D. and Cantorna, A. I. S. (2005). The effect of the implementation of advanced manufacturing technologies on training in the manufacturing sector. *Journal of European Industrial Training* 29(4), 268–280.

Charalambous, G., Fletcher, S. and Webb, P. (2013). Human-automation collaboration in manufacturing: Identifying key implementation factors. In M. Anderson (Ed.), *Contemporary Ergonomics and Human Factors 2013: Proceedings of the International Conference on Ergonomics & Human Factors 2013*, Cambridge, UK, 15–18 April 2013.

Charalambous, G. (2014). *Development of a human factors tool for the successful implementation of industrial human-robot collaboration*. Unpublished doctoral thesis, Cranfield University, UK.

Freedy, E., DeVisser, E., Weltman, G. and Coeyman, N. (2007). Measurement of trust in human-robot collaboration. *International Symposium On Collaborative Technologies and Systems*, CTS 2007, 106–114.

Horiguchi, Y., Sawaragi, T. and Akashi, G. (2000). Naturalistic human-robot collaboration based upon mixed-initiative interactions in teleoperating environments. *International Conference on Systems, Man, and Cybernetics*, 2000 IEEE, 876–881.

ISO: International Organization for Standardization. (2011). ISO 10218-2:2011 *Robots and robotic devices – safety requirements for industrial robots – part 2: Robot systems and integration*. ISO.

Lee, J. D., & Moray, N. (1994). Trust, self-confidence, and operators' adaptation to automation. *International Journal of Human-Computer Studies* 40(1), 153–184.

Lewis, M. W. and Boyer, K. K. (2002). Factors impacting AMT implementation: An integrative and controlled study. *Journal of Engineering and Technology Management* 19(2), 111–130.

McDermott, C. M. and Stock, G. N. (1999). Organizational culture and advanced manufacturing technology implementation. *Journal of Operations Management* 17(5), 521–533.

Muir, B. M. and Moray, N. (1996). Trust in automation. Part II. Experimental studies of trust and human intervention in a process control simulation. *Ergonomics* 39(3), 429–460.

Parasuraman, R. and Riley, V. (1997). Humans and automation: Use, misuse, disuse, abuse. *Human Factors* 39(2), 230–253.

Salmon, P. M., Stanton, N. A., Walker, G. H. and Jenkins, D. P. (2009). *Distributed situation awareness: Theory, measurement and application to teamwork*. Surrey: Ashgate.

Venkatesh, V. and Davis, F. D. (2000). A theoretical extension of the technology acceptance model: Four longitudinal field studies. *Management Science* 46(2), 186–204.

Walton, M., Webb, P. and Poad, M. (2011). Applying a concept for robot-human cooperation to aerospace equipping processes. Technical Paper No. 2011-01-2655. SAE Technical Paper. doi:10.4271/2011-01-2655.

Walton, M., Fletcher, S. and Webb, P. (2012). Developing true cooperation between human operators and robots in assembly work systems: A proxemic study of behaviour, workload and comfort. In: *Advances in Manufacturing Research XXVI Vol. 1: Proceedings of the 10th International Conference on Manufacturing Research* 2012.

Zammuto, R. F. and O'Connor, E. J. (1992). Gaining advanced manufacturing technologies' benefits: The roles of organization design and culture. *Academy of Management Review* 17(4), 701–728.

SEARCH STRATEGIES IN HUMAN VISUAL INSPECTION

Rebecca L. Charles[1], Mitul Tailor[2] & Sarah R. Fletcher[1]

[1]*Industrial Psychology and Human Factors Group, Cranfield University*
[2]*Wolfson School of Mechanical & Manufacturing Engineering, Loughborough University*

Visual inspection is a critical and highly skilled area of work in many high value manufacturing systems. It is currently undertaken by highly trained human operators who are provided with prescribed standards and classifications for assessing defects. However, previous work indicates that operators' inspection methods often deviate from prescribed standards and entail variability in detection rates and strategies. Additionally, an understanding of human skill and acuity in visual inspection will be advantageous when considering the feasibility and appropriation of potential replacement technologies. This paper describes a small initial study investigating human visual search patterns and their effect on defect detection rates.

Introduction

There are various definitions of visual inspection, but from a human perspective it can be simply described as the process of using the eye (aided or unaided) to make judgements about the condition of an item. From a scientific perspective, it is a change in the light's properties after contact with the surface of the item. Whether this change is identified by the human eye, a vision enhancing accessory, or an automatic system does not change the key objective of visual inspection (Drury and Watson, 2002). In high value manufacturing, human visual inspection is critical, particularly for products requiring safety-critical standards compliance and quality assurance such as in the aerospace sector. For this reason organisations produce prescribed defect classifications and provide work standards and related training. However, previous field study work conducted by Cranfield University's Industrial Psychology and Human Factors Group in aerospace manufacturing facilities has found that after training and over time operators often adapt the methods they originally learned and develop idiosyncratic techniques which, particularly in situations of ambiguity, involve a higher degree of subjective interpretation. Without a more in-depth understanding of these individual approaches to defect detection and classification, and their effectiveness, we are currently unable to advocate best practice in recommendations for training/process improvement. Moreover, when considering whether to automate visual inspection it is vital to gain a thorough understanding of the tasks currently undertaken by a human operator to ascertain feasibility of potential replacement technologies.

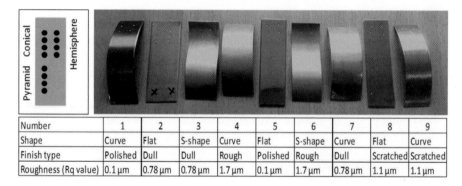

Number		1	2	3	4	5	6	7	8	9
Shape		Curve	Flat	S-shape	Curve	Flat	S-shape	Curve	Flat	Curve
Finish type		Polished	Dull	Dull	Rough	Polished	Rough	Dull	Scratched	Scratched
Roughness (Rq value)		0.1 μm	0.78 μm	0.78 μm	1.7 μm	0.1 μm	1.7 μm	0.78 μm	1.1 μm	1.1 μm

Figure 1.   Test pieces and defect types.

Spitz and Drury (1978) developed a widely recognised two component model of visual inspection in which the 'search' and 'decide' elements are totally independent of one another. Other literature supports that the search aspect and strategies used cannot be carried over to a different task, whereas learned decision making strategies can (Drury and Wang, 1986). The aim of this paper is to describe a small human factors study that focuses on the search aspect of Spitz and Drury's model to identify if patterns emerge through practice, or if optimal strategies exist, and what impacts these. Work conducted at Loughborough University as part of the EPSRC Centre for Innovative Manufacturing in Intelligent Automation has already effectively measured surface defects using a 6 axis industrial robot to provide a robust, repeatable, mathematical solution for automatic defect detection and characterisation (Tailor et al., 2013). This human factors study will provide early analysis to a) initiate wider investigations for the development of visual inspection best practice and b) to integrate human aspects into the intelligent automation solution that is being developed.

Method

Participants

Five participants with no previous industrial inspection experience took part in this study, recruited by opportunity sampling.

Equipment

Nine steel test pieces were used, each with 12 indentation defects (Figure 1).

Eye tracking data were collected using a 'commercial off the shelf' system (SensoMotoric's BeGaze© system). This eye tracking equipment consists of glasses with in-built cameras that track human eye pupil activity while simultaneously recording field of vision.

Procedure

Participants were asked to identify 12 different indentation defects (4 different sizes of conical and hemi-spherical indentations following the Rockwell specification set out in ISO 6508-1, and four different sizes of pyramidal indentation following the Vickers hardness test specifications set out in ISO 6507-1 (Baxevani and Giannakopoulos, 2009)) on metal test pieces of four different surface finishes (Tailor et al., 2013) while wearing eye tracking glasses. The test pieces were assigned as shown in Figure 1 to each participant. They were told to look for indentation defects, and asked to verbally portray how many defects were on each test piece; they were not told how many were on each test piece.

After the tests were completed and glasses removed, each participant was asked a short series of questions using the following schedule:

1. Did you find that task easy or difficult? Why?
2. Did it get easier as the trial progressed? Why?
3. Did you have a strategy? What was it?
4. If the trial was more complex do you think you may have approached it differently? How?

Analysis

Eye tracking data were analysed using the Begaze© software; information on timings, strategies and methods were logged on the software for each participant.

Results

The polished finish led to increased detection of defects, although in four cases participants identified non-defects. The eye tracking data revealed that the participants tilted the test pieces to catch the light in most cases. This technique seemed more prevalent for the curved test pieces where participants would move the piece so that the light followed the curved profile.

There were no clear patterns observed for the length of time taken to identify defects in relation to shape or finish of test piece. There was a general scanning pattern observed that involved detecting the 'obvious' defects first (noted by participants as being 6 on each test piece) then scanning the remainder of the area in a sweeping left to right motion, finishing with counting the defects. All of the participants used touch in addition to visual search to identify defects: this was observed most on the scratched finish. The defects most commonly identified were the conical and hemispherical defects (see figure 1). Most participants felt that the dull finish made the defects harder to see, and that the shape of the test piece made little difference. They also felt that the task got easier as it progressed as the location of the defects became predictable and that the small size of the test pieces meant a set strategy was not necessarily required. If a larger test piece was used, all of the participants stated they would devise a strategy. Suggestions ranged from dividing the piece into sections, to looking for each type of defect (scratch, indent etc.) separately.

Conclusions

This study has demonstrated that strategies may develop through practice, and that systematic approaches were adopted automatically by all participants. However, there were instances of non- and mis-identification which varied with differing surface finishes, and the smallest, pyramidal defects were the least detected. This trial was limited due to the size of the test pieces and the uniformity of the defects, but it is an early study to start understanding the potential benefits of scanning strategies and their effect on detection rates.

Further work will be carried out to include larger test pieces to start to identify the development of search strategies dependent on size of piece, and type of defects. This could potentially lead into the development of training plans, and suggestions for automation. By considering human behaviours and approaches to visual inspection tasks, automation can be optimised, and is more likely to be accepted and trusted by human operators. This is important when considering collaborative environments and should be incorporated into any work package considering automating a human led task.

Acknowledgements

This work was funded by the UK Engineering and Physical Sciences Research Council Centre for Innovative Manufacturing in Intelligent Automation.

References

Baxevani, E. and Giannakopoulos, A. (2009). The modified Rockwell test: a new probe for mechanical properties of metals. *Experimental Mechanics* 49(3), 371–382.

Drury, C. G. and Wang, M. (1986). Are Research Results in Inspection Task Specific? *Proceedings of the Human Factors and Ergonomics Society Annual Meeting, 30,* 476.

Drury, C. G. and Watson, J. (2002). Good practices in visual inspection. *Human factors in aviation maintenance-phase nine, progress report, FAA/Human Factors in Aviation Maintenance.* Retrieved 28/10/14: http://hfskyway.faa.gov.

Spitz, G. and Drury, C. G. (1978). Inspection of sheet materials–test of model predictions. *Human Factors* 20(5), 521–528.

Tailor, M., Phairatt, P., Petzing, J., Jackson, M. and Parkin, R. (2013). Real-Time Surface Defect Detection and Traceable Measurement of Defect in 3D. *11th International Symposium on Measurement and Quality Control, 2013.*

THE CASE FOR THE DEVELOPMENT OF NOVEL HUMAN SKILLS CAPTURE METHODOLOGIES

J.W. Everitt & S.R. Fletcher

Industrial Psychology and Human Factors Group, Cranfield University, UK

As the capabilities of industrial automation are growing, so is the ability to supplement or replace the more tacit, cognitive skills of manual operators. Whilst models have been published within the human factors literature regarding automation implementation, they neglect to discuss the initial capture of the task and automation experts currently lack a formal tool to assess feasibility. The definition of what is meant by "human skill" is discussed and three crucial theoretical underpinnings are proposed for a novel, automation-specific skill capture methodology: emphasis upon procedural rules, emphasis upon action-facilitating factors and taxonomy of skills.

Introduction

The capability of automation to replace or supplement human activity in manufacturing is increasing, however there exists no method to decide what tasks are suitable for intelligent automation and to what extent should they be automated. Often the default strategy taken by system designers is to simply automate all the functions that it is economically viable and technically possible to do so. Typically, the human is still required for task performance involving 'fiddly' manual dexterity or which requires cognitive reasoning and decision-making. However, this often leaves the human operator to 'pick up the slack' and perform the functions that the system designer could not automate; thus the human operators' role is defined not by the optimisation of the entire human-machine system but rather by the creativeness (or lack thereof) of the system designer (Parasuraman, Sheridan and Wickens, 2000). This disregard for the abilities, limitations and needs of the human operator can lead to a range of human factors issues such as loss of skills, operator complacency and reduced situational awareness (Endsley and Kiris, 1995).

A number of papers have been published discussing the process of automation implementation and function allocation with an eye on both operator well-being and system reliability (Endsley and Kaber, 1999; Kaber and Endsley, 2004; Lin, Yen and Yang, 2010; Parasuraman et al, 2000). Notably the initial capture of the task is not discussed by these studies, perhaps under the assumption that traditional methods such as Hierarchical Task Analysis (HTA) are suitable for function allocation (Marsden and Kirby, 2004). Whilst in the past these methods may have been sufficient for the automation of simpler, less cognitively demanding tasks, but more tacit and complex cognitive factors must be accounted for during the implementation of intelligent automation. Effective automation strategy depends upon accurate and sufficiently detailed decomposition of tasks. Thus, one of the challenges of effective intelligent automation in manufacturing is being able to accurately capture both the implicit and tacit skills deployed by human operators during task completion, both physical and cognitive.

From a mechatronic perspective, experts in that field are better suited to determine the feasibility of automation based upon technical possibility and/or cost effectiveness. However, human factors is ideally placed to inform and advise these decisions if it can offer a detailed, accurate account of the physical, perceptual and cognitive processes that facilitate skilled behaviour. A detailed account of the human faculties should assist system designers to identify analogies between human processes and potential automated solutions which mimic human performance, but also devise ways in which automation may circumvent the required faculty.

In brief, human factors and automation engineers currently lack a formal methodology to fully evaluate the suitability of high skill manual tasks for automation. This paper aims to outline why a novel skill capture methodology is needed for intelligent automation implementation and the required theoretical underpinnings. In order to accomplish this we first need to understand what is meant by 'skill'.

Defining skill

Human skill can apply to a broad spectrum of behaviours. For example, both an expert cellist and a grand master chess player could be considered highly skilled in their respective domains. However, playing the cello requires precise motor actions of the fingers and upper limbs, while all the 'skill' which differentiates an expert from a novice chess player is purely cognitive; an expert is no better than a novice at physically moving a piece. Skill is therefore best considered an umbrella term for a range of abilities, and thus a comprehensive understanding of human skill must recognise and effectively discriminate between different types of skilled behaviour as well as consider their interactions.

To understand what is meant by 'skill' is to define skilled behaviour from unskilled behaviour. Proctor and Dutta (1995) defined human skill as 'goal directed', well-organised behaviour that is acquired through practice and performed with economy of effort" (p. 18). This definition posits four characteristics of human performance which define it as skilful; the first characteristic is 'goal directed', it could be argued is not exclusive to skilled performance as unskilled performance can also be goal directed and as such this exists as a pre-condition rather than a defining feature of skilful behaviour. This then places an emphasis on the economical, well organised, and acquired by practice characteristics that define skilful behaviour from unskilful.

Another perspective is offered by Rosenbaum, Carlson and Gilmore (2001), who stated that skill is "an ability that allows a goal to be achieved within some domain with increasing likelihood as a result of practice" (p. 454). As with Proctor and Dutta (1995), goal direction is noted as well as acquisition through practice. However, instead of the emphasis being placed upon the efficiency of actions instead skilled performance is defined by the likelihood of success. This is not a trivial distinction, the latter definition allows inefficient but successful practices to be labelled as skilful whilst the former argues that it is increasing efficiency through practice which defines the acquisition of skill and that the likelihood of success is less relevant.

It is likely that the validity to both assertions is, to varying extents, dependent upon the task. For example, if completion of task by a novice is likely regardless of time taken or errors made then the skilled component of the behaviour is better defined by Proctor and Duttas' (1995) definition based on efficiency. However, if the task is unlikely to be completed by a novice (e.g. trying to defeat an expert chess player) then Rosenbuam et al.'s (2001) definition based upon likelihood of success is more appropriate. Most real world tasks are unlikely to be arbitrarily one or the other; instead most tasks are composed of both characteristics in varying proportions.

Both definitions agree that skill is acquired through practice. Thus, it can be inferred that the improvement of task performance related to practice is the key defining feature of skilful behaviour and that improvement is defined by the context and features of the task, rather than by arbitrary declaration. Although it could be argued that this simpler definition is too reductionist, as discussed, the term skill is an umbrella term for a broad range of behaviour rather than a precise definition due to the vast differences in what can be considered 'skilled'.

Human skills capture

Within a manufacturing context the view of human skills being defined by the practice placed upon them holds extra significance when one considers the nature of manufacturing tasks: being predictable, occurring in stable work environments, and being oft practiced and repeated. Thus it could be speculated that within manufacturing operators rarely have to strategise independently or create novel solutions. Rather, operators' actions are governed by pre-determined procedures, i.e. the tacit skills in manufacturing are more like the cellist's than the chess player's. Thus the key to capturing tacit human skills in manufacturing is to capture the tacit 'unofficial' procedures that govern skilled behaviour. In turn, effective skills capture methodologies should assume a procedural nature of the task and emphasise the capture these tacit procedural rules used by operators in task performance.

It could be argued that current methods of task analysis such as hierarchical task analysis (HTA) (Annett, 2003; Annett and Duncan, 1967) would be sufficient to capture the tacit procedural rules deployed by operators during task performance. HTA's emphasis is upon the identification of task goals and sub-goals to create a nested hierarchy. This nested hierarchy is then furnished with 'operations' and 'plans' to record the actions being undertaken and how these operations are ordered respectively (Stanton, 2006). Whilst HTA has been demonstrated to assist automation implementation (Tan, Duan, Zhang, and Arai, 2008) it does not account for more cognitive elements (Phipps, Meakin and Beatty, 2011) and in particular *action-facilitating factors* like visual judgement and haptic feedback. Whilst these factors may be superfluous for simple tasks, when considering high level, tacit skills these factors must be accounted for in order to achieve a comprehensive account of the task and how optimal performance is achieved.

Finally, Bullock, Ma and Dollar (2013) demonstrated the effectiveness of standardised taxonomy of fine motor movements for the transfer of simple tasks, such as picking up a coin. It is posited that a novel, more comprehensive taxonomy of skills that encompasses all physical, perceptual, and cognitive faculties could be used to identify analogies between human operations and potential automated solutions. The identification and classification of the human faculties that facilitate human performance would allow automation and human factors experts to analyse each individual sub-task for feasibility of automation and assess the impact of automation upon the entire task.

Conclusions

This paper has highlighted three major requirements for an effective human skills capture methodology to assess the feasibility of automation: an emphasis upon capturing the tacit procedures, an emphasis upon capturing the action-facilitating factors, and taxonomy of skill types. The specificity of the requirements are such that novel analytical tools are necessary to keep pace with the growing capabilities of intelligent automation.

At the time writing a novel skills capture methodology is currently under development which seeks to encompass the three major requirements; Discrete Task Analysis (DTA). Instead of replacing traditional methods such as HTA, DTA is a Task Decomposition based theoretical framework intended to supplement HTA by capturing physical manipulations and, perceptual and cognitive action-facilitating factors within the context of specifically devised skills taxonomy. DTA's development has been grounded in industrial case studies in tasks such as aircraft assembly and steel polishing.

References

Annett, J. 2003, Hierarchical task analysis. In Hollnagel, E. (Ed.), *Handbook of cognitive task design,* (CRC Press), 17–35.

Annett, J. and Duncan, K. D. 1967, Task analysis and training design, *Occupational Psychology* 41: 211–221.

Bullock, I. M., Ma, R. R. and Dollar, A. M. 2013, A hand-centric classification of human and robot dexterous manipulation, *IEEE Transactions on Haptics* 6(2): 129–144.

Endsley, M. R. and Kaber, D. B. 1999, Level of automation effects on performance, situation awareness and workload in a dynamic control task, *Ergonomics* 42(3): 462–492.

Endsley, M. R. and Kiris, E. O. 1995, The out-of-the-loop performance problem and level of control in automation, *Human Factors* 37(2): 381–394.

Kaber, D. B. and Endsley, M. R. 2004, The effects of level of automation and adaptive automation on human performance, situation awareness and workload in a dynamic control task, *Theoretical Issues in Ergonomics Science* 5(2): 113–153.

Lin, C. J., Yenn, T. and Yang, C. 2010, Optimizing human–system interface automation design based on a skill-rule-knowledge framework, *Nuclear Engineering and Design* 240(7): 1897–1905.

Parasuraman, R., Sheridan, T. B. and Wickens, C. D. 2000, A model for types and levels of human interaction with automation, *IEEE Transactions on Systems, Man, and Cybernetics Part A:Systems and Humans* 30(3): 286–297.

Phipps, D. L, Meakin, G. H. and Beatty, P. C. W. 2011, Extending hierarchical task analysis to identify cognitive demands and information design requirements, *Applied Ergonomics* 42: 741–748.

Proctor, R. W., and Dutta, A. 1995, *Skill Acquisition and Human Performance.* (Sage Publications, England).

Rosenbaum, D. A., Carlson, R. A., & Gilmore, R. O. 2001, Acquisition of intellectual and perceptual-motor skills. *Annual Review of Psychology* 52(1): 453–470.

Stanton, N. A. 2006, Hierarchical task analysis: Developments, applications, and extensions, *Applied Ergonomics* 37(1): 55–79.

Tan, J. T. C., Duan, F., Zhang, Y., and Arai, T. 2008, Task Decomposition of Cell Production Assembly Operation for Man-Machine Collaboration by HTA, *Proceedings of the IEEE International Conference on Automation and Logistics* 1066–1071.

A COMPARISON OF USER REQUIREMENTS AND EXPECTATIONS FOR CLOUD MANUFACTURING

David Golightly, Hanne Wagner & Sarah Sharples

Human Factors Research Group, University of Nottingham, Nottingham, UK
Horizon Centre for Doctoral Training, University of Nottingham, UK

Cloud manufacturing offers a new paradigm in providing on-demand manufacturing services. This should offer benefits to both large organisations, and small and micro manufacturing service users and providers. It is not clear, however, how user requirements might diverge, or show consistency, depending on the size of the organisation. The paper compares two requirements gathering activities – one with major international organisations, and the other with 'cottage' industries – to compare how their needs, expectations and capabilities will influence their engagement with the manufacturing cloud.

Introduction

The notion of the cloud applied to manufacturing is set to evolve with the emergence of 'cloud manufacturing'. Wu et al. (2013) define cloud manufacturing as "a customer-centric manufacturing model that exploits on-demand access to a shared collection of diversified and distributed manufacturing resources to form temporary, reconfigurable production lines" (Wu et al., 2013). Cloud Manufacturing moves beyond the idea of simply using cloud computing resources within a manufacturing context, proposing the use of remote manufacturing resources, the sharing of a single resource between multiple users, and the notion of delivering manufacturing as a service (MAAS) (Rauschecker et al., 2011). In this manner, manufacturing services, including design, simulation and other knowledge-based process (Tao et al., 2011), can be used on as 'pay-as-you-go' basis. This new paradigm hopes to provide heightened levels of quality and value for consumers of manufacturing services, and allows manufacturing service providers to engage in new, flexible arrangements leading to better utilisation of capabilities. It also allows consumers to use third-party manufacturing services without the capital expenditure costs that might otherwise prove prohibitive.

While the relevance of factors such as trust models and the importance of HCI and user experience are highlighted as areas for research within cloud manufacturing (Xu, 2012; Wu et al., 2013) they have yet been subjected to specific investigation. It is also highly likely that new issues will also come to light, such as the management of knowledge and skills in production and assembly, or the successful integration of new technology into the workplace. These are challenges where the ergonomics/human factors community can contribute. Key to addressing these issues is establishing knowledge of user requirements – what are the potential use cases that are relevant for various stakeholders in cloud manufacturing, what are their expectations and pre-existing knowledge, and what are the constraints that might affect the successful acceptance and adoption of cloud manufacturing?

While much of the emphasis in cloud manufacturing so far has been places on major manufacturing organisations, there is also interest in the utility of the concept for smaller businesses, including micro-sized cottage industries. These small and medium sized businesses make up 99% of all the businesses in the United Kingdom, and is where cloud manufacturing could have particular value, providing manufacturing capabilities that are encapsulated in a cloud service, making the production a 'service' that can be used similarly to an utility. Small businesses could hence use as much production capability as they need, without having to pay for more capability than they need or having to make large up-front investments, giving them the potential to experience a level of growth that would be impossible to attain through their regular production and business processes. This is already being seen in developments such as making high-cost manufacturing simulation services available to occasional users on a pay-as-you-go basis (Taylor et al., 2014),

Little distinction is yet made in the literature, however, between traditional OEMs (e.g. aerospace or automotive) or global ICT providers who might power the manufacturing cloud, and the small and medium sized enterprises (SMEs) or even micro-SMEs of one, two or three people. Despite the potential, it is as yet unknown how well the cloud manufacturing model really fits the cottage industry model. Given prior work in cloud computing that suggests SMEs have different challenges from large organisations in adopting that technology (Werfs et al., 2013), it seems likely that different levels of knowledge and different aims will influence the adoption of cloud manufacturing.

Two independent pieces of work have been conducted to understand organisational and user needs for cloud manufacturing. The first of these studies specifically targeted major organisations that would use or support the manufacturing cloud – aerospace, automotive, and global ICT. The second study investigated the needs of 'cottage industries' – companies of one, two or three people, providing niche products and services. Both studies used specific methods, relevant to the given aims of each research project, containing numerous results. While these results are being prepared as papers in their own right, we take the opportunity in this paper to compare and contrast the two sets of results, in order to identify commonalities and ideosyncracies in the requirements found.

Methods

As the two pieces of work were conducted with difference participant groups, with different constraints on availability as well as on investigator time, different approaches were appropriate for the two studies.

For the study with large organisations, a workshop was arranged, building on a series of interviews with representatives from various companies with an interest in cloud manufacturing. The workshop involved 11 senior industry representatives covering consumers of cloud manufacturing services and providers of the underpinning ICT technology that would deliver the interface between consumers of manufacturing services and providers of services. Within that workshop, a structured elicitation exercise was conducted asking participants to present their needs for cloud manufacturing, either at the short term (12 months) or long-term (5 year) timescale, against a set of pre-defined research areas (listed on the X-axis in Figure 1). Participants were then requested to give post-it note descriptions of their priorities against the pre-determined themes, giving in total 57 detailed comments (mean of 5.2 comments per participant). Discussion then explored the relevance of meaning of these themes to the participants. Post-its were captured and quantified to give a ranking of research priorities, leading to a set of documented requirements and use case priorities.

For the study on the potential for cloud manufacturing usage in cottage industries, one-on-one interviews, either face-to-face or via VoIP application, were conducted and recorded. These interviews were semi-structured, using a combination of questions on the businesses themselves, the owner's background, technology used as well as brief scenarios to test the usability of individual concepts and aspects of cloud manufacturing for micro SMEs. Five individuals participated in the research. Three were sole owners/members of a micro-SME and two where part of a team of three that run a small company together. Products manufactured and/or sold included artisan baking, educational software, crafting components and the development of a platform for sustainable asset tracking. After having been conducted, interviews were transcribed and coded to be analysed through thematic analysis, focusing on business processes and how/if cloud manufacturing could be introduced to them.

Therefore, while both of the studies used different methodologies, both resulted in thematic analyses of the major motivators and barriers to the adoption of cloud manufacturing, which supported an initial comparison.

Results

Large industries

Figure 1 shows a prioritisation of the major themes identified in the large organisation workshop. One immediate impression is that the major themes prioritised by industry had a strong human factors/HCI element. To eleborate on the feedback from participants, cyber physical systems issues included defining the means for human operators to be able to work within a highly automated manufacturing environment. This included effective HCI within smart working environments, and for operators to be able to manage robots safely and with a view to effective resource utilisation on the production line.

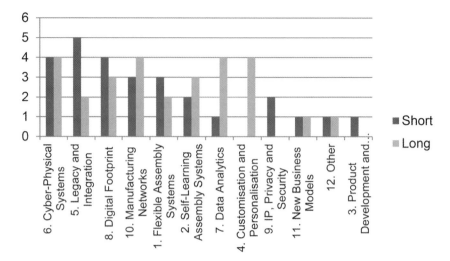

Figure 1.   Prioritisation of major themes from industrial workshop.

Similarily, legacy and integration was raised as a major concern. This concern covered the ability to integrate cloud processes into the manufacturing envrionment in a way that maximises existing technology and, as specifically highlighted by industrial stakeholders, human investment. Operators prior experience of integrating new technology meant that they placed a value on retaining knowledge and expertise, and being able to apply that.

A third aspect was the digital footprint of processes and products. While at first this may seem a technology-driven issue, the comments revealed (1) that issues around the data footprint related to visualisation and the effective use of product and process data within human decision making, and overall system performance (2) that at least some of the use of the data footprint was to capture product and customisation data; in essence, a form of user requirements gathering.

Manufacturing networks related to the organisation of the supply chain, and how suppliers and consumers could be grouped together. This again was in part related to how this process could reflect not just technical capability, but also human and organisational definitions of skills and competence. Also, concerns related to how the automated aspects of supply chain management could be embedded within the decision-making processes of confirming a supply chain that could be trusted to deliver.

Cottage industries

When conducting the thematic analysis, four main themes emerged. These are efficiency, control, dependency on technology and issues regarding growth- and general strategy for the business.

Efficiency was mostly expressed as how efficient a task can or needs to be performed for a business. This was usually put in the context of limited human as well as financial resources, in very small organisations, and time saving. Due to financial restrictions, the ability to invest in new technology was generally felt to be limited and any investment would have to provide excellent value for money. Furthermore, efficiency was also expressed in technology not having to "sing and dance", but instead be good at the task at hand. Efficiency was also thought of as something best achieved through the owner alone, rather than external resources. Yet, if outside capabilities were brought in, they would need to increase efficiency in an obvious value-added way, not achievable by the owner alone. At the same time, there was also the need felt for any new technology to fit into the existing workflow while maintaining customer focus. This customer focus was very high and important overall, and potential further automatization was for some perceived as a threat to what would make their business special.

Related to this is also the need for control over as many aspects of the business as possible, ranging from everything being as 'hands on' and done by the owner himself as possible to having control over potential risks such as logistics and financial matters, which again has to do with the overall limited resources these businesses have. There was also a general interest in being as self-sufficient as possible, as this would allow for more control and was also felt to be more appropriate for the limited resources available.

The role of technology and how important it is for the businesses became also emerged from interviews, yet while it is an important enabler for the micro SMEs, it also creates critical dependancies. As such the quality of service of any technology used is paramount, and even basic issues such as quality and speed of an internet connection can have grave impact on these businesses. Additionally, being mostly self-employed it is a lot more difficult for owners of micro SMEs to keep up with technological developments, perceived as a problem by many.

The last main theme found focused on business growth and strategy. Here it became obvious that most owners either do not have long-term strategies or are planning to sell their business eventually. Therefore big long-term investments were not always seen as something desirable. Also, for those in the 'craft' sector, their business was more perceived as a creative outlet that did not have exponential growth as its main aim, but instead enabled them to do what they love for a living. Finally, small businesses were also seen as way to integrate work or running a business into other aspects of one's life such as family responsibilities or ongoing studies.

Discussion

Having outlined the individual themes relating to requirements for cloud man-ufacturing both settings, we now consider the broader implications for design, development and deployment.

For the large organisations, a major theme is around the integration of new technology, including the integration of new processes into legacy working practices. It is important for them that value is recovered from sunk costs and existing technology, and emphasises that cloud manufacturing should not be viewed purely in terms of 'green field sites'. Similarly, for the micro SMEs, their major issue is efficiency and, coupled with that, control. They too are very concerned with how new technologies can enhance existing workflow. This is interesting as where in theory they might be more agile, in practice the scarcity of resources (time, knowledge etc.) everything has to count.

Cloud manufacturing is proposed as one important means to support customisation and mass customisation (Wu et al, 2013). While lower down the list than might be expected and classed as a long-term aim, this was highlighted by OEMs. Data footprint is a way of tracking a product and thus provides one heavily data centric way to understand customers and their use of products. On the other hand, one of the cited strengths of the micro SME was to be able to provide bespoke products, and owners prided themselves on being able to be close enough to their customers to know their needs. Therefore it is less clear how the value of customer-centricity and customisation that is there for large organisation translates to micro SMEs. It is possible, however, that these micro SMEs become suppliers to the larger organisations and this customer knowledge and agility is therefore harnesssed by the larger organisations.

One area where the manufacturing cloud clearly had advantages for the micro SMEs was in 'functional off-loading'. They are simply too small the operate many functions, and many of the owners cited that they actually preferred to concentrate on their own strengths. Strategically, these owners were in their business for reasons such as creativity, autonomy or being able to balance work with family. Therefore, the manufacturing cloud offers a way to offload non-core functions to other providers and suppliers. Also, micro SME owners are not looking for exponential growth. Many saw their current business as a temporary thing, likely to contrast with the globe spanning strategies of OEMs. Again, this suggests micro SMEs would prefer to use services on a pay as you go basis rather than sinking costs in a way the larger organisations are able to afford.

A major difference between the two groups is likely to come in technical know-how and confidence. IP and security is on the list of concerns for major organisations, but relatively far down. For the micro SMEs, trust comes much higher in their priorities. It is integral to trust and confidence in how their knowledge and data might be treated by others, and they are cautious about how this would be carried out by third parties. While major industries all have dedicated IT departments, micro SMEs depend on their own expertise and trying to stay abreast of the latest developments. Indeed, participants cited examples of areas where they really wanted to improve their functioning, and suspected there might be IT out there to achieve this, but did not have time or technical knowledge to find out more. Some cited that despite the urge to stay ahead, the lack of time and knowledge meant new capabilities could only be found opportunistically. It is also interesting that while both technical proposals for cloud manufacturing (Wu et al., 2013) and discussions with OEMs have highlighted the importance of standards, ontologies and APIs for effective data exchange, this was not even a topic that micro SMEs raised. In part, this may be because standards simply do not exist for the kind of products that micro SMEs and craft providers deal in.

Conclusions

In conclusion, we see some significant differences in how organisations approach their work and therefore may approach manufacturing on the cloud. Indeed, one comment might be that the small, personal nature of cottage SMEs amplifies the diversity in this sector not just in product, but also in knowledge, competence and business stratey/motivation. This has a human factors angle as the ways they use the cloud will affect the kind of knowledge they will bring, their expectations and the kind of tasks they need to support to be reflected in new technology, HMIs, processes etc. Example differences include :-

- Different motivations for using the cloud – large organisations are looking to customise and adapt, whereas smaller organisations are looking for 'functional off-loading'. This means engagement strategies are likely to be different for both, and attempts to pull either organisation into cloud manufacturing may need to be significantly different. For large comp.anes, integration with legacy and meeting the skills and knowledge of the workforce is critical to strategy. For smaller companies, the issues link to technical awareness and aking owners aware of the opportunities that exist.
- Different levels of IT competence going into the cloud and different expectations for knowledge, for example their knowledge of standards, or indeed the applicability or relevance of standards - being able to standardise requests to the manufacturgin cloud is likely to be a major barrier to entry for micro-SMEs, yet routine for OEMs.
- Both types of organisation place an importance on how technology will be integrated into their existing functions, and an importance of cloud fitting with them, not the other way round. However, there are likely to be key differences in how organisations approach their work, and therefore care needs to be taken to understand and model the types of task that different organisation. Customisation and customer-centricity is the appeal for large organisations. Smaller organisation may already possess this customer centricity but place importance on functional off-loading.

Limitations of this work include quite small numbers in both activities, and potentially an over-emphasis on consumers to the manufacturing cloud, rather than supplier to the manufacturing cloud. Also, it is noted that data capture used different methodologies across the two studies which, while similar, did use different elicitation and thematic analysis approaches. Therefore the comparisons shown in this paper serve as a guide to a further, more coordinated, parallel examination of different cloud manufacturing users from different sectors. Future work could more formally verify whether the distinctions identified so far are valid and, if so, what are the implications for the design and implementation of cloud manufacturing. Emphasis on engagement processes may be particularily valuable.

Acknowledgements

Work presented in this paper was funded by EPSRC Cloud Manufacturing project (EP/K014161/1) and Horizon Doctoral Training Centre at the University of Nottingham (RCUK Grant No. EP/G037574/1).

References

Rauschecker, U., & Stohr, M. (2012, June). Using manufacturing service descriptions for flexible integration of production facilities to manufacturing clouds. In *Engineering, Technology and Innovation (ICE), 2012 18th International ICE Conference on* (pp. 1–10). IEEE.

Tao, F., Zhang, L., Venkatesh, V. C., Luo, Y., & Cheng, Y. (2011). Cloud manufacturing: a computing and service-oriented manufacturing model. *Proceedings of the Institution of Mechanical Engineers, Part B: Journal of Engineering Manufacture*, 0954405411405575.

Taylor, S. J., Kiss, T., Terstyanszky, G., Kacsuk, P., & Fantini, N. (2014, April). Cloud computing for simulation in manufacturing and engineering: introducing the CloudSME simulation platform. In Proceedings of the 2014 Annual Simulation Symposium (p. 12). Society for Computer Simulation International.

Werfs, M., Baxter, G., Allison, I., & Sommerville, I. (2013). Migrating software products to the cloud: An adaptive STS perspective. *Journal of International Technology and Information Management*, 22(3), 37–54.

Wu, D., Greer, M. J., Rosen, D. W., & Schaefer, D. (2013). Cloud manufacturing: Strategic vision and state-of-the-art. *J. Manuf. Syst. Available at: http://www. sciencedirect. com/science/article/piiS.*

Xu, X. (2012). From cloud computing to Cloud Manufacturing. *Robotics and computer-integrated manufacturing*, 28(1), 75–86.

THE LIMITATIONS OF USING ONLY CAD AND DHM IN DESIGN RELATING TO HIGH VALUE MANUFACTURING

Teegan L. Johnson & Sarah R. Fletcher

Industrial Psychology and Human Factors Group, Cranfield University

The ergonomics suites available within computer aided design and digital human modelling programs are increasingly being used to predict and prevent ergonomic and human factors risk due to poor design. To further aid the reduction in poor design, it is of importance to understand the need for user input and the limitations of these software programs. These limitations include: the small number of available anthropometric population samples; and the disconnect between what a designer perceives as possible, and what is possible within a manufacturing environment. A method of mitigating these limitations is the use of user input using virtual reality suites, mock-ups and motion capture technology.

Introduction

Poor design for ergonomics has been shown to negatively impact worker health (Broberg, 1997) and workplace performance (Wang and Lau, 2013). As a result the integration of human factors (HF) into the planning process of product development has been identified as a major strategy for mitigating work related injuries and illness (Broberg, 1997). Despite the availability of ergonomic guidelines for engineers (Lehto, Landry and Buck, 2008), ergonomic methods and tools have historically not been available in accessible formats for design engineers, hampering the integration of HF into the design process (Broberg, 1997). The integration of digital human modelling (DHM) and human activity analysis suites into computer aided design (CAD), has given designers access to HF methods and tools (Broberg, 1997). This has led to CAD human modelling tools becoming a common technique for predicting and preventing ergonomic risk within manufacturing (Lockett, Fletcher and Luquet, 2014).

The aim of this paper is to discuss some of the ways HF tools in CAD are used in design and redesign within manufacturing. Some of the issues and limitations that should be considered when using these tools to reduce negative impacts on worker health and performance will be highlighted.

CAD and DHM

CAD can be used in the design and redesign of structures within manufacturing. Structures are the parts and the framework that holds these parts in place. Within manufacturing these structures tend to be large and occlude operators and their postures, making it difficult to accurately assess for musculo-skeletal risk. The DHM and ergonomic assessment capabilities in CAD enables the easy variation of anthropometric and biomechanical parameters of virtual manikins (Paul and Wischniewski, 2012). Helping to improve design and usability of products and work systems (Paul and Wischniewski, 2012) by allowing designers to identify how their structure will be used by individuals of different physiologies. Designers can identify how operators will likely interact with their structure; allowing problem areas to be identified such as those that may cause the development of musculo-skeletal disorders (MSD). This reduces the need for large participant pools and mock-ups in the early stages of the design process when assessing structure accessibility. There are, however, limits in the application of CAD and DHM in practice that should be taken into account.

Designers using DHM to inform their design decisions should be aware of the anthropometric differences between countries (Cavelaars, et al, 2008). CAD programs are limited on the types of pre-set manikins available, which can vary based on the percentiles within the specific populations. However, the number of population groups is limited, meaning that designers looking to develop structures for specific populations not represented in the program will need to identify and interpret the relevant anthropometric data. This requires an awareness of how to manually make changes to the manikins, and create manikins that fit the percentiles of interest; this can be a time consuming process in relation to the ease of use for the preprogramed populations.

When modelling the postures of manikins while interacting with structures, the postures adopted through the task from beginning to end should be modelled. For some activities the working posture may be judged as acceptable by a postural analysis tool, while MSD risk lies in the transition postures. It is therefore important that the entire interaction is modelled to identify problem areas. CAD software has the ability to record frame-by-frame the activity of the manikin and ergonomically assess every frame rapidly (Johnson and Fletcher, 2014). This makes assessment much faster than traditional forms of postural analysis and allows the designer to identify exact problem areas in a design.

Operator/user perception and feedback

CAD can be used at the start of a project to design a structure and design out obvious problems using DHM and HF tools available. However, input should be gained from individuals outside of the design process to identify the more subtle or missed issues with a design. Ideally this would include representative users.

An issue with DHM is that it is possible to make a task look feasible that when translated into a manufacturing environment, can only be accomplished with a high level of discomfort. Physical discomfort is indicative of poor biomechanics (Helander and Khalid, 2012), which can contribute to the development of MSD. The movement of a manikin in a CAD program can be limited by fixing the degrees of freedom and applying collision detection to limit unnatural movements. The discomfort levels while completing the task have to be inferred by the designer or engineer. If possible representative users, such as those that will be directly interacting with the structure, should be consulted about perceived discomfort levels of the structure. Virtual reality (VR) and physical mock-ups can be used to gain this information on the accessibility and comfort associated with a structure.

Virtual reality

A CAD model can be presented in an immersive VR suite or on a screen and the representative user can either point out where they perceive problem areas to be or physically enact how they would interact with the structure. For individuals physically enacting their interaction, a motion capture (MoCap) suit can be used to capture postures and input that data into CAD. This gives the user visual feedback which can inform their movement and can be ergonomically assessed. This data will be limited because interactions are devoid of tactile information; the structures used for support, guidance or as a barrier are unavailable. However problem areas can still be identified, which can aid in refining the design of the structure. A solution to the need for tactile information is to use mock-ups.

Mock-ups

Upon the reduction of problems using DHM and VR suites, the structure can be constructed using inexpensive materials with a degree of assembly flexibility. This gives operators a more realistic idea of what interacting with the structure will involve, helping to inform data on physical discomfort. Gaining the physical movements as well as the qualitative discomfort data involved with interacting with the structure would be advantageous. Discomfort can be used to identify where the structure could be causing stress, which may play a role in the development of MSDs (Gielo-Perczak, Karwowski, Hancick, Marras, Karwowski, and Bonato, 2012). A limitation when using mock-ups to gain an understanding of how operators interact with a structure is that the activity of the individual may be occluded. This can make it difficult to pinpoint the exact problem areas on or in the structure that require redesign. Non-optical MoCap suits, particularly where the data can be fed into the CAD program, would allow the whole posture to be captured and translated to a manikin. Problem areas and structures essential for task completion can be identified with the removal of the structures surfaces within CAD. Postural assessment suites within the CAD program can be applied to the data captured giving postural information about the task.

Conclusion

CAD modelling with DHM can be used at the start of a design process to improve design and usability. However it is limited to the perceptions of those working on the design, their understanding of how a structure will be implemented and used on a shop floor may be limited. Therefore, representative user input is essential to identify problems that may be missed. User input can be gained through the use of VR and mock-ups to identify physical usability and ease of access for structures along with qualitative comfort data. This information can be used to improve design and help to mitigate work related injuries and stress, helping to reduce cost relating to injuries and improve worker performance.

Acknowledgements

This work was supported by the UK EPSRC Centre for Innovative Manufacturing in Intelligent Automation.

References

Broberg, O. (1997). Integrating ergonomics into the product development process. *International Journal of Industrial Ergonomics* 19, 317–327.

Cavelaars, A. E. J. M., Kunst, A. E., Geurts, J. J. M., Crialesi, R., Grötvedt, L., Helmert, U., et al. (2000). Persistent variations in average height between countries and between socio-economic groups: an overview of 10 European countries. *Annals of Human Biology* 27, 407–421.

Gielo-Perczak, K., Karwowski. W., Hancock, P.A., Marras, W.S. and Bonato, P. (2012). Multidisciplinary Concepts in Ergonomic Design and Individual Differences in Performance. *Proceedings of the Human Factors and Ergonomics Society Annual Meeting*, 56, 1034–1038.

Helander M. G. and Khalid, H. M. (2012). Affective Engineering and Design. In In G. Salvendy (Eds.) *Handbook of Human Factors and Ergonomics, Fourth Edition*. New Jersey: John Wiley & Sons, Inc.

Johnson, T. L. and Fletcher, S. R. (2014). A Computer Software Method for Ergonomic Analysis Utilising Non-Optical Motion Capture. In Sharples, S. & Shorrock, S. (Eds.), *Contemporary Ergonomics and Human Factors*. (pp. 93–100). Croydon: CRC Press.

Lehto, M. R., Landry, S. J. and Buck J. (2008). *Introduction to Human Factors and Ergonomics for Engineers*. New York: Lawrence Erlbaum Associates

Lockett, H., Fletcher, S. and Luquet, N. (2014). Applying Design for Assembly Principles in Computer Aided Design to Make Small Changes that Improve the Efficiency of Manual Aircraft Systems Installations. *SAE Int. J. Aerosp.* 7.

Paul, G. and Wischniewski, S. (2012). Standardisation of digital human models. *Ergonomics* 55, 1115–1118.

Wang, L. and Lau, H. Y. K. (2013). Digital Human Modeling for Physiological Factors Evaluation in Work System Design. In V. G. Duffy (Eds.), *Digital Human Modeling and Applications in Health, Safety, Ergonomics, and Risk Management*. Heidelberg: Springer.

ELIMINATION OF NON-VALUE-ADDING OPERATIONS AND ITS EFFECT ON EXPOSURE VARIATION AT AN ORDER-PICKING WORKPLACE

Michael Kelterborn[1], Volker Jeschke[2], Sebastian Meissner[2], Carsten Intra[2] & Willibald A. Günthner[1]

[1]*TU München, Lehrstuhl für Fördertechnik Materialfluss Logistik*
[2]*MAN Truck&Bus AG*

The elimination of non-value-adding operations is a common measure to increase efficiency. However, this can lead to monotonous working procedures and one-sided workload. The aim of the present paper was to examine if, for the example of an order-picking workplace, the elimination of non-value-adding operations leads to a more one-sided workload. For the study a Predetermined Time System (MTM) was combined with a working posture analysis (OWAS). It was shown that non-value-adding operations can provide an opportunity for exposure variation. This aspect should therefore be considered in the ergonomic risk assessment.

Introduction

The reduction of time losses and non-value-adding operations is often achieved by splitting up the work into simple, standardized work tasks. However this can result in fewer opportunities for variation and recovery. It is generally believed that variation in biomechanical exposure is beneficial to musculoskeletal health and well-being (Wells, 2007), though empirical evidence for this assumption is low (Mathiassen, 2006). Visser (2006) showed, that without sufficient recreation periods, muscles can be damaged. In a laboratory study, Sundelin (1993) found out, that short breaks reduce muscle fatigue, even if the total amount of work remains constant. Further studies show that the introduction of repetitive, short cycled work can lead to an increase of musculoskeletal disorders (Fredriksson, 2001; Moreau, 2003). In the present paper the workers body posture at an order-picking workplace was analysed. It is examined whether the elimination of non-value-adding operations leads to a more one-sided workload. Socio-technical aspects (as defined by Cherns, 1987) are not within scope of the study.

Method

For the study a Predetermined Time System (MTM), which is used to describe the temporal order of operations, was combined with the OWAS-method for continuous posture logging. Commonly used workload-assessing-techniques, for instance

**Table 1.   Classification of value-adding and
non-value-adding operations.**

Task	Classification
search for the next box	non-value-adding
walk	non-value-adding
put label on box	value-adding
put box on conveyer belt	value-adding

EAWS (Schaub, 2012) are not suitable here, as they do not consider the temporal order of tasks, or consider only single aspects of physical exposure, for example liftings (Waters, 2007). Workers were video-taped during their regular work at the observed workplace. Afterwards, work-processes were converted into OWAS-codes and described in terms of the MTM-1-system, using the recordings. The worker walks down a corridor with storage racks on both sides and a belt-conveyer in front of one of the storage racks. He has to search the correct box, put an adhesive label on it and then put the box on the belt conveyer. The sequence of operations and classification into value-adding and non-value-adding operations for one pick is presented in Table 1. For the study an exemplary order of 14 picks was analysed.

This workplace is compared with a virtual workplace, at which all non-value-adding operations are eliminated. In the following the original, "real" workplace will be referred to as "workplace 1", the "virtual, optimized" workplace will be "workplace 2".

Results

The following observations could be made for the different body-regions. In the following, the numbers in brackets refer to the OWAS code as defined by Karhu (1977). For workplace 1 the "neutral" back posture "straight (1)" offered opportunities for short, recreational rests for this part of the body. These opportunities were smaller in numbers at workplace 2. Regarding the upper limbs, workplace 2 shows an increased portion of "limbs above shoulder level" compared to workplace 1. Concerning the lower limbs, workplace 1 is mostly characterized by the alternation of "standing (2)" and "walking (7)". Workplace 2 in contrast, is very one-sided as it mostly consists of "standing". The factor "load" was not observed due to the fact, that the weight of the boxes is below 10kg and the OWAS method makes no distinction between 0 and 10 kg.

Discussion

Non-value-adding operations serve as an opportunity for recreational rests between work-phases that require bad working postures. When analysing the temporal distribution of working postures, it was conspicuous that at workplace 2 the block

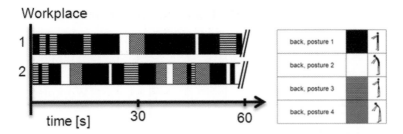

Figure 1.   Example of temporal distribution of body postures (back, 60s).

lengths of single postures were generally more homogenous in length, than at work-place 1 (see fig. 1). As an example, back postures during the first 60s of work cycles at workplaces 1 & 2 are shown in figure 1.

Mathiassen (2006) stated that a reasonable way to quantify exposure variation would be the use of common statistical measures, as this would provide a possibility to compare different studies and workplaces. Thus, a possibility to quantify this aspect of exposure variation would be the coefficient of variation (CV) of the block length for each body-region. The CV is defined as the ratio of the standard deviation σ to the arithmetic mean μ of a distribution. The CV makes data sets with different means comparable, because it is a normalized measure.

In this special case:

$$CV = \frac{\sigma \ (standard\ deviation\ of\ the\ block\ length)}{\nu \qquad (mean\ block\ length)}$$

As a possible interpretation of the CV, the following thoughts can be made. High values of σ, and going along with it high values of CV, stand for high differences in the actual lengths of different postures. In other words, the worker sometimes stays in a certain posture for longer, and sometimes for shorter periods, which can be interpreted as a feature of a workplace with a high posture variation. The other way round, low values of CV can be seen as a characteristic of a highly repetitive workplace. The application of the CV on the here described workplaces shows the following results (*CV for workplace 1, CV for workplace 2*): For the categories *back (0.83, 0.66)* and *upper limbs (1.85, 1.62)* the values of CV were lower for workplace 2 than for workplace 1. For the *lower limbs (0.82, 0.83)* no significant change was observed. Results are shown in figure 2.

We would like to point out that block length variation covers only one aspect of exposure variation. Standing alone, it is not an adequate criterion for the assessment of exposure variation. Regarding posture logging with the OWAS method (as shown in Fig. 1) the following further aspects of variation should be considered. First, the mean block length, as a measure for the time spent in one posture, second, time ratios for all posture categories. For example, this would allow investigating whether a worker has a variation between standing, walking and sitting, or not.

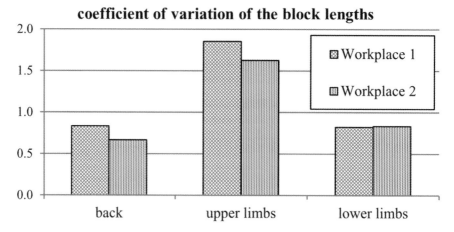

Figure 2.   **Comparison of the CVs of workplace 1 and 2.**

Conclusion

It was shown that the change of the temporal sequence of tasks, particularly the elimination of non-value-adding work, can have a significant influence on the physical workload. Existing ergonomic methods do not consider this aspect. The here presented approach is a contribution to fill this gap. The results for the examined example are promising, yet further validation is necessary. The CV seems to be a useful indicator when analysing exposure variation, but covers only a single aspect of exposure variation. Further research could therefore lead to a set of key figures for the quantification of exposure variation.

References

Cherns, A. B. (1987): Principles of socio-technical design revisited. *Human Relations* 40(3): 153–162.

Fredriksson, K.; Bildt, C.; Hägg, G.; Kilbom, Å. (2001): The impact on musculoskeletal disorders of changing physical and psychosocial work environment conditions in the automobile industry. *International Journal of Industrial Ergonomics* 28(1): 31–45.

Karhu, O.; Kansi, P.; Kuorinka, I. (1977): Correcting working postures in industry: A practical method for analysis. *Applied Ergonomics* 8(4): 199–201.

Mathiassen, S. E. (2006): Diversity and variation in biomechanical exposure: What is it, and why would we like to know? *Applied Ergonomics* 37(4): 419–427.

Moreau, M. 2003, Corporate ergonomics programme at automobiles Peugeot-Sochaux. *Applied Ergonomics* 34(1): 29–34.

Schaub, K.; Caragnano G., Britzke B., Bruder R. (2013): The European Assembly Worksheet. *Theoretical Issues in Ergonomics Science* 14(6): 616–639.

Sundelin, G. (1993): Patterns of electromyographic shoulder muscle fatigue during MTM-paced repetitive arm work with and without pauses. International archives of occupational and environmental health 64(7): 485–493.

Visser, B.; van Dieën, J. H. (2006): Pathophysiology of upper extremity muscle disorders. *Journal of Electromyography and Kinesiology* 16(1): 1–16.

Waters, T. R.; Lu, M.-L.; Occhipinti, E. (2007): New procedure for assessing sequential manual lifting jobs using the revised NIOSH lifting equation. *Ergonomics* 50 (11): 1761–1770.

Wells, R.; Mathiassen, S. E.; Medbo, L.; Winkel, J. (2007): Time—A key issue for musculoskeletal health and manufacturing. *Applied Ergonomics* 38(6): 733–744.

JOB DESIGN FOR MANUFACTURING IN AN ERA OF SUSTAINABILITY

M.A. Sinclair, C.E. Siemieniuch & M.J.D. Henshaw

ESoS Research Group, Loughborough University

The paper explores the changes that are likely to be necessary as the world moves to a more sustainable way of life. When these changes are added to the development of the Internet of Things, in which it is envisaged that devices with some level of embedded intelligence will communicate with each other, as will intelligent services, it appears that our current ways of conducting job design may be found wanting. The principles of socio-technical design will still apply; how these principles will necessarily be extended within the manufacturing domain is the subject of this paper; covering issues such as how to include aspects of sustainability and the need to train for resilience.

Introduction

This paper explores some likely changes over the next decade to job design as practised by Ergonomists/Human Factors Engineers. The context for this paper are the foreseeable changes in the manufacturing domain that will be produced by the global drivers that require a sustainability response (Allwood, Ashby et al. 2011, Sulston 2012, Lavery, Penell et al. 2013). The paper outlines briefly what these global drivers are, the likely response of manufacturing, the outlines of which can already be seen, and the implications for job design in manufacturing. These implications are likely to be apparent in other domains as well.

Some global drivers, for the next 35 years

The global drivers are those that are likely to affect the population of the world as a whole, and whose effect, if not addressed, will likely be unpleasant.

Eight drivers are commonly identified:

- Population demographics (estimated to rise by about 60% by 2050, with consequent demand for resources)
- Food security (agricultural land fixed; crops affected by climate change)
- Energy conservation (emissions are a major cause of climate change)
- Water conservation (potable water is limited in many countries)
- Resource depletion (some key minerals will be difficult to access)
- Emissions and global climate (the main cause of climate change)

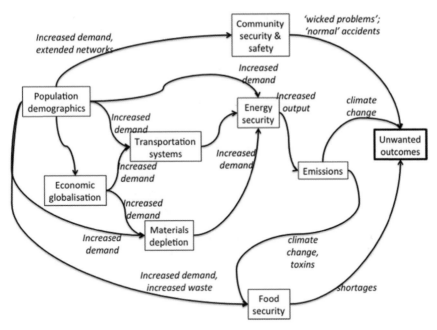

Figure 1a.   Illustration of the high-level interactions between the Global Drivers, leading to unwanted outcomes. Only a few of the interactions are shown, to assist clarity.

- Transportation (also contributing to the emissions problem)
- Globalisation (improving living conditions around the world will increase demand for resources)

There are interactions between these drivers, implying that efforts to mitigate their effects will need to be concerted and co-ordinated. Fig. 1a below indicates some of these interactions, and Fig. 1b indicates some countervailing strategies.

Fig. 1b includes approaches to mitigate the unwanted outcomes that could accrue from the harmful interactions among the Global Drivers (CCS = Carbon capture and storage.)

Three conclusions may be drawn about these Global Drivers:

- Each of the Global Drivers, operating on its own, could have very significant effects on the world as a whole. Together, they pose a significant threat to the health and well-being of all of us on this planet
- Mitigating this set of drivers necessitates a connected, comprehensive approach; it is evident that tackling one, or another, is unlikely to have much impact by itself.
- A combination of political persuasion and technology will be required to reach any satisfactory conclusion; a comprehensive socio-technical solution will be necessary.

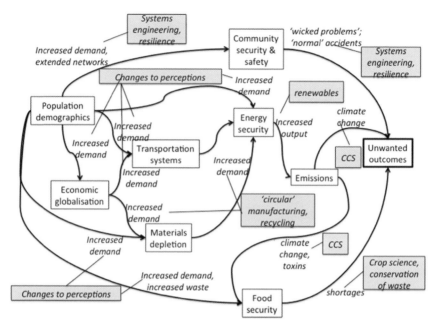

Figure 1b.   Global Drivers, including approaches to mitigate the unwanted outcomes that could accrue from the harmful interactions of the Global Drivers (CCS = Carbon capture and storage).

Turning now to manufacturing, it is clear that this domain has a massive part to play in reaching a sustainable state:

- Manufacturing creates the main demand for the exploitation of resources, and is a major producer of emissions and of landfill.
- Manufacturing will create the devices and the processes that will enable the world to reach a sustainable state.
- Re-engineering manufacturing processes to be more efficient will be necessary to reduce the emissions and other waste. 'Circular manufacturing' will have to become a dominant philosophy (Kagermann, Wahlster et al. 2013, Lavery/Penell 2014).
- Re-engineering the products of manufacturing to minimise waste and to enable the recycling implicit in 'Circular Manufacturing' will have to become a necessity. Legislation is already providing manufacturing push; for example in the UK there are the End-of-Life Vehicles (ELVs) Regulations 2003 and the ELVs (Producer Responsibility) Regulations 2005; in the EU27 there is the ECODESIGN Directive 2009/125/EC, and in Japan the 'Top-runner' scheme has been in existence for many years (Siderius and Nakagami 2007).

Many new jobs will be created by a move to a circular economy, replacing those lost in the extractive industries, and most manufacturing jobs will require some degree of redesign; the latter caused by the re-engineering points listed above.

The re-engineering of manufacturing

The philosophy of manufacturing is changing markedly. Some trends are already evident, as is indicated below:

- Manufacturing is moving from a major-plant-centric view to a networked view, as exemplified in the 'Industrie 4.0' agenda adopted by the German government; first there was the introduction of steam and mechanical production (a technical revolution), followed by the move to standardised parts, mass production and task specialisation (an organisational revolution), then the introduction of information technology (technical again), to be followed by the 4th, networking revolution (organisational again), involving the 'Internet of Things' (IoT) which includes data, services and cyber-physical systems (e.g. robots and other devices with embedded intelligence), including intelligent machine-to-machine communications, perhaps with significant autonomy. The network, not the factory, becomes the core of manufacturing.
- The demands of circular manufacturing imply change to materials, machines, products and processes; all of which will impact job design
- Because of the interlinked nature of the global drivers and their impact on local communities and the manufacturing organisations within these communities, issues such as local ecology/biodiversity and corporate social responsibility are intertwined with more traditional manufacturing concerns. This is for two reasons; legislative push, and wide recognition that governments and the tax-payer by themselves cannot produce the mitigations required.

Figure 2 below is an illustration of what is meant by 'Circular Manufacturing'.

These trends have implication for an organisation's policies, and their decomposition into jobs. This is shown, for example, by Nestlé™, a major company in the food industry, which has adopted six policy principles within their 'New Accelerated Model', in addition to the usual 'faster, better, cheaper' mantra, widespread in manufacturing (NMR/2deg 2014):

- Energy – transition to low energy sites
- Water – optimised water withdrawal & use across sites
- Waste – transition to low waste sites
- Biodiversity – recognised for promoting & developing biodiversity
- Value chain – reduction in environmental impact across the value chain
- Community & people – recognised as a 'good corporate citizen' and adding value to local communities

At a site where these principles were first piloted, the company has been able to achieve, over a period of 6 years, savings of about 22% in green-house gas (GHG) emissions, about 30% in energy consumption, and about 22% in water usage per tonne of product. This site has become a 'zero waste' site (implying the creation of a value chain for waste) and has made contributions to 'butterfly meadows'.

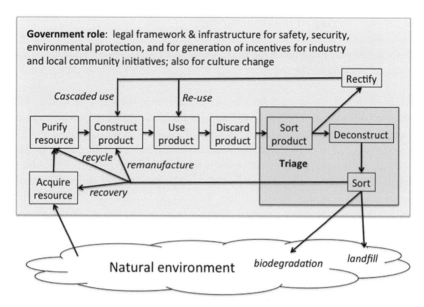

Figure 2.   Illustration of 'Circular Manufacturing', as an instance of the 'Circular Economy'. Note that the 'use product' box may involve considerable iterated re-use by a sequence of owners.

Similarly, Toyota[TM] in their UK operations over a period of 20 years have reductions per vehicle of 74% for energy, 69% for water, and 60% for waste. Both of these companies have emphasised that these improvements necessitated the interpretation of these principles into the jobs that their employees do, accompanied by continuous efforts to gain acceptance, not just acquiescence, of these re-defined roles and responsibilities.

Re-emphases for job design in manufacturing

The first point to make is that the core, socio-technical principles for the design of human jobs and roles within organisations as enunciated for example by Cherns (Cherns 1976, Cherns 1987) and subsequently elaborated by Eason, Clegg and Doherty (Eason 1988, Clegg 2000, Doherty 2014) remain the same. As an aside, as cyber-physical systems within the IoT become more intelligent with more autonomy, these principles (adapted) may be necessary for them, too.

But networking manufacturing via the IoT will bring some important changes to jobs; firstly, there are cultural implications; while large assembly plants will still exist (because physical components must be brought into contact to create products), much of the manufacturing of components, particularly components that can exist as software, will be in different, smaller enterprises that may be widely distributed. 'Company culture' may become local, with enterprise culture becoming more of a mosaic, predicated on trust in performance and with an emphasis on 'good citizenship', built into job design and support.

Security and confidentiality issues are already prevalent in the internet. However, it brings a number of issues with it, as outlined below:

- 'Informed Consent'; for example, you give consent to a request from the network to use your current location data. It is unlikely that you know, or ever could know, who or what has access to that data, what meaning(s) will be inferred from the data, for how long this access will be enabled, how your data will be combined with other data, and so on. Solutions are not immediately evident.
- 'Informed Command', as given, for instance, by a human controller involved in a networked process in which some devices may have some degree of behavioural autonomy. The UK Ministry of Defence has a rule of for this; summarised as 'whoever gives the last command is responsible for the outcomes.' This is also embodied in International Humanitarian Law. The implication is that whoever gives a command must be able to anticipate the likely outcomes and side-effects of any command. Again, solutions are not immediately evident; nevertheless, this problem sits at the heart of systems of systems/SoS ergonomics.
- Identity. As a Foresight document has pointed out, each of us has many identities, and some of these are not constructed by us but exist on the internet (Foresight-FFI 2013). There is a question of who owns and who can use these identities and for what purpose, especially those we have not constructed ourselves. This has implications for the notions on informed Consent and Informed Command, above, especially if some of these identities have been created or utilised with criminal intent.
- Autonomy and learning. Not all cyber-physical systems need to be given the capability to learn, though some must; robots, for instance. Current thinking suggests that autonomy should be constrained to level 6 or below on Sheridan's scale (Parasuraman, Sheridan et al. 2000): 'The [robot] allows the human a restricted time to veto before automatic execution'; however, this returns the problem to 'Informed Consent' and 'Informed Command'; the rule assumes that the command-giver is in an appropriate 'informed' state.

However these issues are resolved, there are a few things of which we may be sure; networked manufacturing will produce unexpected problems of varying extent and severity on a frequent basis, both in 'normal' operations and when changes are introduced. These will need to be solved by those involved (as implied by Clegg and other authors), but over the network; an important part of job design will be problem-solving via the network, involving an understanding of how the IoT works and can be fixed, how cultures interact, and the chances of miscommunication across cultures, disciplines, ontologies and legal frameworks.

Secondly, there must be concentration on training and education for resilience; this may obviate some of the problems above, and will certainly improve local operations, both with regard to speed and reliability.

Thirdly, the IoT and the networking of processes may result in a much more labile approach to jobs; there may be much more redefinition of jobs and roles, leading to more churning of jobs and hence of people who do these jobs. This raises issues of training for employability and for career progression; issues well-addressed by Zink in his recent review paper (Zink 2013).

Finally, there is the necessity of turning the Nestlé™ principles (for example) into the tasks and roles that people undertake. Firstly, the principles must be translated into Key Performance Indicators (KPIs) appropriate to the performance of the organisation in relation to these policies, and then cascading these down into the activities (i.e. Activity Performance Indicators – APIs) that people undertake within their roles; for example, they may have to become part of the Minimal Critical Specification for a role, following Cherns and others. Given this, the role-holder will require training and authority over resources to be able to achieve these APIs. Given the range that these APIs may take – consider support in the job for biodiversity, for example – the organisation may well have to have very good links into local education and training facilities, professional services and the like in order to fulfil its obligations to the role-holder.

Conclusions regarding Ergonomics/Human Factors Engineering

As things stand, it is difficult to see how an Ergonomist/Human Factors Engineer would be able to design jobs for this new, networking model of manufacturing within the constraints and imperatives of a sustainable world, based on current approaches such as Cognitive Work Analysis, Hierarchical Task Analysis and Situation Awareness. There appears to be a knowledge gap, requiring some effort to fill. Einstein's famous quotation is relevant here:

> "The world as we have created it is a process of our thinking. It cannot be changed without changing our thinking."

We have some work to do.

References

Allwood, J. M., M. F. Ashby, T. G. Gutowski and E. Worrell (2011). Material efficiency: a white paper. *Resources, Conservation and Recycling* **55**: 362–381.

Cherns, A. B. (1976). Principles of socio-technical design. *Human Relations* **29**: 783–792.

Cherns, A. B. (1987). Principles of socio-technical design revisited. *Human Relations* **40**(3): 153–162.

Clegg, C. W. (2000). Socio-technical principles for system design. *Applied Ergonomics* **31**: 463–477.

Doherty, N. F. (2014). The role of socio-technical principles in leveraging meaningful benefits from IT investments. *Applied Ergonomics* **45**: 181–187.

Eason, K. D. (1988). *Information technology and organisational change*. London, Taylor & Francis.

Foresight-FFI (2013). *Foresight Future Identities (executive summary)*. L. The Govenment Office for Science. London, The Government office for Science.

Kagermann, H., W. Wahlster and J. Helbig (2013). *Recommendations for implementing the strategic initiative INDUSTRIE 4.0*, acatech: National Academy of Science and Technology.

Lavery, G., N. Penell, S. Brown and S. Evans (2013). *The next manufacturing revolution: non-labour resource productivity and its potential for UK manufacturing*. Cambridge, UK, Institute for Manufacturing, Cambridge University: 164.

Lavery/Penell (2014). *The new industrial model*. London, Lavery/Pennell.

NMR/2deg (2014). *2degrees' Resource Efficiency Summit 2014: Learn how Toyota and Nestlé are tackling resource efficiency*. London, Next Manufacturing Revolution: 42 minutes.

Parasuraman, R., T. B. Sheridan and C. D. Wickens (2000). A model for types and levels of human interaction with automation. *IEEE Transactions on Systems, Man & Cybernetics* **30**(3): 286–297.

Siderius, P. J. S. and H. Nakagami (2007). Top Runner in Europe? Inspiration from Japan for EU ecodesign implementing measures. *Summer Study Series*. Stokholm, European Council for an Energy Efficient Economy: 1119–1126.

Sulston, J. (2012). *People and the planet*. London, The Royal Society.

Zink, K. (2013). Designing sustainable work systems: The need for a systems approach. *Applied Ergonomics* **45**: 126–132.

FIAT CHRYSLER AUTOMOBILES ERGONOMICS APPROACH IN DEVELOPING NEW CARS: VIRTUAL SIMULATIONS AND PHYSICAL VALIDATION

Stefania Spada[1], Danila Germanà[1], Fabrizio Sessa[2] & Lidia Ghibaudo[1]

[1]*Fiat Chrysler Automobiles, Manufacturing Engineering – Ergonomics, Torino, Italy*
[2]*Fiat Chrysler Automobiles, Manufacturing Engineering Southern Italy – Digital Manufacturing and Ergonomics, Pomigliano d'Arco (NA), Italy*

Car manufacturing is subject to radical change due to demand for new models. Original equipment manufacturers and suppliers have to develop more flexible assembly chains, manufacturing services and methods for job planning. This requires new concepts for process design and for production: the human centered approach to improve manual assembly. In Fiat Chrysler Automobiles, digital manufacturing and immersive virtual reality have been used in the design phase. Holistic ergonomics assessment methods EAWS and ErgoUAS methods have been used for ergonomic optimisation of assembly tasks and for optimal line balancing. Parallel physical assessments on prototypes have been run to validate design/virtual solutions.

Introduction

During recent years, the global car market has undergone to radical change, becoming more competitive and unstable due to several factors (legal, financial, environmental, etc.). The most important automotive original equipment manufacturers (OEMs) have been obliged to react. The continuous request to produce different new models every few years obliges OEMs to develop more flexible assembly lines and better methods for job planning on the same lines. Safety norms and product quality requirements also demand improvements in working conditions according to international standards. Moreover, the assembly line is seeing increased levels of robotics and hybrid human-automation decision making and this requires new concepts for process design and for production: the human centered approach to improve manual assembly operations, even assisted by automation.

Therefore, ergonomics and work organization concepts are undergoing strong development and new organisational models have been created using ergonomics guidelines to organize and optimise production in the plants. Important examples include world class manufacturing (WCM) for production theories and the European assembly worksheet (EAWS) for ergonomics and work organisation (Ergo-UAS). These concepts allow the achievement of the most important benefits only when they can be applied in the design phase of a new car's manufacturing process, because, in this case, the whole production system can be changed and optimised. For this reason, Fiat Chrysler Automobiles (FCA) has created the 'digital manufacturing' (DM) team project whose final goal is to create simulation tool and procedure that allow to apply ergonomics principles and work organisation methods in the early phases of a new project development for all manual assembly operations.

In order to show how to achieve this result, it important to remember that, generally speaking, ergonomic analyses in this context aim to improve different aspects of manual work: postures, forces, manual material handling, etc. The DM approach has been based on a process to adapt these ergonomics methods to the 'virtual plant' environment in order to get, in the design phase, a high correlation between the results obtained by the virtual simulation with those obtained on the physical analysis of the work tasks. Thanks to this process, it becomes possible to use ergonomic risk assessment indices in design phase in order to change and improve project solutions and to well balance the work load between workers.

Moreover, the use of immersive reality allows an effective idea of visibility and reachability aspects. However, virtual simulations do not always allow the solution of all the typical problems related to workstation design, because experimental data and physical validations on real components are necessary (for example to evaluate forces, flexible objects behaviours, etc). For this reason, FCA set up an ergonomics laboratory for experimental data evaluation. In this lab, tools and equipment as well as assembly methods are tested and measured using systems commercially available to study physical factors influencing ergonomics, quality and efficiency of manufacturing process. Through advanced instruments and specific methodologies of analysis, detailed data are measured and gathered. The information is then analysed and validated by field test in order to arrive at final design solutions.

Digital manufacturing

One of the most important aspects of the DM project is the information's 'active management' that allows designers and engineers to make detailed virtual simulations in order to improve the car's manufacturing process. This concept involves an important development on the methods used by FCA manufacturing engineering to manage all the information related to the production process: technological data are available in a unique simulation environment that contains virtual models of the car components and of the production plant (robots, tools, equipment, etc.). The key factor to improve DM simulations has been to insert digital human models and human centred simulation methods in the virtual environment in order to perform certified simulations for all the manual operations that are designed for the production process in the future plant.

Digital human models are now a standard in simulation software and many commercial products (e.g. "Jack" by Siemens), and are available on the market (Stephens, 2006). The digital manufacturing allows the creation of a virtual plant where virtual

Figure 1.   A virtual model of a production plant.

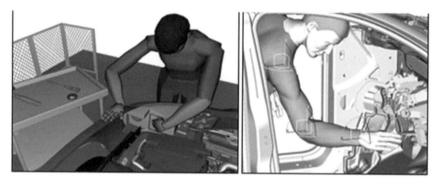

Figure 2.   An example of assembly tasks.

mannequins interact with digital models of car's components, equipment, containers, etc., in order to simulate and improve working conditions with many benefits regarding ergonomics, safety, final product quality, work organisation and general production costs.

3D immersive virtual reality

Mathematics can be used to create a 3D image of car and components. The operator, wearing glasses, can be set in the immersive environment and interact with objects and tools. This system enables operators and designers to simulate all assembly operations, giving a realistic perception of size and space. In this way, it is possible to highlight critical issues related to visibility and accessibility at the working point, while ergonomics specialists can check the possible risks related to the workstation directly with the operator or using the information provided by the dummy virtual system.

Figure 3. An example of a working team using 3D immersive reality and an operator using the system.

Ergonomics method

The fundamental development on which FCA has based the digital manufacturing project for human simulations has been the effort to improve standard digital human modelling software. This is in order to be compliant with FGA's design standards (based on international technical norms – ISO/EN) and FGA's production standards that are based on European/Italian legal specifications for safety and ergonomics. In this way, designers have not only a virtual representation of the future plant but also the virtual tools to validate design solutions. Reference frames on these mannequins have been created in order to have specific points on which evaluate body angles, distances, etc. according to the most important ergonomics methods requested by ISO/EN standards (ISO 11226) and by FGA's standards (like OCRA NIOSH, Snook & Ciriello, EAWS).

EAWS especially is the ergonomics risk assessment method applied in the design phase as an ergonomic screening tool. It links corrective and proactive ergonomics, points out ergonomic problems and offers design solutions to overcome them. The EAWS consists of four sections for the evaluation of

- working postures and movements with low additional physical effort,
- action forces of the whole body or hand-finger system,
- manual materials handling, and
- repetitive loads on the upper limbs.

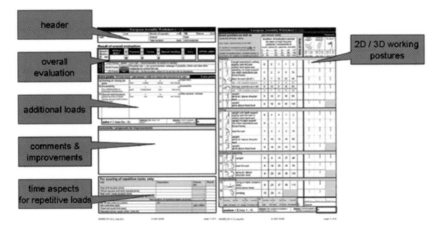

Figure 4.   EAWS structure and sections (pp. 1–2).

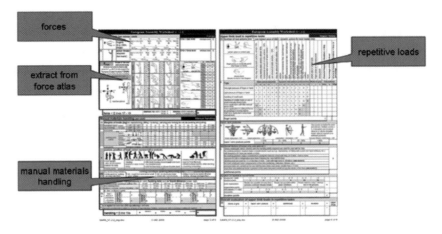

Figure 5.   EAWS structure and sections (pp. 3–4).

The evaluation, of mechanical load request by the workplace (considering product and process together) also in the design phase, has to be carried out by the measurement of physical parameters on prototypes or similar models.

Physical assessment

The main analysis concerned postural aspects and effort by operators during working tasks and experimental measurement of physical parameters involving equipment and tool use. In the Mirafiori Plant, in the Manufacturing Engineering area, a new ergonomic laboratory was established (ErgoLab). In ErgoLab, the

Figure 6. Examples of forces measurement (pp. 3–4).

Figure 7. Acquisition of joints angles.

workplace of production units has been recreated: Assembly and Painting are provided by a car hook that can accommodate car models of different segments; Body in White is provided by a welding system and a dedicated preparation workbench; Handling is provided by different container types, LCA systems and Ro-Ro; a mechanic workshop is provided by an adaptive workbench to simulate preparation workstations.

The laboratory is grounded on experience gained over the years from collaboration with FCA Research Centre on several issues: the physical parameters measurement; the ergonomic design of the workplace; support during the design phase; and selection and training in correct use of pieces of equipment according to national and international legislation requirements. The aim of ErgoLab is to support plants, the manufacturing engineering area and designers with reliable information and data to design technological systems and ergonomic workstations. The main purpose is to make measurement procedures more accessible and to recreate working conditions without interfering with the working process.

The main field of laboratory activities concern:

- measurement of force, localised and distributed pressure, reaction force couple, hand/tool interaction;
- acceleration;
- angles assumed by the main joints.

All measurements focus on the physical characterisation of the following three main areas: tools, equipment and work methods. The acquisitions are made using the most advanced commercial instruments and sensors and the experimental procedures are well defined, documented and repeatable.

Conclusions

In this paper, we have shown the approach used in FCA based on simulation methods and experimental facilities to analyse ergonomics aspects of future workcells.

The approach is based on virtual and physical methods that can be quickly used by project designers and plant ergonomics experts. Using these tools it becomes possible to do a preliminary ergonomic analysis of the future workcell according to the most important ergonomics indexes (like OCRA and EAWS) in order to get a preliminary ergonomics optimisation of workcells during the initial phases of a new product/process.

References

Di Pardo, M., Riccio, A., Sessa, F., Naddeo, A. and Talamo, L. *Methodology Development for Ergonomic Analysis of Work-Cells in Virtual Environment*, SAE Paper 2008-01-1481

Stephens, A. and Godin, C. 2006, *The truck that jack built: digital human models and their role in the design of work cells and product design*, SAE Paper 2006-01-2314

ISO 2008, International Standard ISO 7250-1, *Basic human body measurements for technological design – Part 1: Body measurement definitions and landmarks.*

ISO 2010, International Standard ISO/TR 7250-2010, *Basic human body measurements for technological design – Part 2: Statistical summaries of body measurements from individual ISO populations.*

ISO 2000, International Standard ISO 11226:2000, *Ergonomics – Evaluation of static working postures.*

MOBILE

FAME OR FUNCTION? HOW WEBCOMIC ARTISTS CHOOSE WHERE TO SHARE

Liz Dowthwaite[1,2], Robert Houghton[1] & Richard Mortier[3]

[1] *Human Factors Research Group, Faculty of Engineering, University of Nottingham*
[2] *Horizon Centre for Doctoral Training, University of Nottingham*
[3] *Network Systems Group, University of Nottingham*

Online social networks are complex systems that can be variously construed as websites, platforms and communities but relatively rarely are they considered from the perspective of a goal-directed user. A mixed methods study consisting of questionnaires and in-depth interviews was carried out in order to examine the use of social networking within the online cartoonist community. We consider this subset of users, who rely on the use of social networking for specific work-related activities, because they form a group of motivated innovators who explore the uses and abuses of different services in terms of both feature set and community. A complex picture emerges of the strategic combination and interplay between platforms to co-optimise function and reach.

Introduction

Online social networks are complex systems that are traditionally discussed in several different ways, often revolving around either the technology and design of the platforms themselves, or the behaviour and social interactions of the communities that use them. These two areas are often studied as separate topics, but in reality how people interact online is affected and restricted by the functions of the sites they use. Social networking is becoming an important tool for both businesses and creative individuals to find audiences and customers, and it is important to understand how contrasting factors are perceived to affect each other. From a systems ergonomics perspective (Wilson, 2014), how and why people in a particular industry or with a specific goal make use of existing platforms can help to improve the functionality of future designs.

In this paper we study an online community consisting of creators and consumers to examine the use of different social media platforms and the ways in which their interactions may be restricted or facilitated by differing platform dynamics. We chose this community because the use of social media is an integral part of the creators' business models, and social networking is very much a necessary part of their work. The majority of users of social media platforms are performing non-directed activities, browsing with no particular aim in mind, but our creators must act in a goal-directed way to maintain their communities. As such, our creators are an expert group of users and can therefore provide insights which will help other communities, as well as suggesting issues to be considered in future platform design. We now give a brief overview of this community.

Webcomics are comics that are distributed through the internet by an independent creator with no corporate sponsorship (Fenty et al., 2005). Conservative estimates for the number of titles online at any time place the number around 15 to 20 thousand (Walters, 2009) although this varies considerably. Most webcomics do not make a large income, however there are an increasing number of creators who can support themselves full-time. Internet technologies are vital to these creators as the costs of production and distribution are much lower; cheap website hosting, free blogging software, and access to large audiences make it easy for an artist to display their work and get it seen.

The importance of the relationship between comics' creators and their readers has long been recognised, and with webcomics in particular, the use of technology in this relationship is vital (McCloud, 2000). Creators are able to build up meaningful relationships with readers over time and through many different avenues. The internet negates geography, which for webcomics means that they can maintain a large reader base whilst still catering to a niche audience (Guigar et al., 2011). It takes time and dedication to build up a reliable community, but in response creators often end up with a group of readers willing to spend their time and money to support the artist. Most financially successful webcomics artists cite close relationships with their readers as a major factor (e.g. Watson, 2012). In order to cultivate relationships, many creators encourage dialogue, and include blogs with their comics. Readers can post comments, and visit forums or social media pages to engage with creators and other fans; they can be involved in every stage of comic production, with artists posting ideas, concept sketches, and works-in-progress, and livestreaming their process. They make use of many different websites for varying purposes. For example, if they wish to reach as large an audience as possible they may post predominantly on 'famous' sites such as Facebook and Tumblr, but if they are more concerned with reaching a particular audience they may focus on comic-specific sites such as Comic Fury, or particular communities such as gamers. Alternatively, creators may choose sites based on what they wish to achieve by posting work online; this leads to choosing sites based on their particular features, for example using an image-focused site such as Tumblr or Deviant Art as a form of online portfolio. It is most likely that artists will use a combination of 'fame' and 'function' to decide where to post their work, and will use several different combinations of sites to maximize their reach online.

This brief summary shows that webcomics creators are motivated to carefully consider their usage of social networking; their businesses depend on rich interactions and sharing content and so they must use these sites more often and with more consideration than other groups might. Often, part of their working day is dedicated to social media (Guigar, 2013), and they have particular mental models about how such sites work. Their entire working life is visible online, and although their business models only exist due to the power of social networks, their livelihoods may also be threatened by what is available and how they manage their internet presence. This paper highlights some of the issues and opportunities faced by these creators, which in turn may affect the more casual user and other communities.

Methodology

A mixed methods approach was taken in this study, in order to build a compre-hensive picture of how people within the webcomics community make use of the internet, particularly social media. A multiple-choice questionnaire was distributed online, followed by a series of semi-structured interviews with 6 full-time and 5 part-time webcomic creators (7 males and 4 females). Questionnaire participants were recruited through social media, webcomic sites, direct emails to creators, and creators sharing the study with their readers. As an online industry with a large emphasis on interaction, recruiting in this way was felt to be appropriate, particu-larly with the wide range of avenues utilised. All 209 questionnaire respondents were webcomics readers, with 92 also being creators. Almost half were aged between 26 and 35 (46.4%), most were male (53.6%), and American (43.1%) or British (20.1%). Creators had typically created one comic (43.5%), for up to two years (48.9%) although a substantial number had been updating their current comic for more than five years (19.6%). Comics are typically updated once a week (44.6%), and receive fewer than 5,000 unique visitors per week (84.8%); only six creators considered themselves to make a living from their webcomic work. Interviewees were recruited via email; all webcomics creators listed to attend Thought Bubble Comic Art Festival in 2013 were contacted. 19 creators indicated they would be happy to be interviewed, but due to time constraints only 11 interviews were fully completed, with 9 UK-based and 2 US-based creators. Whilst this sample was somewhat opportunistic, Thought Bubble is the largest gathering of independent creators in the UK and therefore it was the best place to gain a diverse group of artists. Questions were kept very open to allow creators to say that they wanted without constraint, and to prevent guiding opinions. Interviews lasted for between 10 and 50 minutes and creators were asked to discuss how they use social media and the internet to interact with their readers. Interviews were transcribed fully and an inductive, iterative, and grounded approach was taken to analysis in order to code for recurring themes. This included grouping comments by keyword, and coding for positive and negative comments. Quotes used in this paper were chosen as representative of the theme.

Results – *Fame*

As can be seen from Table 1, most artists maintain a dedicated site for their web-comics; the following seven most used sites are social media (indicated by *),whilst the final two of the top ten are comic-related or webcomics-hosting sites (#).

Table 1. Top ten websites for webcomics content (Position in brackets).

Website	% Post Comic	% Access Comic	% Post Additional Content	% Access Additional Content
Any	100.0	100.0	83.7	75.6
Dedicated site#	84.8 (1)	96.8 (1)	48.9 (1)	67.5 (1)
Twitter*	54.3 (2)	34.8 (2)	31.5 (4)	38.8 (2)
Facebook*	52.2 (3)	22.2 (3)	33.7 (3)	22.5 (4)
Tumblr*	46.7 (4)	14.5 (4)	42.4 (2)	24.9 (3)
Reddit*	15.2 (5)	5.8 (5)	1.1	1.4 (6)
Google+*	12.0 (6)	3.4 (6)	3.3 (=6)	1.0 (=7)
Deviant Art*	9.8 (7)	1.0 (=9)	15.2 (5)	2.9 (5)
Pinterest*	5.4 (8)	0.0	0.0	0.0
Smack Jeeves#	5.4 (9)	1.0 (=9)	0.0	0.0
Comic Fury#	3.3 (10)	1.9 (8)	2.2 (=9)	0.5 (=9)
Instagram*	0.0	0.0	3.3 (=6)	0.0
Blogger	0.0	0.0	3.3 (=6)	0.0
Webcomic Underdogs#	1.1	1.0 (=9)	2.2 (=9)	1.0 (=7)
Comic Rocket#	1.1	2.4 (7)	1.1	0.5 (=9)

This generally matched with where people read webcomics, although readers much prefer the dedicated site rather than reading elsewhere; this may be because creators often simply post links directing people to their main site on Facebook or Tumblr. Posting and accessing additional content, such as concept art and blogs, differs considerably, although the top five sites were also exclusively social media. Readers also appear to have a very slightly higher preference for comic-related and webcomics-hosting sites; 27 of the 46 other sites mentioned were comic-related, and were only used by one or two people. Generally speaking, the sites with the most users are also the most popular: Facebook boasts 1.32 billion users (Facebook Newsroom, 2014) and Twitter has 255 million (Digital Market Ramblings, 2014), whilst Comic Fury has nearly 45 thousand (ComicFury, 2014) and Tapastic has 1 million (Digital Market Ramblings, 2014).

Interview data backs up these numbers, with all of those interviewed showing preference for Facebook, Tumblr, and Twitter. The most obvious reason was that they reach more people: *"I'm still small, I'm building up my career and I need as many people as I can to see my work"* (P8). Users will share content that they like with others, increasing the work's audience; this is particularly useful when introducing a new comic or seeking the viral effect: *"you need to get into their feed, you need to get into what they look at every single day"* (P8). Posting on social media can be very successful: *"there's some people that do follow me on Facebook that will buy pretty much anything that I put out"* (P1). However, large numbers of followers does not necessarily transfer into more success, either in terms of dedicated readership or merchandise sales: *"Tumblr is more sort of, well kids really, who you know, clicking is free and they'll look at a thing but they don't, they can't reach into their pockets or anything"* (P6). In some cases it may be better to find a smaller, more appreciative audience elsewhere: P2 found most success posting to gaming communities rather than social media.

Getting lost in the crowd was also of concern to those interviewed, for example P1 and P10 felt that Tumblr could becoming confusing in terms of who created a post, whilst P8 thought it was *"flooded with lots of things like little gifs"*. P2 felt that keeping up with social media was often too time consuming. P9 in particular had some issues with large social media sites. *"Once everybody's on social media just honking their horn non-stop it's too much noise and people aren't interested. [. . .] As soon as every voice is heard with equal volume, nobody can possibly benefit the same as they did when it first started out."* (P9). He had more success with fewer followers, which he feels is because people follow so many people that they do not have time to read everything, or click every link. He believes the sheer volume and speed of posts also makes it harder for the artist to engage with their readers: *"I can't try and keep up with replying to comments, if you reply to everything you don't reply to anything properly."*

So whilst webcomic artists mainly post to large social networking sites, corresponding to where content is read, it appears that the opportunity to reach more people may not be the main or only reason for engaging with social media. Creators show clear preferences for particular social media sites, even if they post to all of them; additionally, outside large social media sites, where content is posted and where it is read varies a lot, implying that creators may not simply 'go where the readers are'.

Results – *Function*

This section looks at possible reasons for choosing one site over another, based on three major function-types derived from the interview themes.

News feed

Everyone interviewed indicated that they use social media to post every time they have an update to their comic, or a piece of news to share. P2 felt that readers tend to use *"social media as a trawl to find out interesting things happening"*. Twitter and Facebook were the most preferred; creators were roughly evenly split between the two, with two artists favouring them equally. Updates are usually in the form of links on both Twitter and Facebook, driving traffic towards the creators' own sites. Twitter was often used for automatic posting of links to new comics, although most creators also liked to post more personal, funny, or interesting tweets in between these links. The idea of using Twitter just for automatic links was not well liked: *"I like reading the little life-titbits of people I follow. [. . .]Under those circumstances, I may as well use an RSS feed"* (P11), and it was only used as such by one creator who had multiple Twitter accounts. Facebook tended to be used for more extensive, less frequent news posts between comic updates. *"If someone's subscribing to you on Twitter they kind of expect that sort of junk and terrible jokes and things. If I clog up somebody's newsfeed [on Facebook] I think they're more likely to just defriend me"* (P3). Tumblr was generally not felt to be suited to this type of content: *"I do cross-post all my updates there, but I don't think my audience is primarily reading it on Tumblr"* (P5).

Interaction

Cultivating a loyal readership is a major goal for webcomics creators: *"there's a lot more value in the long-term fan"* (P5). Two-way interaction between creators and readers helps to achieve this, and is a major reason for using social media. It was felt that *"people like to know there's a person behind a webcomic"* (P8) and rather than simply driving traffic towards their sites, creators see social media as *"a way of building engagement with the audience"* (P3). This leads to increased reader loyalty, which may translate into not only more site visits and merchandise sales, but also support when things go wrong; readers often report art theft and copyright violations to creators through social media, and help them to campaign against it. Other creators do this too, and through social media artists are able to form tight-knit communities who provide each other with advice and encouragement. *"It's that kind of nice connectivity of knowing other artists and having a little gathering of artistic minds"* (P1); *"We're constantly chatting with each other, because it's nice to build those bridges. [. . .] it's massively useful to chat with other people doing this for a living."* (P6).

Of the eight creators who expressed a preference with regards to interaction, six of them preferred Twitter. *"I use [Twitter] to kind of help readers to get to know me [. . .] I vent the frustrations of my daily life in a way that I think is going to be amusing for people"* (P5). A potential reason for this preference is that Twitter provides an instant 'chat'-like element to the interaction; another is that it provides slightly more privacy. Several artists were not comfortable with readers being able to see where they lived, who their family was, and so on, and on Twitter these things are easier to hide. Facebook Pages and Groups were preferred however when creators needed a permanent base for group discussions, longer bits of news, and opinions that were not practical to post in 140 characters. *"I was very keen to sort of build a kind of fan club around the comic so that people would regularly tune in"* (P8). Other suggestions for finding this interactivity were through allowing comments on the main comic's website, or encouraging discussion in dedicated forums.

Portfolio/Art sharing

The third major theme emerging with regards to functionality is the posting of images. Many creators like to showcase their work outside the individual pages of comics that they post. Everyone interviewed preferred Tumblr or Instagram for this. *"Tumblr is mostly just the one panels, and when I do a bit of promotional art like a post card I put it on there and sometimes I ask for feedback."* (P6); *"Instagram is a good community, that is, it feels like the other social networks did at first. [...] The quality of work that you see on there is higher than you get on any other social network"* (P9).

Facebook is used by some, but P10 felt *"it seems like not the done thing, like to put a load of, like an album of drawings on Facebook"*. Twitter is also seen as rather awkward for posting images, as most of the time the user has to follow a link to see the picture properly. However, some did link to their Tumblr posts on Twitter, to reach a wider audience. Deviant Art was occasionally used, although P8 found that the *"submission process was too long winded and archaic for me to update daily"* so it wasn't really suited to a webcomic; P10 felt that this was the case for Flickr also and it was *"not really for sharing things around"*. Additionally, P11 thought that Deviant Art had become an *"art harvesting platform"* and always made sure to add a watermark to her work.

A separate portfolio or art blog is also an option for this functionality. P10 in particular made use of several different art sites for different purposes. *"[M]y portfolio site, that's like the final, main things that I want to show off [...] and then Instagram is literally just anything, just work-in-progress study really [...] and then Tumblr, I can do more, cos I do like animated gifs and little drawings, and also more finished things."*

Conclusions

Webcomics creators post to a wide range of sites, but the most popular are the large social media sites, particularly Facebook, Twitter and Tumblr. These are also the sites where they can reach the most readers. However, interview data shows that creators do have concerns about using these sites, particularly getting lost in the crowd and reaching the right demographic. Often it is more important that the sites they post on have the functionality that they need; their main requirements are news posts and links, interactivity, and displaying work. All three of the preferred social media sites can be used for these purposes, but each one is optimised for a particular thing in the view of creators. It is important for platform designers to consider different types of users and their requirements. Most creators *"use the different mediums for what [they] see as their intended purposes."* (P11), and in order to make the most of what is on offer, creators most often prefer to use a combination of Facebook and Twitter, or Tumblr and Twitter. It is important for these creators to be able to appropriate the functionality they need from each site, and to use them compatibly with other sites.

Creators are generally very aware of how they can find the biggest combination of reach (fame) and convenience (function): *"it changes every year which [social media sites] are the main ones, and which are the best ones, and you just have to kind of stay on top of it"* (P10). Particularly in the case of posting their artwork, creators will choose the site that performs the best for them, and then link to it on the bigger sites to encourage readers to visit them there.

A complex picture of social media use has emerged from this study. Time and effort is needed to create a useful online network, and creators must be highly aware of how they can make the most of the tools available. It is clear that webcomic creators are experimenters, who take different parts of each platform and combine them to form a network that works for them in the way that they need it to. Whilst the structure of a platform dictates how it is used, these creators work around any issues in a sophisticated and strategic way, reaching as wide an audience as they can on some networks and directing them to content on others. It would be beneficial to look at other communities who make use of social networking for various reasons, and how the different functionalities affect interactions of different levels.

Acknowledgements

This work is supported by Horizon Digital Economy Research, RCUK grant EP/G065802/1; and by CREATe, the Centre for Copyright and New Business Models, RCUK grant AH/K000179/1.

References

ComicFury (2014) "ComicFury webcomic hosting". Retrieved 2nd September, 2014, from http://comicfury.com/

Digital Market Ramblings (2014) "How many people use Facebook, Pinterest, Twitter, Gmail and 700 of the top social media and digital services". Retrieved 2nd September, 2014, from http://expandedramblings.com/index.php/resource-how-many-people-use-the-top-social-media

Facebook Newsroom (2014) "Company Info". Retrieved 2nd September, 2014, from http://newsroom.fb.com/company-info/

Fenty, S., Houp, T. & Taylor, L. (2005) "Webcomics: The influence and continuation of the comix revolution." *ImageTexT: Interdisciplinary Comics Studies.* 1(2). Retrieved 30th July, 2014, from: http://www.english.ufl.edu/imagetext/archives/v1_2/group/index.shtml

Guigar, B. (2013) *The Webcomics Handbook: the cartoonists guide to working in the digital age* (Greystone Inn Comics, Philadelphia, PA).

Guigar, B., Kellet, D., Kurtz, S. & Straub, K. (2011) *How to Make Webcomics,* (Image Comics, Berkeley, CA).

McCloud, S. (2000) *Reinventing Comics,* (HarperPerennial, New York, NY).

Walters, M. (2009) "What's up with webcomics? Visual and technological advances in comics." *Interface: The Journal of Education, Community, and Values.* 9(2). Retrieved 30th July, 2014, from: http://commons.pacificu.edu/ cgi/viewcontent.cgi?article=1008&context=inter09

Watson, J. (2012) "Four year experimentaversary, five year anniversary & more experimenting." *HiJiNKS Ensue – A Geek Comic.* Retrieved 30th July, 2014, from: http://hijinksensue.com/2012/05/07/4-year-experimentiversary-5-year-anniversary-more-experimenting/

Wilson, J. R. (2014) "Fundamentals of systems ergonomics/human factors." *Applied Ergonomics.* **45**(1), 5–13.

FROM DIALLING TO TAPPING: ATTITUDES OF YOUNG USERS TO MOBILE PHONES

J. Fowler & J. Noyes

School of Experimental Psychology, University of Bristol, UK

A survey of 168 children, aged 11–14 years, is reported. Three questionnaires were developed to find out about mobile phone use with follow up interviews. It was found that the main purpose of using a mobile phone for 11–14 year olds is communication. Texting and calling have particular advantages of usage, for example, texting allows time for deliberation and social control whereas calling allows a meaningful interaction due to the communication through the voice. The mobile phone is an integral tool in the everyday life of an 11–14 year old and there are many benefits of ownership and use. However, a level of caution in use is recommended due to concerns expressed by participants.

Introduction

As mobile phones become more ubiquitous, their use for various communication modes continues to increase. The mobile is the most popularly owned electronic gadget (Pew Internet and American Life Project, 2011). Ownership and use are most prevalent in young people (Cotten, 2009, Pew Internet and Life Project, 2011). A survey of American teens (12–17 year olds) showed that 37% owned a smart phone (Pew Research Internet Project, 2013). Predictions from Emarketer for 2014 identified the mobile phone rate as being highest in the 17–24 year old age group followed closely by the 12–17 year old age group at 87% (Emarketer, 2013).

The reasons for the technological and social revolution of the mobile phone are due to its portability, multi-purpose use and constant accessibility and reachability. The latter has led to terms such as "always on" (Baron, 2008) and "perpetual contact" (Katz and Aakhus, 2002) as a way of describing the way users interact with this technology. Brown (2011, p. 1) endorsed the impact mobile technology has had on people's lives as "permanently changing the way we work, live and love". Mobile phones are "redefining careers, the family unit and social intercourse" (Arbitron Inc and Jacobs Media, 2011, p. 1). The mobile phone facilitates people in the organisation of their lives. Ling and Ytrri (2002) introduced the terms, micro-coordination and hyper-coordination, to help understand the way users organise their lives through using their mobile phone. Micro-coordination refers to the flexibility that can occur with a mobile phone when arranging meetings; it is possible to change and adapt the agreement as the need arises. Hyperco-ordination refers to micro-coordination and the expressive use of the mobile for social and emotional communication. This connectivity from hyper-coordination refers to work, family and personal relationships.

Communication in mobile phone use

Text messaging was introduced into the domain of communication technology almost by accident. In 1982, a voice mobile telephone system that would work throughout Europe was created by a multinational European initiative known as Group Special Mobile, or GSM. This came into operation by 1992. A piece of leftover bandwidth was made available so that users could create short messages on the keypad by tapping the number keys between one and four times to produce alphabetic characters. This was SMS texting (Short Message System) which very quickly became popular amongst young adults and teenagers.

The creation of a mobile phone communication culture amongst teenagers has been recognised in many studies (De Gournay, 2002, Skog, 2002). Madell and Muncer (2004) found that the most important uses for mobile phones amongst teenagers were for making and receiving calls and texting. Geser (2006) documented the ever-increasing use of the mobile phone by younger and younger users and the importance of the text messaging culture. Reasons for texting as a preferred method for communicating have been put forward; for example, texting allows user time to deliberate over the message being sent and also gives the user a level of social control.

Concerns in everyday use

One of the key factors that have been observed in many studies of mobile phone use is that the very function that people enjoy from using their mobile phone is also the very function that has a negative aspect. For example, a major benefit of the mobile phone is the freedom offered by being able to be contacted at any time, but this constant connectivity and accessibility brings with it the problem of being always available leading to a sense of dependence, sleep disturbance (Thomee, 2011) and in extreme cases, addiction. Negative characteristics range from "disruptive sleep to carpel tunnel syndrome (from constant texting), distraction from the cognitive or social task at hand, or clinical addiction" (Baron, 2009).

Distraction from the task at hand can be a problem when users carry out another activity and use the phone at the same time. The experiments of Hyman et al. (2009) showed that attention was affected when students were walking and using their phones. It was found that mobile phone users were less than half as likely to notice an unusual activity (a unicycling clown) and it was suggested that inattentional blindness occurred as a result of a simple activity that required few cognitive resources.

The purpose of this study is to continue the exploration of this conundrum in teenagers (11–14 year olds) and to find out what young people's perceptions of mobile phone are. Drawing on studies of teenager's use of mobile phones (Madell and Muncer, 2007, Ling and Yttri, 2002), a survey was conducted to find out how teenagers are using their mobile phones and also to find out what their attitudes are. Most previous mobile phone studies have considered empirical usage data. This study, while considering usage issues through questionnaire data, also looked at concerns users themselves have about their phones. The latter is carried out through open-ended interview questions. What 'contradictory pulls' do teenagers experience when they use their mobile phone and what do they think about this? Previous studies have shown communication, in particular texting, to be the most important mobile phone use (Baron, 2009). The first hypothesis is that communication, in particular texting, is the most important use of young mobile phone users (11–14 year olds). Texting, as discussed earlier, was an unexpected area of technological development, largely spurred on by teenage youth. Research Question 1: Is communication, in particular texting, the most important use for young mobile phone users?

To understand young user's attitudes towards mobile phones, the second research question asked users if they preferred to call or text and what influences their decision to text or call. Research Question 2: Do young users prefer to call or text and what influences their decision to text or call? This was broken down into three questions so users were asked, do you prefer to call or text? This was followed by the question, 'What makes you decide to text rather than call?' and 'What makes you decide to call rather than text?

The third research question asked about the kinds of concerns young users have about mobile phones. Research Question 3: What kinds of concerns do young users express about mobile phones? This was carried out through a range of open-ended questions. The first question was a word association task; participants were asked, what are the first three words you think of when you think of mobile phones? To find out further about attitudes, participants were also asked questions about what they liked most about mobile phones and what they liked least. Further questions asked about their opinions of mobile phones and views on why they thought mobile phones had become such an important part of everyday life.

Method

Questionnaires about mobile phone use were completed by 11–14 year olds in small groups in four different schools in Somerset, UK in 2012. There were 168 participants with an equal balance of girls and boys. All participants who carried out the questionnaire were included in the study. The questionnaires involved open ended questions and a range of closed questions. Some questions had a likert scale which was either 3 or 5 point depending on the type of question. 21 participants took part in follow-up interviews. Selection for interview was random. Ethical procedures were followed and respondents had the opportunity to ask questions and make comments. The questionnaires took about 20 minutes to complete.

Table 1.   Three main themes of the IPA.

Super-ordinate themes	Emergent themes	Phone function
Evaluative issues	Useful for homework	Texting, reminders, storing contacts, calculations Physical Internet (music, games)
Communication	Positive: being able to keep in contact with family and friends for micro-coordination and hyper-coordination Negative: being woken up in the night	Texting, Calling, Internet, Snapchat, Facebook
Physical Attributes/functions	Entertainment (music/games)	Internet (music, games)

Results

Questionnaire results show that nearly all 11–14 year olds owned a mobile phone (98%). The most popular use was texting. Over three quarters of participants said they liked to play games on their phone (75.6%) and over three quarters liked to listen to music (84.5%). Responses revealed that over three quarters (77%) access the internet through their mobile phone.

Five themes were identified from an Interpretative Phenomenological Analysis (IPA) of the interviews. These were Communication, Physical Attributes and Functions, Evaluative issues, Cost and Safety Issues. The three main themes of the IPA are shown in Table 1.

Other themes were issues around safety (18.5%) and concerns to do with cost (13.7%). Some respondents made no comment (2.4%). Many valued the safety and security that the phone affords them: "In an emergency, if you want to get hold of someone." "I think it's good because they can help you if you're lost or need something, then it saves time, you can just call someone." "If someone took you, you could record it so you could play it back and listen to it." "It makes me feel safer."

Cost was mentioned in the interviews as a reason for why they would text rather than call. "Texting is cheaper and it doesn't use up your money so much." "Calling takes a bit more credit than texting." "It's always more expensive to call people."

Analysis of the first three words

Analysis of first word responses showed that nearly half of young users mentioned texting (46.4%) whereas only a few referred to calling (7.1%). The words, communication (6.5%) and contact (4.2%) were reported by a few participants. Perceived

Table 2. Aspects that were liked most.

Communication	63.7%
Physical	26.2%
Evaluative issues	10.1%

Table 3. Aspects that were liked least.

Communication	16.7%
Physical	20.8%
Evaluative issues	25.6%

ease of use is an important consideration for users and with no prior information or prompting, the word, easy, was mentioned by 7.1%.

The like most/like least responses

Tables 2 and 3 show the major components of the coding scheme created to classify responses for the like most/like least question.

Communication was the main aspect liked most by users with nearly two thirds of responses (63.7%). The most significant 'like least' factors are those to do with 'evaluative issues'; the most significant factor being the stress that can be caused by dependency on the phone.

Discussion

Questionnaire results show that 98% of 11–14 year olds own a mobile phone. This is a high number of teens compared to the Pew Research Internet Project (2013) where it was found that 78% of teens (12–17 year olds) owned a mobile phone.

The purpose of the mobile for communication

The IPA analysis of the interviews showed that nearly two-thirds of participants saw communication as the characteristic they liked most about their phone. Previous studies identified communication through talking and texting as the main use of the mobile phone. Now that the mobile phone has become such a multi-purpose device, it is interesting that communication is perceived as the most significant use. Results showed that the most popular use of the mobile phone was texting. Nearly all users identified sending and receiving texts (94%) as the most popular use, followed closely by making and receiving calls (80%). Other studies (Geser, 2006, Madell and Muncer, 2004) found that texting was the most popular use amongst teenagers. Phones are "good for keeping in contact with people" and "good for relationships".

To talk or text

When asked whether they would choose to talk face to face with each other, talk on the phone or text, teens chose face-to-face contact (97%) over talking (89.3%) and texting (91.1%). However, texting and talking for communication is highly valued by participants.

"It is easier to text". Texting information allows users time "to choose what information to include" and also a time when it is convenient to them. It also saves them from become embroiled in extended conversations. Other reasons for texting rather than calling were to do with cost, noisy surroundings or poor reception, as evidenced by these participants: "It depends what the noise is like round you because sometimes it's really noisy, you might not be able to hear them if you ring them". "If you call someone on the phone it might cut out because the range isn't that good".

Calling is used for micro-coordination when an instant response is needed; for example, one participant said she used calling if she needed "to know something really quickly and I haven't got time and I know they'll answer" and "I would prefer to call, just because it's quite hard, not hard, but, it takes a long time to text but in a phone call you can get everything done much quicker".

Calling however mainly facilitates hyper-coordination as the preferred way of communicating because teens like to "hear the tone or expression in people's voices". Listening to the human voice is active but texting is passive. Receiving a call and hearing the human voice captures "the essence of the person" (Baron, Radio 4, April 14, 2014) The interviews also indicate that when participants want a more 'meaningful' contact, they prefer to call because "you can get a sense of how the other person is feeling". "I think I only want to ring if it's a really personal issue or if someone's upset about something. And I want to talk to them to find out how they're feeling because with texting you don't get to know how they're feeling. It's completely different. They could be making up that they're re upset or they are fine when they're not." Another participant commented on the length of time texting would take if she were trying to have a conversation, "Just because you can have a conversation, like an hour's conversation".

Everyday concerns

Users expressed a range of everyday concerns. Comments from the interviews show concerns from 11–14 year olds. "I think they could get quite addictive. If you have a routine of going home or texting your friends, listening to music and playing games and um, it's something to do when you're bored as well, you can just, yeah, go on your phone". With reference to dependency and possible addiction, it was found that a high number of teens slept with a phone next to their bed (82%) and many left their phone on when they went to bed (72%). This means that sleep disturbance from calls or texts during the night is likely. The mobile phone has become such an integral part of individual's lives that many do not wish to part with the phone even when sleeping. However, sleep disturbance from mobile phones has become a problem for young people (Thomee, 2011) and in extreme cases, stress and addiction to the mobile phone has occurred (Arbitron Inc and Jacobs Media, 2011).

The findings of this study are useful for identifying the main trends in mobile phone use in this age group. They are important for many reasons and for many organizations, including educational establishments, mobile phone operators and networks providers. If, for example, it is known the reasons why individuals choose to text rather than talk or talk rather than text, then educators. mobile phone companies and network providers can plan future developments with this in mind.

References

Arbitron Inc and Jacobs Media. (2011) [Guest Blog]: The impact of smart-phones on American life, 1–3. Retrieved 11th January 2011 from http://redcrowblog.blogspot.com/2010/01/guest-blog-impact-of-smartphones-

Baron, N. S. (2014) 14 April 2014, 16:30. BBC Radio 4, Digital Human, Series 5, Voice and Text in the Digital World, -Codec Nova, codec-nova.com/programmes/b040hjy6

Baron, N. S. (2011) Concerns about mobile phones: A cross national study, *First Monday*, **16**(8), 1–28. Retrieved January 12th 2011 from: http://firstmonday.org/htbin/ojs/index.php/fm/article/view/article/3335/30

Baron, N. S. (2009) Talk about texting: Attitudes towards Mobile Phones. In Proceedings, London Workshop of Writing, University of London.

Baron, N. S. (2008) *Always On: Language in an online and mobile world.* New York: Oxford University Press.

Brown, A. (2011) Cell phone usage. Retrieved February 10th 2011 from http://www.compukiss.com/articles/cell-phone-usages.html.

Cotten, S. (2009) Old wine in a new technology, or a different type of digital divide? *New Media Society*, **11**(7), 1163–1186.

De Gournay, C. (2002) Pretence of intimacy in France. In Katz, J. E. and Aakhus, M. (Eds) *Perpetual contact: Mobile communication, private talk, public performance.* (University Press, Cambridge).

Emarketer, (2013) April 24, 2013, UK Teens Far Outshine US Counterparts in Smartphone Usage, Retrieved October 9th 2014, eMarketer www. Emar-keter.com/Article/UK Teens-Far

Geser, T. (2006) Pre teen cell phone adoption: consequences of later patterns of usage and involvement, sociology of the mobile phone, Retrieved January 11th 2011 from http://socio.ch/mobile/t_geser2.htm

Hyman, I. E., Boss, S. M. Wise, K. E. & Caggiano, M. (2009) Did you see the unicycling clown? Inattentional blindness while walking and talking on a cell phone. *Applied Cognitive Psychology*, **24**(5), 597–607.

Katz, J. E. & Aakhus, M. (Eds) (2002) *Perpetual contact: Mobile communication, private talk, public performance.* Cambridge: Cambridge University Press.

Ling, R. & Yttri, B. (2002) Hyperco-ordination via mobile phones in Norway. In Katz, J. E. and Aakhus, M. (Eds) *Perpetual contact: Mobile communication, private talk, public performance*. Cambridge: Cambridge University Press.

Madell D. & Muncer, S. J. (2004) Back from the beach but hanging on the telephone? English adolescents' attitudes and experiences of mobile phones and the Internet. *Cyberpsychology and Behaviour*, **7**(3), 359–367.

Madell, D. & Muncer, S. J. (2007) Control over Social Interactions: An important reason for Young People's Use of the Internet and Mobile Phones for Communication, *Cyberpsychology and Behaviour*, 10,1,137–140. Doi:10.1089/cpb.

Pew Internet and American Life Project, (2011) Smartphone Adoption and Usage, Retrieved 11th January from http://itcnetwork.org/ 2011/11/Smartphone adoption and usage-2011/

Pew Research Internet Project (2013) Teens and Technology 2013, Retrieved June 10th 2013 from http://www.pewintenet.org/2013/13/teens and technology-2013/

Skog, B. (2002) Mobiles and the Norwegian Teen: identity, gender and class In Katz, J. E. and Aakhus, M. (Eds.), *Perpetual contact: Mobile communication, private talk, public performance*. (University Press, Cambridge).

Smith, J. A. & Osborn, M. (2003) Interpretative Phenomenological Analysis. In J. Smith (Ed). *Qualitative Psychology*. (pp. 51–80). London: Sage Publications.

Thomee, S., Harenstam, A. & Hagberg, M. (2011) Mobile Phone Use and Stress, sleep disturbances, and symptoms of depression among young adults – a prospective cohort study, *BMC Pubic Health*, 11.66 doi.10:1186/1471 2458-11-6

THE IMPACT OF TASK WORKFLOW DESIGN ON CITIZEN SCIENCE USERS AND RESULTS

James Sprinks[1], Robert Houghton[2], Steven Bamford[3] & Jeremy Morley[1]

[1]*Nottingham Geospatial Institute, University of Nottingham, UK*
[2]*Human Factors Research Group, University of Nottingham, UK*
[3]*School of Physics and Astronomy, University of Nottingham, UK*

Virtual citizen science platforms allow non-scientists to take part in scientific research across a range of disciplines. In addition to this, what they ask of participants varies considerably in terms of task type, user judgement required and user freedom. A study was performed, involving the Planet Four platform, to investigate the effect of task workflow design on both the user and the scientific results. Results showed that participants found the autonomous interface with more variety faster to learn ($p < 0.05$) and easier to use ($p < 0.05$) than other interfaces, while more rigid and less complex interfaces resulted in more time being spent on each analysis and more results being collected ($p < 0.01$).

Introduction

Citizen science, also known as *"public participation in scientific research"* (Hand, 2010), can be described as research conducted, in whole or in part, by amateur or nonprofessional participants often through crowd sourcing techniques. Citizen science projects exist that require the participant to either act as a sensor – collecting data in the wild using an array of mobile technologies, or utilise Virtual Citizen Science (VCS) platforms (Reed *et al.*, 2012) – usually analysing previously collected data through a website platform. Launched in 2009, the Zooniverse is home to some of the internet's most popular VCS projects, and the scientific research addressed is wide-ranging, with participants asked to (for example) classify different types of galaxies from photographs taken by telescopes, transcribe historical ships logs and weather readings, or marking craters found on images of planetary surfaces. Due to citizen science being a new area of work, although there has been research into interface HCI and functionality (Prestopnik & Crowston, 2012), there has been relatively little attention paid specifically to human factors issues. This is perhaps ironic given the importance of the 'citizen' part of the discipline, especially as the effectiveness of a citizen science venture is related to its ability to attract and retain engaged users, both to analyse the large amount of data required, and to ensure the quality of the data collected (Prather *et al.*, 2013). At the present time, citizen science platforms tend to require the user to carry out tasks in a very repetitious manner, the design of which are arguably driven more by the 'science case' (analogous to a 'business case') rather than any consideration of the experience of the citizen scientist. In the study reported here we make a first step in considering how virtual citizen science systems can be better designed for the needs of the citizen scientists by exploring whether manipulating task flow would affect time spent on task and number of features indicated, as well as user ratings on difficulty and usability issues.

Some studies have considered motivation amongst citizen science volunteers (Raddick *et al.,* 2010, Reed *et al.,* 2013) but not considered in any depth the form of work activity itself. This may be considered remiss as over thirty years of human factors research has identified a relationship between motivation, satisfaction and work design. Hackman & Oldham (1975) developed the 'Job Diagnostic Survey' in order to better understand jobs and how they could be re-designed to improve motivation and productivity. Factors such as task variety, complexity and autonomy were identified as key to this process, all of which can be influenced in VCS design. Building on these findings, further research has found a positive correlation between motivation and task complexity (Gerhart, 1987, Chung-Yan, 2010), task autonomy (Dubinsky & Skinner, 1984, Chung-Yan, 2010) and variety (Ghani & Deshpande, 1994, Dubinsky & Skinner, 1984).

This paper describes the current active Zooniverse site *Planet Four,* with over 100,000 registered users as the research context, presenting the results of three different iterations of the site that differ in task workflow design, in terms of user experience/satisfaction and scientific output. Finally the impact of task workflow design on these results, and the implications for VCS platforms and other online mechanisms, are discussed.

Methodology

In order to investigate the effect of task workflow design on user experience and scientific results, a new version of the Zooniverse's Planet Four platform has been developed. The new site allows users to mark craters on the planetary surface, and an initial study has been carried out to both consider task workflow factors and also act as a technical test, identifying any general functionality and usability issues that might be present.

Participants

A total of 30 participants took part in the pilot study of the crater counting platform between January and March 2014. There were no specific prerequisites for participation in order to reflect the open-access nature of citizen science web platforms.

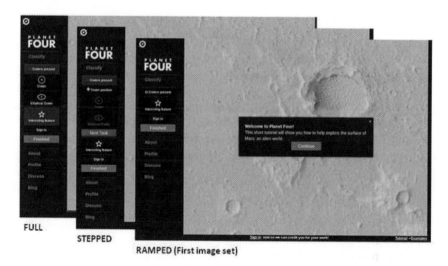

Figure 1.   Screenshots of planet four study interfaces.

Materials

For the study, image G05_020119_1895_XN_09N198W taken by the context camera (CTX) from NASA's Mars Reconnaissance Orbiter mission was used to be analysed by the participants. It was chosen because it contains a variety of landscapes common to the Martian surface, and has also been previously studied by the scientific community, therefore allowing comparisons to be made between citizen scientist results and those measured by planetary science experts. Before being uploaded to the platform, the image was 'sliced' into a number of smaller images that can be more easily handled. Original NASA imagery is often gigabytes in size, making it time consuming to upload to a website display. A total of 78 smaller image 'slices' were created, measuring 840×648 pixels with an included overlap of 100 pixels to ensure features on the edges are adequately displayed.

The platform has been developed to include three different interfaces for marking craters that vary in task type, number of tasks available to the user and user freedom. Figure 1 shows screen shots of these three interfaces, which include:

- **FULL**: Users have access to all the tools and can complete all crater marking tasks for each image in any chosen order
- **STEPPED**: All tools are made available to the user and all tasks completed in a predefined order (increasing in complexity) for each image, which cannot be diverged from
- **RAMPED**: Users have access to one tool and complete one crater marking task for a set number of images, then use another tool and complete another task (increasing in complexity) for the next set of images etc.

Design & procedure

The study took place in the same room and was carried out on the same laptop for each participant to ensure factors such as lighting conditions and screen setup remained constant, as differences could otherwise influence an image analysis task. Participants used each of the interfaces in turn to mark craters on a set number of the image slices, with the interface order differed (same number of participants doing each order) to account for bias caused by learning of the system. The order in which image slices were displayed to each participant was also randomised, to prevent bias being caused by image content (images with little or no craters appearing in the same interface each time etc.). Before using each interface, each participant completed an online tutorial learning how to use the tools, marking craters on a separate example image.

After using each interface, each participant completed a questionnaire asking them to rate their agreement, via a nine-point Likert scale, with a number of different statements across themes including *design & usability*, *tasks & tools* and *imagery*. At the end of each section, 'free text' boxes were available for participants to give additional comments and opinions.

Results

Participant questionnaire results

Participant responses to a number of statements included in the *design & usability* and *imagery* sections of the questionnaire showed no significant difference between each interface, and as such will not be discussed in this paper. This is as expected as many of the design features and the image format are constant across each interface.

When considering the visual attractiveness of the site and image quality, participant responses were positively skewed. Over 80% of all raw scores across each interface lie between 6 and 9 (very attractive/high quality). Likewise when considering how demanding the task of marking craters is, over 70% of all scores across each interface lie between 6 and 9 (low demand). In terms of crater tool 'ease of use', again there is no significant difference in scores between each interface. Participant responses however showed no clear preference, with the FULL, STEPPED and RAMPED interfaces having mean scores of 6.00, 5.70 and 5.77 respectively.

Differences between each interface do exist when considering how quickly participants learnt to use them. A repeated measures ANOVA shows a statistically significant difference in participant scores between interfaces ($F(1.297, 37.621) = 6.232, p < 0.05$), and Post hoc tests using the least significant difference correction revealed that participants indicated that they learnt the FULL interface more quickly than the STEPPED (mean score of 7.90 ± 1.54 vs. 6.87 ± 2.06, $p < 0.002$). Likewise when considering how easy to use the participants found each interface, a statistically significant difference is shown ($F(1.817, 52.684) = 4.957$, $p < 0.05$). Participants scored the FULL interface easier to use than the STEPPED

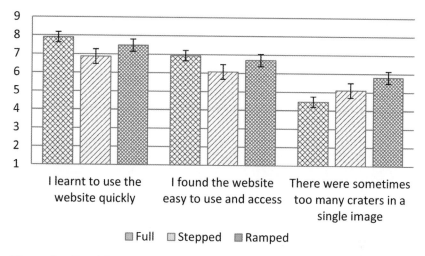

Figure 2.   Participant questionnaire results (with standard error shown).

(mean score of 6.93 ± 1.68 vs. 6.07 ± 2.18, $p < 0.02$) and slightly easier than the RAMPED (6.70 ± 1.80), although the difference is not statistically significant ($p = 0.118$).

In terms of the number of craters in each image, there is also a difference in participant responses across the interfaces ($F(1.811, 52.507) = 5.184$, $p < 0.05$). Participants felt that there were sometimes too many craters in an image with the RAMPED interface compared to the FULL (mean score of 5.80 ± 2.20 vs. 4.47 ± 2.05, $p < 0.005$) and slightly more than the STEPPED (5.10 ± 2.16), although the difference is not statistically significant ($p = 0.053$). Figure 2 shows the average participant level of agreement to these statements for each interface, along with standard error.

Crater marking results

Participant crater marking behaviour has been compared across each interface in terms of percentage of participants who marked craters per image, number of markings per image and time on each image. As explained in the methodology, the RAMPED interface requires participants to use one tool over a number of images, and then another tool etc. therefore RAMPED (position) represents results where participants only mark the centre of craters and RAMPED (mark) represents results where participants mark the shape. The comparatively large values of standard deviation can be explained by image variation, with some images having no craters present and others having several dozen.

When considering the percentage of participants that marked craters per image, a repeated measures ANOVA shows a statistically significant difference between

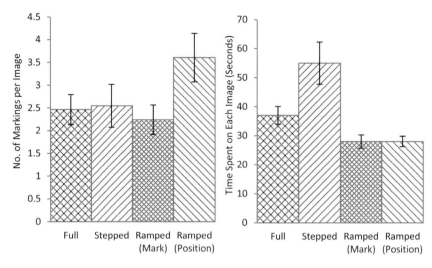

Figure 3.   Crater marking results (with standard error shown).

each interface ($F(2.685,\ 204.055) = 2.709$, $p < 0.05$). Post hoc tests using the least significant difference correction revealed that participants were more likely to mark craters using the RAMPED (position) interface compared to the STEPPED ($63.24 \pm 33.76\%$ vs. $53.67 \pm 33.21\%$ of users marking craters per image) although the difference is not statistically significant ($p = 0.098$). The percentage of participants that marked craters per image for the FULL and RAMPED (mark) are $58.24 \pm 35.45\%$ and $57.94 \pm 35.02\%$ respectively.

In terms of the number of crater markings per image, again a statistically significant difference is shown ($F(2.656,\ 201.83) = 7.416$, $p < 0.0005$), where the RAMPED (position) interface resulted in a greater number of markings (3.61 ± 4.67) compared to the FULL (2.46 ± 2.93, $p < 0.002$), STEPPED (2.55 ± 4.17, $p < 0.005$) and RAMPED (mark) (2.24 ± 2.85, $p < 0.002$) interfaces. Finally when considering the amount of time spent on each image, a statistically significant difference again exists across interfaces ($F(1.570,\ 119.290) = 12.755$, $p < 0.0005$). Participants spent more time using the STEPPED interface compared to the RAMPED interfaces (55 ± 64 seconds vs. 28 ± 16 seconds, $p < 0.002$) and more time compared to the FULL (37 ± 27 seconds, $p < 0.02$). Figure 3 shows the average number of crater markings and the average time spent on each image using each interface, along with standard error.

Conclusion

The Planet Four task design study found that altering the task workflow design of the interface can have an effect both on the user experience and on the resulting scientific data. Participants found the FULL interface design, which has the greatest user freedom and variety, both quicker to learn and easier to use than the other interfaces. This is in agreement with Hackman & Oldham's (1975) findings which suggest that both variety and autonomy are important in ensuring greater job satisfaction. Participants found the RAMPED interface, which requires completing only one task per image, was more of an issue with large numbers of craters, again highlighting the importance of task variety. When considering scientific output, again task workflow design can have an effect. The RAMPED interface resulted in both a greater percentage of images with craters marks, and a greater number of craters marked per image than the other interfaces. This is an important result, as reducing the number of null returns (images with no markings) would in turn reduce the time spent on data reduction by the science team. Participants using the STEPPED interface spent more time on each image than the other interfaces, suggesting that when users are forced to complete the task in a set, non-autonomous manner they are perhaps more considered in their analysis and are less likely to miss an important step.

The effect task workflow design has on both user experience and scientific results could have major implications for future VCS platform design. Although this study agrees with existing literature regarding the importance of variety, autonomy and complexity in user satisfaction, it also suggests that designs involving more rigid, non-autonomous workflows or repetitive tasks with less complexity can result in more considered, complete data or a greater data density respectively. When considering task workflow design, future VCS platform science teams and developers will need to perform a balancing act, weighing up the importance of user satisfaction and community engagement against the data needs of the science case and the resources that can be committed to data reduction, more than likely on a case by case basis.

While the present paper is concerned with volunteer work, it could be argued many similar forms of work are being carried out on a paid basis, via services such as Amazon's Mechanical Turk. With access to online services and data likely to become more widespread in the future, these work mechanisms could become more commonplace, and as such the impact of task workflow design on both their success and popularity will become an important consideration.

Acknowledgement

The first author is supported by the Horizon Centre for Doctoral Training at the University of Nottingham (RCUK Grant No. EP/G037574/1) and by the RCUK's Horizon Digital Economy Research Institute (RCUK Grant No. EP/G065802/1). Special thanks to Brian Carstensen, web developer and Michael Parrish, software developer based at the Adler Planetarium, Chicago for their support in developing the Planet Four interfaces. Special thanks also to Jenny Taylor, planetary seismologist at the University of Bristol, for developing the crater counting science case and identifying the required imagery.

References

Chung-Yan, G. A. (2010) The Nonlinear Effects of Job Complexity and Autonomy on Job Satisfaction, Turnover, and Psychological Well-Being. *Journal of Occupational Health Psychology* **15**(3): 237–251.

Dubinsky, A. J. & Skinner, S. J. (1984) Impact of Job Characteristics on Retail Salespeople's Reactions to Their Jobs. *Journal of Retailing* **60**(2): 35–62.

Gerhart, B. 1987, How Important are Dispositional Factors as Determinants of Job Satisfaction? Implications for Job Design and Other Personnel Programs. *Journal of Applied Psychology* **72**(3): 366–373.

Ghani, J. A. & Deshpande, S. P. (1994) Task Characteristics and the Experience of Optimal Flow in Human-Computer Interaction. *The Journal of Psychology* **128**(4): 381–391.

Hackman, J. R. & Oldham, G. R. (1975) Development of the Job Diagnostic Survey. *Journal of Applied Psychology* **60**(2): 159–170.

Hand, E. (2010) Citizen Science: People power. *Nature* **466**(7307): 685–687.

Pelli, D. G. & Farell, B. (2010) Psychological Methods. *Handbook of Optics* **3**(3): 3.1–3.12.

Prather, E. E., Cormier, S., Wallace, C. S., Lintott, C., Raddick, M. J. & Smith, A. (2013) Measuring the Conceptual Understandings of Citizen Scientists Participating in Zooniverse Projects: A First Approach. *Astronomy Education Review* **12**(1): 010109.

Prestopnik, N. R. & Crowston, K. (2012) Citizen Science System Assemblages: Understanding the Technologies that Support Crowdsourced Science. *Proceedings of the 2012 iConference* 168–176.

Raddick, M. J., Bracey, G., Gay, P. L., Lintott, C. J., Murray, P., Schawinski, K., Szalay, A. S. & Vandenberg, J. (2010) Galaxy Zoo: Exploring the Motivations of Citizen Science Volunteers. *Astronomy Education Review* **9**: 010103-1.

Reed, J., Raddick, M. J., Lardner, A. & Carney, K. (2013) An Exploratory Factor Analysis of Motivations for Participating in Zooniverse, a Collection of Virtual Citizen Science Projects. *Proceedings of the 2013 46th Hawaii International Conference on System Sciences* 610–619.

Reed, J., Rodriguez, W. & Rickhoff, A. (2012) A framework for defining and describing key design features of virtual citizen science projects. *Proceedings of the 2012 iConference* 623–625.

MOBILE TECHNOLOGY AND NEW WAYS OF WORKING – IS IT HELPING OR HINDERING HEALTH AND PRODUCTIVITY?

Jim Taylour[1] & Patrick W. Jordan[2]

[1]*Orangebox Limited, Parc Nantgarw, Cardiff*
[2]*Loughborough University, Loughborough*

Background

Rapid changes are occurring in people's work environments and the devices that they use such as tablets, smartphones, laptops and emerging wearable technology. This raises issues for how to mitigate work-related discomfort including new strains of potential ill health brought on by new ways of working. The current Display Screen Equipment (DSE) regulations have not been updated since 2002 and have little or nothing to say about either these devices or mobile working practices.

In addition, flexible working practices and the changing nature of the office mean that more and more people spend a significant proportion of their working time away from a fixed workstation. A recent study by the authors investigated the pros and cons of this, in terms of productivity and wellbeing. Results suggest that flexible working is popular, increases wellbeing and is more productive. There are drawbacks too, including expectations of longer working hours and lack of visibility within an organisation. Meanwhile offices themselves need to be able to support a variety of types of work, devices and spaces, something that many do not currently do well.

We want to debate some of the issues raised in our paper proposals at the conference and generate interest in the subject of health issues surrounding new technology and new ways of working. Is the CIEHF behind on this topic and is there more in depth research required? We think so but we also think there's some interesting initiatives already in place that need debating.

Purpose

The purpose of the workshop is to highlight these emerging workplace trends, and debate the role of the ergonomics community (both academic and commercial) in measuring the potential health issues, challenging existing guidance where necessary and developing and communicating best practice at home, the workplace and in schools.

STRESS

ETHNICITY AND WORK-RELATED STRESS: MIGRANT WORKERS IN SOUTHERN ITALY

Roberto Capasso & Maria Clelia Zurlo

University of Naples "Federico II", Italy

This paper reports a study of ethnicity and work-related stress that takes the Demands-Resources-Individual-Effects (DRIVE) model (Mark & Smith, 2008) as a framework of reference. The model combines individual differences, job characteristics, ethnicity dimensions and appraisals as independent variables in the prediction of psychophysical health outcomes and it was applied in a sample of 900 workers, differing in ethnicity, in Southern Italy. Results confirmed the hypotheses predicted by the suggested model, showed different profiles of associations and the main effects of the independent variables on the psychophysical health outcomes. The importance of including ethnicity dimensions in work-related stress research requires further investigation.

Introduction

This research proposes and tests an ethnicity and work stress model (taking a cue from the DRIVE model, Mark and Smith, 2008) that would combine aspects of traditional job stress models (the DCS and ERI) with individual differences in the forms of coping styles and personality behaviours, appraisals (job satisfaction/stress) and all the ethnicity variables in the prediction of psychophysical health outcomes. The DRIVE model was considered as a framework for this new model because some of its key relationships propose that work demands, individual differences, and work resources have main effects on anxiety, depression, and job satisfaction and that work resources and individual differences may moderate the relationship between work demands and health outcomes, while perceived job stress may mediate this relationship.

In accordance with the literature on stress and wellbeing at work, ethnicity and culture represent critical issues because there has not been enough consideration of them in studies of work stress. Measures of work characteristics and/or work stress have been developed largely within single ethnic group data sets. Most of studies measure ethnicity as a descriptor of the population studied or as an objective category (i.e. country of birth, nationality, language, skin colour, origin, racial group) and associate some descriptive aspects of ethnicity with psychophysical health conditions (Karlsen et al., 2005), work characteristics or appraisals

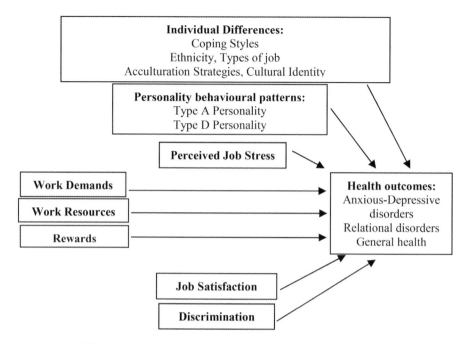

Figure 1.   An ethnicity and work-related stress model.

(Nazroo, 2003; Szczepura et al., 2004) as single associations rather than in a general model that integrates ethnicity and work-related aspects in a transactional perspective. Cultural dimensions are very complex because they are related to three core concepts: acculturation (Berry, 1997), racial discrimination (Noh & Kaspar, 2003) and ethnic identity (Phinney, 1992). For this reason, this study considers all these ethnicity dimensions and their relationships with individual differences, work characteristics, appraisals and health outcomes. The suggested model can be seen below (Figure 1) and testing it formed the basis for the research. All the dimensions involved in this model, namely individual differences, work characteristics, appraisals and ethnicity (independent variables) were evaluated.

The above figure shows that work demands, individual differences and personality behavioural patterns, work resources/rewards, appraisals (perceived job satisfaction/stress) and ethnicity dimensions (acculturation/cultural identity behaviours and racial discrimination) are all proposed to have main effects on anxious-depressive symptoms, relational disorders and physical problems. For this reason the aim of the present study was to test a model of stress that integrates all the ethnicity aspects with work-related stress dimensions and investigates the associations between individual differences, work characteristics, appraisals, ethnicity and health outcomes in sample of varying in ethnicity.

Method

Sample and materials

Multi-stage sampling was used in the selection of the study sample considering the following inclusion criteria on the basis of: largest ethnic communities around Naples and Caserta (Eastern European, Moroccan and Ghanaian), employment, aged between 25–45 years and language capacities. We paid close attention to job type in particular in order to reduce the confounding factors related to different work environments while ensuring that each ethnic group had a similar profile of associations between ethnicity, work characteristics and health. A total of 1061 questionnaires were distributed, with 900 returned and considered valid (response rate = 84.8%) and the sample of the study consisted of: Eastern European care workers (N = 250), Moroccan factory workers (N = 250), Ghanaian masons (N = 200), Italian factory workers (N = 100) and Italian masons (N = 100).

This questionnaire consisted of six sections.

- Section 1: respondent's personal and biographical details (e.g., gender, age, nationality, education) and job characteristics (e.g., employment, type of contract, number of hours worked).
- Section 2: individual characteristics; Coping Style Inventory (Cooper, Sloan, & Williams, 1988), Bortner's Type A Behavioural Style Inventory (Bortner, 1969), Type D Personality (Denollet, 2005).
- Section 3: cultural dimensions; Multigroup Ethnic Identity Measure (MEIM, Phinney, 1992), Berry's measures of acculturation (1997) and racial discrimination (a single item reported discrimination at work on the basis of race or ethnicity – Smith et al., 2000).
- Section 4: work characteristics; Job Content Questionnaire (JCQ; Karasek, 1985), Effort-Reward Imbalance (ERI test; Siegrist, 1996).
- Section 5: psychophysical health; Symptom Checklist 90 R (SCL-90-R, Derogatis, 1994) and a single item asking "Over the past 12 months, how would you say your general health has been?" (Smith et al., 2000).
- Section 6: appraisals (perceived job satisfaction/stress); Job Satisfaction Scale (Warr, Cook & Wall, 1979), a single item asking "In general, how do you find your job?" (Smith et al., 2000).

The principal component analyses (PCA) of the subscales for each dimension reported in the questionnaire above were run to reduce the huge numbers of variables. All the factors extracted for this study were split at the median into low and high groups and in three categories for the acculturation dimensions taking into account the different recoding for these cultural factors due to the presence of the missing values corresponding to the Italian sample (Capasso, thesis, in preparation). The statistical analyses carried out for the whole sample (using SPSS-X software) were: descriptive analyses for gender, age, ethnicity, marital status, education, type of job, work status and type of contract; and forward LR logistic regression analyses between the psychophysical health outcomes and independent variables.

Table 1.   Multi-variable associations of significant independent variables with anxious-depressive disorders.

Variables	OR	CI
Nationality/Jobs		
Italian masons	1.00	
Eastern European care workers	2.955	1.573–5.552
Moroccan factory workers	.460	.285–.918
Ghanaian masons	.314	.155–.698
Italian factory workers	5.735	2.520–13.051
Work Demands		
Low	1.00	
High	3.275	2.107–6.456
Objective Coping		
Low	1.00	
High	.371	.189–.775
Type A behaviour		
Low	1.00	
High	3.516	2.223–7.466
Social Inhibition		
Low	1.00	
High	3.383	2.158–6.808
Work Stress		
Low	1.00	
High	1.919	1.361–3.971
Affirmation/Maintenance Culture		
Low	1.00	
High	.404	.213–.852
Perceived discrimination		
Low	1.00	
High	2.422	1.344–4.856

Results

Among the whole sample of workers, 68.1% were men, 27.8% were from Eastern Europe, the same percentage were Moroccan and 22.2% were Ghanaian (same percentage were Italians). Moreover 38.8% worked as factory workers, 33.3% as masons and 27.7% as care workers for the elderly. Most of them were married (95.5%) and 50.1% had a high school education compared to 47.3% who had a middle school education. Finally, 62.6% worked part-time and 48.2% with a fixed-term contract. Logistic regression analyses were carried out to determine the main effects in the whole sample of workers (see tables 1, 2, 3).

Table 1 shows the significant multivariable associations of individual differences, work characteristics, ethnicity dimensions and appraisals on Anxious-Depressive disorders. The workers with perceptions of high work demands and work stress, with high recourse to social inhibition and Type A behaviours and who had experienced racial discrimination were more likely to suffer Anxious-Depressive

Table 2. Multi-variable associations of significant independent variables with relational disorders.

Variables	OR	CI
Nationality/Jobs		
Italian masons	1.00	
Eastern European care workers	.362	.155–.732
Moroccan factory workers	3.230	1.589–6.565
Ghanaian masons	3.134	1.507–6.518
Italian factory workers	2.814	1.375–5.774
Work Demands		
Low	1.00	
High	1.939	1.121–3.834
Rewards		
Low	1.00	
High	.497	.301–.919
Objective Coping		
Low	1.00	
High	.475	.286–.892
Total Job Satisfaction		
Low	1.00	
High	.276	.142–.569
Search identity/Adoption of the host culture		
Low	1.00	
High	.505	.319–.935
Perceived discrimination		
Low	1.00	
High	2.525	1.994–4.879

disorders. Moreover those who used high levels of objective coping and an affirmation/maintenance culture were less likely to suffer these psychological disorders.

Table 2 shows that the group of workers who perceived high levels of work demands and who had experienced racial discrimination were more likely to suffer relational disorders. Moreover the workers who perceived high levels of rewards and job satisfaction and who used objective coping and a search identity/adoption of the host culture were less likely to suffer this psychological disorder.

Finally data reported in table 3 show that the group of workers with high perception of work demands and with high recourse to Type A behaviours were more likely to report poorer general health, while those who perceived high levels of rewards and who favoured objective coping and an affirmation/maintenance culture were less likely to suffer poor health. Further analyses examined interactions between variables but these were largely absent. Effects of nationality are considered elsewhere (Capasso, in preparation).

Table 3. Multi-variable associations of significant independent variables with general health.

Variables	OR	CI
Nationality/Jobs		
Italian masons	1.00	
Eastern European care workers	.462	.232–.919
Moroccan factory workers	.137	.065–.289
Ghanaian masons	.095	.046–.197
Italian factory workers	2.497	1.008–6.188
Work Demands		
Low	1.00	
High	2.956	1.924–5.966
Rewards		
Low	1.00	
High	.403	.201–.804
Objective Coping		
Low	1.00	
High	.477	.311–.907
Type A behavior		
Low	1.00	
High	3.235	2.127–6.317
Affirmation/Maintenance culture		
Low	1.00	
High	.351	.162–.721

Conclusion

Findings confirmed the main effects of the dimensions related to occupational stress on the health outcomes and the importance of considering each aspect of ethnicity and its potential relationships with these variables. In the prediction of Anxious-Depressive disorders, work demands and work stress, the personality behavioural patterns of social inhibition and Type A and racial discrimination were associated with higher risk of anxiety and depression, while objective coping and an affirmation/maintenance culture seem to reduce this risk. Moreover high levels of work demands and racial discrimination were associated with higher risk to report relational disorders, while rewards, the objective coping strategy, the cultural dimensions of search identity/adoption of the host culture and perceived job satisfaction appear to be associated with a lower risk of this psychological outcome. Finally with respect to general health, high recourse to Type A behaviours and high perception of work demands were associated with poorer general health, while rewards, objective coping and affirmation/maintenance culture seems to reduce the risk of reporting physical problems.

These results confirmed the main effects suggested by the proposed model, the associations of coping strategies and work characteristics with anxiety and depression in accordance with some studies on the application of the DRIVE model

(Mark & Smith, 2012a, b) and the associations reported in literature between Type A personality and physical diseases (Gallo et al., 2001).

With respect to the ethnicity dimensions, the adoption of the host culture could be considered a behavioural pattern that reduces the risk of marginalisation and suffering relational disorders, while affirmation/maintenance culture seems to reduce the risk of physical problems and anxious-depressive symptoms. Results on perceived discrimination were in accordance with the literature on this subject (Troxel et al., 2003; Roberts et al., 2004; Smith et al. 2005). Furthermore language barriers are associated with acculturation stress dimensions and they have a relevant impact in the workplace. Considering the overall sample, the language difficulties could depend on the level of education, the traditional language used by migrants, the type of job that is related to the kind of communication with colleagues and supervisors. In further research on each ethnic group we will pay close attention to this aspect, to the factors that influenced it and to the association with the stress dimensions.

The different ethnic groups had different jobs and future research must examine ethnicity in samples with similar jobs. Moreover it could be important to investigate whether the effects of work demands and ethnicity dimensions on health outcomes could be independent or combined. These aspects will be helpful in the development of psychological interventions to support migrant workers.

References

Berry, J. W. 1997. Immigration, acculturation, and adaptation. *Applied Psychology* 46(1): 5–34.

Bortner, R. W. 1969. A short rating scale as a potential measure of Pattern A behaviour. *Journal of Chronic Diseases* 22: 87–91.

Cooper, C. L., Sloan, S. J. and Williams, S. 1988. *Occupational stress indicator.* (NFER Nelson, Windsor, UK).

Denollet, J. 2005. "DS14: standard assessment of negative affectivity, social inhibition, and Type D personality". *Psychosomatic Medicine* 67(1): 89–97.

Derogatis, L. R. 1994. *Symptom Checklist 90–R: Administration, scoring, and procedures manual* (3rd ed.). (National Computer Systems. Minneapolis, MN).

Gallo, L. C. et al. 2001, Personality traits as risk factors for physical illness. In A. Baum, T. Revenson, J. (ed.) *Handbook of Health Psychology,* (Hillsdale, NJ: Erlbaum) 139–72.

Karasek, R. A. 1985. *Job Content Questionnaire and user's guide.* (University of Massachusetts, Lowell, MA).

Karlsen, S., Nazroo, J.Y., McKenzie, K., Bhui, K. and Weich, S. 2005. Racism, psychosis and common mental disorder among ethnic minority groups in England. *Psychological Medicine*, 35(12): 1795–1803.

Mark, G. M. and Smith, A. P. 2008. Stress models: a review and suggested new direction. In Houdmont, J. and Leka, S. (eds.) *Occupational Health Psychology,* (Nottingham University Press, Nottingham), 111–144.

Mark, G. and Smith, A. P. 2012a. Effects of occupational stress, job characteristics, coping, and attributional style on the mental health and job satisfaction of university employees. *Anxiety, Stress & Coping* 25(1): 63–78.

Mark, G. and Smith, A. P. 2012b. Occupational stress, job characteristics, coping, and the mental health of nurses. *British Journal of Health Psychology* 17(3), 505–521.

Nazroo, J. Y. 2003. The structuring of ethnic inequalities in health: Economic position, racial discrimination and racism. *American Journal of Public Health* 93: 277–284.

Noh, S. and Kaspar V. 2003. Perceived Discrimination and Depression: Moderating Effects of Coping, Acculturation, and Ethnic Support, *American Journal of Public Health* 93(2): 232–238.

Phinney, J. 1992. The multigroup ethnic identity measure: A new scale for use with diverse groups. *Journal of Adolescent Research* 7: 156–176.

Roberts, R. K., Swanson, N. G. and Murphy, L. R. 2004. Discrimination and occupational mental health. *Journal of Mental Health* 13: 129–142.

Siegrist, J. 1996. The effort-reward imbalance model. *Journal of Occupational Health Psychology* 1: 27–41.

Smith, A., Johal, S., Wadsworth, E. et al., 2000. *The Scale and Impact of Occupational Stress: the Bristol Stress and Health at Work Study*. (Health & Safety Executive Research Report 265. HSE Books, London, UK).

Smith, A., Wadsworth, E., Shaw, C., Stansfeld, S., Bhui, K. and Dhillon, K. 2005. *Ethnicity, work characteristics, stress and health*. (Health & Safety Executive Research Report 308. HSE Books, London, UK).

Szczepura, A., Gumber, A., Clay, D., Davies, R., Elias, P., Johnson, M., Walker, I. and Owen, D. 2004. *Review of the occupational health and safety of Britain's ethnic minorities*. (Health & Safety Executive Research Report 221. HSE Books, London, UK).

Troxel, W. M., Matthews, K. A., Bromberger, J. T., and Sutton-Tyrrell, K. 2003. Chronic stress burden, discrimination, and subclinical carotid artery disease in African American and Caucasian women. *Health Psychology* 22, 300–309.

Warr, P., Cook, J. and Wall, T. 1979. Scales for the measurement of some work attitudes and aspects of psychological well-being. *The Journal of Occupational Psychology* 52: 129–148.

RELIGIOUS BEHAVIOUR AS A BUFFER AGAINST OCCUPATIONAL STRESS

Jonathan Fonberg

School of Psychology, Cardiff University

The relationship between religious behaviour (organized and unorganized religious behaviour activity), occupational stress (Robert Karasek's demand-control model and Johannes Siegrist's effort-reward imbalance model) and subjective well-being (life satisfaction) was investigated. Contrary to predictions made on the basis of past research, organized and unorganized religious activity didn't appear to buffer against the negative effects of occupational stress. Future research should focus on more comprehensive models of occupational stress and place a greater emphasis on individual differences.

Introduction

Despite the long-standing debate regarding the benevolent nature of religious behaviour, one that predates modern academia, recent research indicates that it is generally associated with positive life outcomes. Several recent meta-analyses have found that, on the whole, higher levels of religious behaviour are associated with a variety of different positive life outcomes. For example, Moreira-Almeida, Lotufo Neto, and Koenig (2006) reviewed 100 studies that examined the relationship between religious behaviour and subjective well-being. They found that 79 of them reported a significant positive relationship while only a single study reported a negative correlation. Similarly, Hackney and Sanders (2003) found significant inverse correlations between various conceptualizations of religious behaviour and psychological distress. They also found positive associations between the same conceptualizations of religious behaviour, life satisfaction and self-actualization. Generally speaking, it appears that positive outcomes are associated with only certain conceptualizations of religious behaviour. Based on this, it seems plausible that at least part of the debate that has plagued the academic study of religious behaviour is rooted in these varying conceptualizations. One of the most common distinctions is between extrinsic and intrinsic religious behaviour. The former refers to both organized (e.g., church, religious services) and unorganized (e.g., prayer) religious activity while the latter refers to internal commitment and belief. Understanding such distinction appears to be crucial to making sense of past results and it must be kept it in mind for future investigation. From here, the more practical benefits of religious behaviour come into question.

Religious behaviour has been found to serve as a coping mechanism in a variety of different contexts, with both positive and negative outcomes depending on the method used (Ano & Vasconcelles, 2005). Occupation has been noted to be a significant source of stress. Given that, it is somewhat surprising that the research into the relationship between religious behaviour and occupational stress has been minimal: rarely has it been the primary focus of investigation. Given the ubiquity of occupational stress, furthering our understanding of this relationship could be of great practical value.

There have been a variety of different theories put forward that attempt to explain occupational stress. Mark and Smith (2008) make note of two models that, in particular, serve as a good starting point for examining the relationship between occupational stress and religious behaviour: the demand-control (Karasek et al., 1998) and effort-reward-imbalance models (Johannes Siegrist, 1996), both of which are well-established and relevant to today's workforce. Karasek's model focuses on job demands and control: stress is seen as a function of the relationship between these two variables. Job demands are the psychological stressors that occur in the workplace while job control refers to the degree of discretion and authority that an individual has in completing their tasks. When there is an imbalance between these variables an individual will experience stress. Research indicates that negative outcomes, such as cardiovascular problems, are most strongly associated with high demands and low control (Kristensen, 1995). Conversely, the same outcomes were least common in those individuals who experienced moderate, and even high, demands so long as they also had high levels of control (Mark & Smith, 2008).

In Siegrist's model stress is a function of the relationship between effort and reward: it occurs when the former exceeds the latter. Here, rewards are defined as esteem, career opportunities and job security (Johannes Siegrist, 1996). Imbalance between these two factors has been found to be associated with poor self-reported health (Niedhammer, Tek, Starke & Siegrist, 2004) and depression (J Siegrist, 2013) amongst other negative outcomes. An important aspect of the model is the emphasis it places on the subjective experience of stress: imbalance is thought to be subjective and defined by the individual.

Furthering this point, Mark and Smith (2008) note that stress "it is a dynamic process that occurs as an individual interacts with their environment." If the experience of stress is unique to the individual then the same might be true of the methods that are used to cope with it. Given this, it is important to note that there are many different tactics that can be used to manage stress. (DeLongis & Holtzman, 2005). One such method is religion: various aspects of religious behaviour can be used as coping mechanisms.

As there are many different facets of religious behaviour it should come as little surprise there are a variety of different forms of religious coping. Pargament, Koenig, and Perez (2000) found that that religious coping can result in both positive and negative outcomes, depending on the particular method used. This comes as little surprise when one considers the variance in the effectiveness of coping strategies in general (DeLongis & Holtzman, 2005). Given that, a more interesting and unexplored line of inquiry is whether or not general religious behaviour is capable of buffering against occupational stress. Perhaps the best way to delineate between these types of behaviours is to separate those that are driven by the social aspects of religious behaviour (organized religious activity) and those that are motivated by internal belief (unorganized religious activity). Examining how these behaviours relate to occupational stress could potentially provide information pertaining to the function of general religious activity.

As implied above, research into occupational stress typically deals with practical outcomes, such as objective health. Thus, subjective variables are typically ignored in this relationship. In particular, subjective well-being, as it relates to occupational stress, has received little empirical attention. Given this it is unsurprising that there has been minimal investigation into life satisfaction (the cognitive component of subjective well-being) as it relates to occupational stress. Life satisfaction, beyond being a desirable goal in itself, is associated with a variety of different positive life outcomes such as increased longevity (Koivumaa-Honkanen et al., 2000), better self-reported health, overall physical health and the absence of long-term limiting health conditions (Siahpush, Spittal & Singh, 2008). Given that, there is a great deal of practical benefit in increasing the understanding of subjective well-being and the factors that relate to it. The lack of data on the relationship between occupational stress, religious behaviour and life satisfaction represents a gap in the literature. This point is of particular importance when stress is viewed as a subjective experience: a subjective outcome variable might be able to provide different information than an objective one.

Thus, the goal of this research is to assess how the frequency of participation in both organized and unorganized religious activity interacts with occupational stress to impact life satisfaction. As past research has demonstrated the value of religion as a coping mechanism it is hypothesized that religious behaviour will interact with occupational stress to positively influence subjective well-being.

Method

Participants

In total data was collected from 254 participants: Christians (n = 67), Muslim (n = 42), Hindu (n = 73) and Atheist (n = 72). Participants were chosen based on opportunity sampling. There were a similar number of male (n = 138) and female (n = 116) participants. Age ranged from 19 to 63 years (M = 31.67, SD = 9.66). Participants came from the United States (n = 119) and India (n = 135). The majority of the Christian (n = 45) participants and all of the Atheist (n = 72) participants resided in the United States. All of the Hindu (n = 73) participants and the majority of Muslim (n = 40) participants resided in India.

Recruitment

Participants were recruited through Mechancial Turk; an online crowdsourcing website. While relatively new in the field of psychological research, the merits of

Mechanical Turk as a vehicle for the recruitment of participants has already been noted by several authors; for a detailed review see Paolacci, Chandler, and Ipeirotis (2010) and Buhrmester, Kwang, and Gosling (2011). Participants were linked to the Qualtrics website to complete the questionnaire.

Materials

Data pertaining to the demand-control and effort-reward-imbalance models was collected using single item measures developed by Williams and Smith (2012). The items, which use a 10-point likert type response scale, include sample items from the longer questionnaires from which they were developed [e.g.: I find it difficult to withdraw from my work obligations; work is always on my mind; I find it difficult to relax when I get home from work; people close to me say I sacrifice too much for my job]. These models were selected because of the extensive empirical investigation that they have undergone and the need for brevity in this paper. Religious behaviour data was collected using the Duke University Religiosity Index (DUREL), which measures both organized religious behaviour [How often do you attend church or other religious meetings?] and unorganized religious behaviour [How often do you spend time in private religious activities, such as prayer, meditation or Bible study?] (Koenig & Büssing, 2010). The two DUREL questions use 6-point likert type response scales. The cognitive component of subjective well-being was measured using the Satisfaction with Life Scale (SWLS) (Diener, Emmons, Larsen & Griffin, 1985). The SWLS contains five questions and uses 7-point likert type response scales.

Procedure

The data was analysed using a series of univariate general linear ANOVA models. In order to do so median splits of the occupational stress and religious behaviour variables were created. The analyses were structured around the two stress models described above: Karasek's demand-control model and Siegrist's effort-reward-imbalance model. Both occupational stress models were analysed with both religious behaviour variables in isolation due to sample size restrictions. Each analysis used life satisfaction as the outcome variable. Due to the large number of analyses the significance level was set at $p < .01$.

Results

Each of the independent variables (demands, control, efforts, rewards, organized religious activity, unorganized religious activity) had significant main effects on life satisfaction when analysed in isolation. When run in combination, there was only a single significant interaction at the $p < .01$ level. This interaction was between demands, control and organized religious activity $(F(1,246) = 7.399,$

Table 1. Demands-Control, Organized Religious Activity, Life Satisfaction.

Job Demands	Job Control	Organized Religious Activity	Mean
Low	Low	Low	4.192
		High	5.069
	High	Low	4.941
		High	5.176
High	Low	Low	4.933
		High	4.676
	High	Low	4.400
		High	5.364

$p = .007$). The detailed results of this analysis can be seen in Table 1. There were no significant interactions between demands, control and unorganized religious activity ($F(1,246) = 4.631, p = .032$), efforts, rewards and organized religious activity ($F(1,246) = 6.376, p = .012$) or efforts, rewards and unorganized religious activity ($F(1,246) = .577, p = .448$).

Discussion

When analysed in isolation the relationships between the occupational variables and life satisfaction came as little surprise: high demands, high effort, low control and low rewards were negatively associated with subjective well-being. The relationship between both organized and unorganized religious activity was also positive, a finding that was in line with past research, as noted above.

Overall there was little evidence to support our hypothesis that religious behaviour would help buffer against the negative effects of occupational stress. While life satisfaction means were almost universally higher for those individuals who reported frequent participation in religious activities, both organized and unorganized, the differences were statistically non-significant.

However, as noted, there was a significant higher order interaction between job demands, control and organized religious activity. As was the case with the other analyses, individuals who more frequently participated in organized religious activity generally reported higher levels of life satisfaction. This was particularly evident when both demands and control were high and low: in both instances individuals who more frequently participated in organized religious activity reported higher life satisfaction.

However, these are not the conditions that were of primary interest. The purpose of this research was to determine whether or not religious behaviour could provide a buffering effect against occupational stress. According to the demands-control model, stress occurs when demands are high and control is low. When these conditions were met, as can be seen in table 1, individuals who were high in organized religious activity reported less life satisfaction ($M = 4.676$) than those who were low in organized religious activity ($M = 4.933$).

As we investigated general religious activity and not religious coping one possible explanation is that those individuals who experienced higher levels of occupational stress more frequently engaged in organized religious activity in an attempt to cope with it. If this was the case then the data indicates that organized religious behavior is generally not an effective method of coping with occupational stress. The general nature of this statement is important because past research has shown that certain facets of religious coping, such as seeking social support, are associated with positive life outcomes (Pargament et al., 2000). Thus individuals may be using religious behavior ineffectively as a coping mechanism.

With that in mind there was some support for our initial hypothesis: we generally found that life satisfaction was higher amongst those individuals that more frequently engaged in both organized and unorganized religious activity. The non-significance of these results demonstrates one of the potential problems with the models that were used: both have been studied with objective outcomes, such as health-based measures, in mind. Given this, there might be additional aspects of occupational stress that religious behaviour is more capable of buffering against when subjective well-being is of concern. One model which is particularly promising is the Demands, Resources, and Individual Effects model (DRIVE) proposed by Mark and Smith (2008). As the DRIVE model takes a much more holistic approach it seems likely that it could provide a greater understanding of the relationship between occupational stress, religious behaviour and subjective well-being.

One important limitation of this research was the failure to take alternate coping mechanisms into account. Individuals who don't engage in organized religious activity might be coping using alternative methods that are similarly effective in buffering against occupational stress.

Beyond that, religious behaviour may only serve as a coping mechanism for individuals that value it. Life satisfaction is defined by its subjective nature: individuals can use whatever criteria they wish in the cognitive judgment process of evaluating their life. A logical extension of this point is that individuals are likely to use criteria that they place greater value on when making such an appraisal. Perhaps religious behavior does act as a coping mechanism; but only for those people that put stock into it. Individuals who do not engage in religious behavior may find the same type of support through different avenues. This would mean that while religious behaviour is capable of serving as a coping mechanism it does not provide any unique benefits. Thus both religious and non-religious individuals may be using coping mechanisms that differ only in terms of form, not function.

This leads to the question of why some individuals adopt religious coping methods. Perhaps there are significant individual differences in the effectiveness of coping styles: the more an individual values a particular method the more effective it will be for them. Research (DeLongis & Holtzman, 2005) has shown that there are significant differences between individuals in terms of both chosen coping strategies and their effectiveness. While the work of DeLongis and Holtzman (2005) focused on personality traits, the effectiveness of religious coping could perhaps be better attributed to differences in personal values. As noted above, individuals are likely to determine their life satisfaction based on what is important to them. Perhaps a similar pattern might be observed in methods of coping. Exploring this possibility is an important next step in furthering the understanding of the relationship between religious behaviour and occupational stress.

Conclusion

This paper failed to provide evidence that religious behaviour is capable of buffering against the negative effects of occupational stress. Linear trends, despite supporting our initial hypothesis, were non-significant. However, there are several possible explanations for these less than promising results. First, the non-significance may have been the result of incomplete stress models. It is also possible that the lack of significant differences could be a consequence of alternate coping methods that were not controlled for. Extending this point, there may be individual differences in values that impact both chosen coping styles and their effectiveness. Given the positive outcomes associated with life satisfaction, understanding more about this relationship is of great practical importance. Future investigation should focus on more comprehensive models of occupational stress and individual differences in values, both of which are likely to provide additional information about the role that religious behaviour can play as a coping mechanism.

References

Ano, G. G., & Vasconcelles, E. B. (2005). Religious coping and psychological adjustment to stress: A meta-analysis. *Journal of clinical psychology, 61*(4), 461–480.

Buhrmester, M., Kwang, T., & Gosling, S. D. (2011). Amazon's Mechanical Turk A New Source of Inexpensive, Yet High-Quality, Data? *Perspectives on Psychological Science, 6*(1), 3–5.

DeLongis, A., & Holtzman, S. (2005). Coping in context: The role of stress, social support, and personality in coping. *Journal of Personality, 73*(6), 1633–1656.

Diener, E., Emmons, R. A., Larsen, R. J., & Griffin, S. (1985). The satisfaction with life scale. *Journal of personality assessment, 49*(1), 71–75.

Hackney, C. H., & Sanders, G. S. (2003). Religiosity and mental health: A Meta–Analysis of recent studies. *Journal for the scientific study of religion, 42*(1), 43–55.

Karasek, R., Brisson, C., Kawakami, N., Houtman, I., Bongers, P., & Amick, B. (1998). The Job Content Questionnaire (JCQ): an instrument for internationally comparative assessments of psychosocial job characteristics. *Journal of occupational health psychology, 3*(4), 322.

Koenig, H. G., & Büssing, A. (2010). The Duke University Religion Index (DUREL): A Five-Item Measure for Use in Epidemological Studies. *Religions, 1*(1), 78–85.

Koivumaa-Honkanen, H., Honkanen, R., Viinamäki, H., Heikkilä, K., Kaprio, J., & Koskenvuo, M. (2000). Self-reported life satisfaction and 20-year mortality in healthy Finnish adults. *American Journal of Epidemiology, 152*(10), 983–991.

Kristensen, T. (1995). The demand-control-support model: Methodological challenges for future research. *Stress medicine, 11*(1), 17–26.

Mark, G. M., & Smith, A. P. (2008). Stress models: A review and suggested new direction. *Nottingham University Press, 3*, 111–144.

Moreira-Almeida, A., Lotufo Neto, F., & Koenig, H. G. (2006). Religiousness and mental health: a review. *Revista Brasileira de Psiquiatria, 28*(3), 242–250.

Niedhammer, I., Tek, M.-L., Starke, D., & Siegrist, J. (2004). Effort–reward imbalance model and self-reported health: cross-sectional and prospective findings from the GAZEL cohort. *Soc Sci Med, 58*(8), 1531–1541.

Paolacci, G., Chandler, J., & Ipeirotis, P. (2010). Running experiments on amazon mechanical turk. *Judgment and Decision Making, 5*(5), 411–419.

Pargament, K. I., Koenig, H. G., & Perez, L. M. (2000). The many methods of religious coping: Development and initial validation of the RCOPE. *Journal of clinical psychology, 56*(4), 519–543.

Siahpush, M., Spittal, M., & Singh, G. K. (2008). Happiness and life satisfaction prospectively predict self-rated health, physical health, and the presence of limiting, long-term health conditions. *Am J Health Promot, 23*(1), 18–26. doi: 10.4278/ajhp.061023137

Siegrist, J. (1996). Adverse health effects of high-effort/low-reward conditions. *Journal of occupational health psychology, 1*(1), 27.

Siegrist, J. (2013). [Effort-reward imbalance at work and depression: current research evidence]. *Der Nervenarzt, 84*(1), 33–37.

STRESS, HEALTH AND WELLBEING IN CALL CENTRE EMPLOYEES

Helen McFarlane, Rich Neil & Karianne Backx

Cardiff School of Sport, Cardiff Metropolitan University, UK

Call centre staff are at risk of stress, poor psychological wellbeing and physical health problems. The current research, comprising four studies, assesses wellbeing in a call centre in more depth than has previously been attempted. Study 1 is a longitudinal study of psychological wellbeing. Study 2 investigates health outcomes. Study 3 is a qualitative study of wellbeing and Study 4 reviews staff support services. Results show that one in three employees experiences high work stress. High rates of depression, anxiety and obesity and low levels of physical activity were also identified. This reinforces previous research findings suggesting that poor health and wellbeing in call centres is a cause for concern. Future publications will report further findings from the four studies.

Introduction

Staff working in call centres report low levels of wellbeing in comparison to many other occupations (e.g. Johnson, Cooper, Cartwright, Donald, Taylor & Millett, 2005). Problems identified among call centre staff include high levels of work-related stress (Holdsworth & Cartwright, 2004), poor psychological wellbeing (Sprigg, Smith & Lane, 2003), weight gain and obesity (Boyce, Boone, Cioci & Lee, 2008), a high risk of physical health problems and low job satisfaction (Johnson et al., 2005). The poor health and wellbeing among call centre staff leads to higher than average absence from work (CIPD, 2013) and high staff turnover (CfA, 2012). A number of risk factors for poor wellbeing among call centre staff have been identified. These include working in a larger call centre, following a strict script, high workload, lack of clarity (Sprigg et al., 2003), lack of role variety, lack of support from supervisors and colleagues (Deery, Iverson & Walsh, 2002), the degree to which calls are monitored (Sprigg et al., 2003) and the clarity and usefulness of the feedback received from monitored calls (Kochan, 1989). However, it is difficult to draw any conclusions about causality from these findings on wellbeing in call centres due to a number of methodological weaknesses in the existing evidence, including a lack of longitudinal studies, a focus on limited numbers of variables and an over-reliance on questionnaire-based studies.

The lack of longitudinal studies of wellbeing in call centres makes it difficult to draw any conclusions about possible causal pathways linking the identified risk factors and low levels of wellbeing among call centre staff. Employees working in call centres have been identified as being likely to fall into one of three groups: young women with low education levels, older workers who have been made redundant from other sectors and young students or graduates who plan to work in the sector for only a short time (CfA, 2012). It may be that this group of workers is intrinsically at a greater risk of psychological and physical health problems and that the low wellbeing of call centre staff is not in fact an outcome of stressful working conditions. Longitudinal studies are therefore needed in order to establish whether exposure to risk factors precedes the onset of negative outcomes. It is also important to measure life stress, demographic factors and individual differences which could be alternative explanations for the low levels of wellbeing found in call centre employees. However, most studies fail to adequately control for these variables.

As well as failing to control for life stress and variables associated with the individual, studies often focus on only a small number of workplace variables. Many studies of stress and wellbeing at work have used popular theories of workplace stress such as the Demand-Control-Support (DCS) Model (Karasek & Theorell, 1990), Effort-Reward Imbalance (ERI) Model (Siegrist, 1996) and the Job Demand-Resources (JD-R) Model (Demerouti, Bakker, Nachreiner & Schaufeli, 2001). These theories attempt to predict wellbeing outcomes using a small number of variables. Whilst they do so quite successfully, there is little understanding of the cumulative effects of risk factors. This is despite growing evidence that the effects of demands, control and support, hypothesised to interact, are in fact largely additive (Van der Doef & Maes, 1999). To the authors' knowledge, there is no study of the cumulative effects of work factors on wellbeing in call centres and therefore there is a need for research in this area. In addition, some variables appear to have been ignored altogether. For example, many employers offer support to staff via services such as occupational health and employee assistance programmes whereas studies which measure support tend to focus only on colleagues and line managers.

Studies of wellbeing in call centres have not only relied on a limited number of factors but have also tended to rely on a limited number of methods. In particular, many studies use only questionnaire data. Whilst questionnaires may be a convenient way to gather evidence, there is the possibility that the relationships between risk factors and outcomes may have been over-estimated due to common-method variance. There is therefore a need for more research using different methodologies, including mixed- and multi-method approaches.

The present research attempts to address some of the limitations in the current research on wellbeing in call centre staff by providing a more in-depth understanding of wellbeing in call centres. The research uses a multi-method approach made up of four studies (one of them longitudinal) in order to address the identified limitations in the literature. This paper will provide a brief overview of the entire body of research and present some of the early results with a focus on the prevalence of stress and other health and wellbeing outcomes.

Methodology

The research project is made up of four studies. Study 1 is a longitudinal study (4 time points over 2 years) investigating psychological wellbeing using a brief questionnaire. Study 2 is an investigation of health outcomes among call centre staff using both physiological and self-report measures. Study 3 is an in-depth qualitative study of wellbeing among call centre staff and Study 4 reviews the support services which are available to staff. The research was conducted at a large call centre (employing between 800 and 1000 people) in South Wales.

Study 1: Longitudinal Study of Psychological Wellbeing

A longitudinal study of psychological wellbeing was conducted in order to develop an understanding of possible causal pathways between risk factors and wellbeing outcomes. The study measured a broad range of predictive factors in order to look at cumulative effects. Factors such as life stress and individual differences were measured in addition to work factors in order to account for influences on wellbeing which stem from outside the workplace.

All staff working in the call centre were invited to complete an electronic questionnaire measuring psychological wellbeing (work stress, life stress, anxiety, depression, positive affect, negative affect, job satisfaction, life satisfaction, flourishing), work factors (demands, control, colleague support, supervisor relationship, effort, reward, overcommitment, consultation about change), individual differences (extraversion, emotional stability, self-esteem, self-efficacy, optimism), coping styles (self-blame, avoidance, wishful thinking) and life variables (hassles, uplifts). In addition, measures of sickness absence, stress-related absence and presenteeism were included. Four measurement points are planned spanning a two year period. Two questionnaires have been completed with further surveys planned for November 2014 and May 2015. At each time point all staff were invited to take part, allowing new employees to take part.

The Wellbeing Process Questionnaire (WPQ; Williams & Smith, 2012) was used to measure predictors and wellbeing outcomes. This questionnaire uses single item measures in order to measure a large number of variables within a short questionnaire. Multi-item questionnaires are generally seen as more reliable than single item measures (Wanous & Reichers, 1996), however, single items can show good predictive validity (Bergkvist & Rossiter, 2007) and reliability (Wanous & Reichers, 1996). In call centres, the time available for staff to complete questionnaires is very limited and so long questionnaires are impractical. A measure made up of single items therefore provides a useful alternative to full length questionnaires.

Questionnaires were administered electronically via the organisation's intranet. All employees were sent an email inviting them to take part in the research. An attached information sheet explained the nature of the research and the procedures in place to protect anonymity. Longitudinal data were matched using staff numbers as identifiers.

Study 2: Study of Health Outcomes

Study 2 investigated health outcomes among call centre staff, with a focus on cardiovascular risk factors. Cardiovascular health was chosen since stress and poor psychological wellbeing are well established risk factors for cardiovascular disease (Backé, Seidler, Latza, Rossnagel & Schumann, 2011) as are weight gain and obesity which are also problems among call centre staff (Boyce et al., 2008). However, to the authors' knowledge, no study to date has investigated cardiovascular risk among this group of workers.

All employees who completed the questionnaire for Study 1 at time 2 (387 staff members) were asked whether they would be interested in attending a Health assessment. Those who stated that they might be interested in attending were sent further information about Study 2 by email and invited to attend a Health assessment. The Health assessment took place in a private room in the call centre and included a number of health tests. Participants' height and weight were measured and their body mass index (BMI) calculated. Waist circumference was then measured as an indicator of central body fat. Blood pressure was measured using a Mobilograph which also measured arterial stiffness. Blood glucose and overall cholesterol were measured using a portable Accutrend device. Lung function was measured using a spirometer. Prior to the MOT, participants were also asked to complete a questionnaire by hand which included a cardiovascular screening tool (AHA/ACSM Health/Fitness Preparticipation Screening Questionnaire), a measure of physical activity (International Physical Activity Questionnaire [IPAQ]), a measure of alcohol use (FAST [Audit] Alcohol Use Questionnaire) and a questionnaire asking about physical symptoms (Physical Symptoms Inventory [PSI]). Health assessments lasted approximately 30 to 40 minutes including time to complete the questionnaire. Employees were given feedback on their results following the assessment and were sent a written report within 5 days of attending. Those whose results fell outside healthy limits were advised to make an appointment with their GP.

Study 3: In depth study

Study 3 is planned for January 2015. This qualitative study will use daily diaries and interviews in order to understand the experiences of employees working in the call centre in greater depth. Employees who took part in Study 1 will be purposively sampled based on their wellbeing scores. Ten participants who report high wellbeing and ten who report low wellbeing will be chosen and their experiences compared. Diaries will record daily events, stress levels and mood. The immediate and lagged effects of events for staff with high and low levels of wellbeing will be compared. Following completion of daily diaries, employees will take part in a semi-structured interview which will explore the staff members' experiences of working in the call centre and the influences on their wellbeing in more depth. Interviews will be analysed using thematic analysis.

Study 4: Review of existing staff support services

Study 4 reviewed the existing services providing staff support and facilities which impact on staff wellbeing. These include an occupational health service, employee assistance programme, canteen and gym (located at a different site). All staff at the call centre were invited by email to take part in the study. Invitation emails included an information sheet detailing the research. The questionnaire was completed electronically and was administered by a central department outside of the call centre. Questions on support services assessed awareness, acceptability, accessibility, use and usefulness of services. Questions on facilities assessed their use, ease of making healthy choices, cost and barriers to use. Questions were developed specifically for the survey.

Results

Some early results from Study 1 (2 time points) and Study 2 are presented. Due to space constraints, this paper will report findings on the levels of wellbeing in the call centre including the prevalence of stress and other health and wellbeing outcomes.

Study 1: Longitudinal Study of Psychological Wellbeing

At time 1, 397 employees completed the wellbeing questionnaire (49%) and at time 2, 389 employees completed the questionnaire (45%). In total, 186 employees completed the questionnaire at both time points (48% of the original sample). Demographically, respondents were broadly representative of the call centre as a whole with around 60% being female (63% at time 1 and 60% at time 2) and approximately half being under the age of 30 (50% at time 1 and 49% at time 2). However, the higher job grades were under-represented (92% of respondents at time 1 and 95% at time 2 were of the basic call handler grade versus 86% of the call centre as a whole).

Each wellbeing outcome was broken down into three categories. For depression and anxiety, the categories correspond with the anxiety and depression ratings on the Hospital Anxiety and Depression Scale (Williams & Smith, 2013). These ratings are severe, moderate and normal to mild. The other variables were categorised into high, moderate and low where scores of 7 or above were regarded as high and scores of 3 or less categorised as low. The proportions of employees experiencing high and low wellbeing is summarised in Table 1. Approximately 1 in 3 employees in the call centre reported high levels of work stress. This was consistent between times 1 (33%) and 2 (34%). Smaller numbers of employees reported high levels of life stress, although this increased slightly between times 1 and 2 from 26% to 30%. Work stress showed low correlations with life stress (Time 1: $r = .06$, n.s.; Time 2: $r = .20$, $p < .001$). Rates of depression and anxiety were higher than would be expected in the general population (Crawford, Henry, Crombie & Taylor, 2001) and increased

Table 1. Prevalence of low wellbeing.

Outcome		Time 1	Time 2
Work stress	High	33.2%	34.1%
	Moderate	42.8%	46.3%
	Low	23.9%	19.6%
Life stress	High	25.7%	29.9%
	Moderate	33.5%	33.8%
	Low	40.8%	36.3%
Depression	Severe	3.5%	4.9%
	Moderate	21.4%	28.4%
	Normal or mild	75.1%	66.8%
Anxiety	Severe	6.8%	9.5%
	Moderate	23.4%	30.5%
	Normal or mild	69.8%	59.5%
Job satisfaction	High	48.6%	49.7%
	Moderate	34%	30.7%
	Low	17.4%	19.6%

between times 1 and 2, with 40% of employees reporting moderate to severe anxiety symptoms and a third reporting moderate to severe depression symptoms at time 2. However, in contrast to previous studies, around half of respondents reported high levels of job satisfaction, with less than 20% reporting low job satisfaction.

Study 2: Study of Health Outcomes

Health assessments were carried out with 99 staff members. Forty-eight were female and 51 were male, meaning that males were over-represented (making up 39% of the call centre as a whole). Fifty-seven percent were under 30 compared to 48% of the call centre as a whole, meaning that older workers were under-represented. The prevalence of some health risks identified in the Health assessments is summarised in Table 2. Obesity rates were higher than the general population (35% of call centre staff versus 22% of the Welsh population). Whilst 43% of employees met recommended levels of physical activity (higher than the Welsh average of 34%), a high proportion reported very low activity levels (37.4% of call centre employees in comparison to 22% of the Welsh population). Blood pressure and cholesterol levels were broadly in line with the population as a whole with 19% having high blood pressure and 45% having high cholesterol (Welsh Government, 2012).

Conclusion

These findings on the prevalence of stress, low psychological wellbeing and health risks in a call centre reinforce the findings of previous studies which suggest that the health and wellbeing of call centre staff is a cause for concern. Further analysis

Table 2.   **Prevalence of health risks.**

Outcome	Category	Percentage of employees
BMI	Obese	35.4%
	Overweight	31.3%
	Healthy	33.3%
Blood pressure	High	19.2%
	Borderline	27.3%
	Normal	52.5%
Cholesterol	Very high	7.1%
	High	38.4%
	Normal	52.5%
Activity levels	High	30.3%
	Moderate	32.3%
	Low	37.4%

of these data will provide greater understanding of the causes of poor health and wellbeing in this group of workers and the relationships of the predictor variables to outcomes. The research is ongoing and future papers will report on the relationships between predictor variables and outcomes over time, possible causal pathways, relationships between stress and physical health outcomes, experiences of staff working in the call centre and the usefulness of support services. A more in-depth understanding of the health and wellbeing of this group of workers may allow interventions to be developed to improve the health and wellbeing of call centre staff and make this type of work more sustainable in the long term.

References

Backé, E. M., Seidler, A., Latza, U., Rossnagel, K. & Schumann, B. (2011) The role of psychosocial stress at work for the development of cardiovascular diseases: a systematic review. International Archives of Occupational and Environmental Health. Jan; **85**(1): 67–79.

Bergkvist, L. and Rossiter, J. R. (2007) The predictive validity of multiple-item versus single-item measures of the same constructs. *Journal of Marketing Research* 44(2): 175–184.

Boyce, R. W., Boone, E. L., Cioci, B. W. and Lee, A. H. (2008) Physical activity, weight gain, and occupational health among call centre employees. *Occupational Medicine* 58: 238–244.

CfA. (2012) *Contact Centre Operation Labour Market Report 2012,* (CfA Business Skills @ Work, London).

Chalykoff, J. and Kochan, T. (1989) Computer-aided monitoring: its influence on employee job satisfaction and turnover. *Personnel Psychology* 42(2): 807–834.

CIPD (Chartered Institute of Personnel and Development). (2013) *Absence Management: Annual Survey Report 2013,* (CIPD, London).

Crawford, J. R., Henry, J. D., Crombie, C. & Taylor, E. P. (2001) Normative data for the HADS from a large non-clinical sample. *British Journal of Clinical Psychology* 40: 429–434.

Deery, S. J., Iverson, R. D. & Walsh, J. T. (2002) Work relationships in telephone call centres: understanding emotional exhaustion and employee withdrawal. *Journal of Management Studies* 39(4): 471–496.

Demerouti, E., Bakker, A. B., Nachreiner, F. & Schaufeli, W. B. (2001) The Job Demands-Resources Model of Burnout. *Journal of Applied Psychology* 86(3): 499–512.

Holdsworth, L. & Cartwright, S. (2003) Empowerment, stress and satisfaction: an exploratory study of a call centre. *Leadership and Organization Development Journal* 24(3): 131–140.

Johnson, S., Cooper, C., Cartwright, S., Donald, I., Taylor, P. & Millett, C. (2005) The experience of work-related stress across occupations. *Journal of Managerial Psychology* 20(2): 178–187.

Karasek, R. A. & Theorell, T. (1990) *Healthy work: Stress, productivity and the reconstruction of working life,* (Basic Books, New York).

Siegrist, J. (1996) Adverse health effects of high-effort/low-reward conditions. *Journal of Occupational Health Psychology* 1(1): 27–41.

Sprigg, C. A., Smith, P. R. & Jackson, P. R. (2003) *Psychosocial risk factors in call centres: An evaluation of work design and well-being.* Norwich: HSE Books.

Van der Doef, M. & Maes, S. (1999) The job demand-control (-support) model and psychological well-being: A review of 20 years of empirical research. *Work and Stress* 13: 87–114.

Wanous J. P. & Reichers, A. E. (1996) Estimating the reliability of a single-item measure. *Psychological Reports* 78: 631–634.

Welsh Government. (2012) *Welsh Health Survey 2012.* Cardiff: National Statistics.

Williams, G. and Smith, A. (2012) A holistic approach to stress and wellbeing. Part 6: The Wellbeing Process Questionnaire (WPQ short-form). *Occupational Health [at Work]* 9(1): 29–31.

Williams, G. M. & Smith, A. P. (2013) Measuring wellbeing in the workplace: Single-item scales of depression and anxiety. In M. Anderson (ed.). *Contemporary Ergonomics and Human Factors 2013* (Taylor and Francis, London), 87–94.

UNDERSTANDING POLICE OCCUPATIONAL STRESS AND ITS CONSEQUENCES

Kenisha V. Nelson

Centre for Occupational Health and Psychology, Cardiff University, UK

Occupational stress among police officers has been widely studied over the past four decades. Yet, despite the progress that has been made, there are still unanswered questions and some areas that require further elaboration. This paper presents a review of the literature on police stress, with particular attention on the uniqueness of the police stress experience, sources of police stress, and a brief discussion on the consequences of these stressors. The paper concludes with comments on the limitations in the police stress literature and suggestions for future research.

Introduction

Sources and consequences of stress among police personnel have been well documented. However, despite the attention given to stress and its related outcomes and the efforts made to mitigate stress in policing, there has been little change in the stress experience of police officers and ill-health has reportedly worsened (Collins & Gibbs, 2003). This may suggest that stress in policing is not being adequately addressed or the target of intervention and stress management programmes may be misguided. Perusal of the literature on police stress and its correlates presents a substantial amount of information that is at times inconclusive and contradictory and can prove somewhat challenging to navigate through. This paper aims to organise some of the salient areas of enquiry in the police literature. Specifically, it addresses the following three questions: (1) is police work inherently stressful relative to other occupations? (2) What are the sources of police stress? and (3) what is the evidence linking police stress to physical, psychological, and behavioural consequences. The paper concludes with suggestions for future research in the context of the current police stress literature.

Is police stress unique?

The question of how stressful police work is and whether it is more stressful than other occupations remains a source of debate. Whereas there is a body of evidence that suggests police work is one of the most stressful occupations, particularly because of the potential of danger and working with unpredictable situations

(Anshel, Robertson, & Caputi, 1997; Liberman, Best, Metzler, Fagan, Weiss & Marmar, 2002), more recent surveys of the police suggest otherwise. An increasing body of research has provided robust evidence suggesting that police work is no more stressful and police officers are no more prone to mental health problems compared to other occupational groups (Deschamps, Paganon-Badiner, Marchand, & Merle, 2003; Hart & Cotton, 2002; Zhao, He & Lovrich, 2002). In some cases police officers show less severe symptoms of stress when compared with other groups (Berg, Hem, Lau & Ekeberg, 2006; Kop, Euwema & Schaufeli, 1999). Furthermore, in most studies police officers are more likely to endorse stressors related to the organisational climate and managerial practices as more troublesome rather than frontline duties (Biggam, Power, Macdonald, Carcary & Moodie, 1997; Brown & Campbell, 1990; Buker & Weicko, 2007; Collins & Gibbs 2003; Coman & Evans, 1991; Crank & Caldero, 1991; Gershon, Barocas, Canton, Li & Vlahov, 2009; Kop and Euwema, 2001; Liberman, et al., 2002; Morash, Haarr & Kwak, 2006; Storch & Panzarella, 1996). This has given credence to the argument that the stress experience of police officers is comparable with other occupational groups who are just as likely to report organisational factors as their primary source of stress. In short, judging from current evidence it is difficult to say that police is work is inherently more stressful than other occupations.

Nevertheless, while there is little doubt that police work shares many of the stressors that affect other occupations, it would be remiss not to acknowledge that police officers face a number of specific additional challenges that other occupations are unlikely to encounter. The added frustration accrued from some of these unique situations can erode the employee's tolerance for other routine tasks and vice versa which may result in increased levels of stress and negative outcomes (Suresh, Anantharaman, Angusamy & Ganesan, 2013). All things considered, it would be negligent to deny the aggregation of factors contributing to the experience of stress in police work (Webster, 2013), but this has been given little research attention.

In general, it seems less meaningful to consider whether police work is more or less stressful than other occupations. Importantly, the focus should be on identifying sources of stress in policing and related consequences with the intention of devising preventative measures and interventions. Particularly, because police stress can have negative consequences not just for the individual officer but also have the potential for public harm, perhaps more so than other occupations (Webb & Smith, 1980).

Source of police stress

Sources of stress encountered by police officers have been distinguished in a number of ways, from identifying specific individual stressors to defining broad categories. Over the intervening years researchers have agreed on two broad categories of police stressors: operational and organisational. No doubt when one thinks of police work, the obvious focus is on the operational duties. Operational stressors result from factors inherent to police work, typically seen as acute and episodic in nature and sometimes involving critical/traumatic incidences (Stinchcomb, 2004). Incidents involving shooting someone, witnessing the death of a fellow officer, death of a child, attending domestic disputes, unpredictable situations, confronting someone with a weapon and arresting a violent individual have all been identified as major sources of operational stressors among police officers (Anshel, et al., 1997; Biggam et al. 1997; Brown & Campbell, 1990; Comans & Evans, 1991; Violanti & Aron, 1994). In a review of the police stress literature, Abdollahi (2002) classified police operational stressors into six major categories: (1) dealing with the judicial system; (2) public scrutiny and media coverage; (3) officer involved in shootings; (4) encountering victims of crime and fatalities; (5) community relations; and (6) encountering violent/unpredictable situations. Although some researchers contend that stress associated with operational events has been overstated (Hart & Cotton, 2002; Zhao et al., 2002), others document police officers' ratings of these events as highly stressful (Berg, Hem, Lau, Haseth & Ekeberg, 2005; Brown, Fielding & Grover, 1999; Gershon, et al., 2009; Violanti & Aron, 1994. While they have the potential to induce significant stress, under reporting of the intensity of these stressors may be due to the infrequent nature of operational incidents. It may also be argued that police officers come to expect that these activities are a necessary part of their job and are better prepared to handle these stressors. Based on current evidence, attention has shifted to the more routine organisation related stressors in more recent years.

Organisational stressors emanate from the organisational climate and managerial practices, and are commonly experienced in other occupations. They are routinely encountered by the officer, and often times involve situations over which they have limited control (Stinchcomb, 2004). Abdollahi (2002) in her review summarised the most cited organisational stressors as shift work, inadequate supervision and poor relationship, lack of input into policy and decision-making, lack of recognition and insufficient administrative support, excessive paperwork, insufficient pay and poor resources, role conflict and ambiguity, isolation and boredom, and internal discipline structure. While there seems to be enough evidence to suggest that the routine organisation duties are a major source of problems for police officers, stress is still treated as an "individual officer" problem rather than an organisational problem. Hence, prevention and intervention programmes are still focused on treating symptoms rather than the causes of stress (Stinchcomb, 2004).

It is important to note that it is probably overly simplistic to assume homogeneity in stressors experienced by police officers. Stress in policing is dynamic and complex in nature and may not operate uniformly among police officers. This makes it difficult not only to generalise but to effectively evaluate the dimensions of the police officer's stress experience (Dabney et al., 2013). Therefore it would not be surprising if officers in different job roles, ranks, and at different stages of their careers reported varying stress experiences. For example, some researchers report differences in stress exposure, where senior officers are more likely to report more frequent exposure to organisational stressors while lower ranked officers report more frequent exposure to operational stressors (Brown and Campbell, 1990; Biggam et al., 1997; Gudjonsson and Adlam, 1983).

Consequences of police stress

Police stress has been linked to adverse outcomes including a variety of physical and psychological problems. Police officers report physical complaints such as chronic back pain, foot problems, headaches/migraines, and musculoskeletal problems which are linked to work stress (Berg, et al., 2006; Gershon, et al., 2009). Research also suggests that police officers may be at risk for more serious physical disease and higher rates of cardiovascular disease and mortality rates because of stress (Berg et al., 2006; Franke, Collins & Hinz, 1998; Gershon et al., 2009).

Similarly, the taxing nature of work stress in policing has been linked to psychological dysfunction often in the form of depression and anxiety (Gabarino, Cuomo, Chiorri, & Magnavita, 2013; Gershon, et al., 2009). Post-traumatic stress disorder (PTSD) and burnout have also received considerable attention. Organisational variables, rather than operational factors, have been linked to depression and anxiety. However, critical incidents are likely to be stronger predictors of post-traumatic stress symptoms (Liberman, et al., 2002; Komarovskaya, et al., 2011; Wang, et al., 2010). Researchers have also found that organisational hassles such as overtime work, work conflicts, and lack of support from supervisors and co-workers are significant predictors of burnout (Martinussen, Richardson, & Burke, 2007; Kohan & Mazmanian, 2003). Other commonly reported problems are sickness absence, poor sleep quality, and suicide (Biggam et al., 1997; Gabarino, et al., 2013; Gershon et al., 2009; Komarovskaya et al., 2011; Martinussen et al., 2007; Violanti, 1995; Wang et al., 2010). Personal problems such as family discord and high divorce rates have also been documented (Alexander & Walker, 1997; Dechamps et al., 2003; Gershon et al., 2009).

An under investigated area of police work is the influence of stress and related psychological dysfunction on police-citizen relations. Manolias (1983) (in Brown and Campbell, 1994) argue that it is important to study police stress for the police as well as the public because: "the consequences of police stress may have an adverse effect on the development and maintenance of good police relations with the public" and "there exists the possibility that police officers under stress can, in certain situations, constitute a real threat to their own safety, that of their fellow officer, the offenders they deal with, and indeed the general public".

A handful of studies have tested the hypothesis of police stress and its potential to negatively affect police-citizen relations. Work-related stress and burnout have been correlated with negative attitude towards citizens and aggressive tactics such as use of force (Cheong & Yun, 2011; Gershon, et al., 2009; Kop & Euwema, 2001; Kop, et al., 1999). For example, Kop and Euwema (2001) examined the relationship between burnout, attitudes towards citizens, attitude towards use of force, and actual use of force among Dutch patrol police officers. The authors found that all three scales of burnout were associated with negative attitudes towards citizens. Depersonalisation and emotional exhaustion were found to be associated with actual use of force, and officers who scored higher on the scale of depersonalisation had a positive attitude towards use of force. It is hypothesised that police officers experience significant strain which evokes negative emotions such as anger and frustration which then leads to maladaptive behaviours such as aggression and use of force (Arter, 2008). This is an area that certainly warrants further exploration.

Limitations of the literature

This paper would be incomplete if it did not address some of the limitations in the police stress literature. Careful examination of the literature shows that research on police stress is limited in three major ways: (1) the lack of a theoretical foundation, which limits our scope in identifying and explaining the sources and consequences of police stress; (2) methodological problems including almost exclusive attention to cross-sectional research, and unrepresentative samples which limit generalisation of findings; and (3) inconsistent measurement which makes it difficult to make comparisons across studies (Webster, 2013; Abdolahi, 2002). To advance what we know about police stress and its consequences, it is important that future research focuses on designing studies that can account for alternative explanations when trying to establish cause and effect relationships.

There seems to be an underlying acceptance that police stress is homogeneous or universal, with limited consideration of differences in individual and organisational characteristics. Essentially, the police stress research requires a less homogenous approach, taking into consideration variances due to occupational role and individual differences (Malloy and Mays, 1984). Future research should therefore focus on the variances that might be explained by such variables. The extent to which the role of these variables is understood will provide some direction for the development and implementation of policy, interventions, and stress management programmes that will lessen work-related stressors and illnesses. In addition, for the most part, a large portion of the literature on police stress is confined to developed countries, particularly the United States and United Kingdom. Only a limited number of published studies have examined the experience of stress and its effects on police officers in other countries particularly in lesser developed regions. It is therefore debatable whether sources of stress and the overall stress experience of police officers in other countries with different cultures, working environments and conditions, crime rates, and firearms policies would be comparable. More research on the police stress experience across nations may help to determine how much can be generalised about work-related stress in policing.

Finally, one under-investigated area in the literature is the link between police stress, well-being, and attitude and behaviour towards citizens. Due to the nature of policing, police officers need to be functioning at optimal levels to effectively "serve and protect" citizens. A handful of studies have showed that impaired functioning because of stress and burnout can result in negative attitudes and aggressive tactics such as use of force. However, this is an area that requires further elaboration.

Statement of relevance: This paper is relevant to researchers, academics, police administrators and practitioners who are interested in the police stress experience and the potential for adverse outcomes.

References

Abdollahi, M. K. 2002. Understanding police stress research. *Journal of Forensic Psychology Practice* 2: 1–24.

Alexander, D. A. and Walker, L. 1994. A study of methods used by Scottish police officers to cope with work-induced stress. *Stress Medicine* 10: 131–138.

Anshel, M. H., Robertson, M. and Caputi, P. 1997. Sources of acute stress and their appraisals and reappraisals among Australian police as a function of previous experience. *Journal of Occupational and Organizational Psychology* 70: 337–356.

Arter, M. L. 2008. Stress and deviance in policing. *Deviant Behavior* 29: 43–69.

Berg, A. M., Hem, E., Lau, B. and Ekeberg, Ø. 2006. An exploration of job stress and health in the Norwegian police service: a cross sectional study. *Journal of Occupational Medicine and Toxicology* 1: 26.

Berg, A. M., Hem, E., Lau, B., Håseth, K. and Ekeberg, Ø. 2005. Stress in the Norwegian police service. *Occupational medicine* 55(2): 113–120.

Biggam, F. H., Power, K. G., MacDonald, R. R., Carcary, W. B. and Moodie, E. 1997. Self-perceived occupational stress and distress in a Scottish police force. *Work and Stress* 11(2): 118–133.

Brown, J. M. and Campbell, E. A. 1990. Sources of occupational stress in the police. *Work and Stress* 4(4): 305–318.

Brown, J. M. and Campbell, E. A. 1994. *Stress and policing: Sources and strategies.* Chichester: Wiley.

Brown, J. M., Fielding, J. and Grover, J. 1999. Distinguishing traumatic, vicarious, and routing operational stressor exposure and attendant adverse consequences in a sample of police officers. *Work and Stress* 13(4): 312–325.

Buker, H. and Weicko, F. 2007. Are causes of police stress global? Testing the effects of common police stressors on the Turkish National police. *Policing: An International Journal of Police Strategies and Management* 30: 291–309.

Cheong, J. and Yun, I. 2011. Victimization, stress and use of force among South Korean police officers. *Policing: An International Journal of Police Strategies and Management* 34(4): 606–624.

Collins, P. A. and Gibbs, A. C. 2003. Stress in police officers: a study of the origins, prevalence, and severity of stress-related symptoms within a county police force. *Occupational Medicine* 53: 256–264.

Coman, G. J. and Evans, B. J. 1991. Stressors facing Australian police in the 1990's. *Police Studies* 14: 153–165.

Crank, J. P. and Caldero, M. 1991. The production of occupational stress in medium-sized police agencies: A survey of line officers in eight municipal departments. *Journal of Criminal Justice* 19: 339–349.

Dabney, D. A., Copes, H., Tewksbury, R.and Hawk-Tourtelot, S. R. 2013. A qualitative assessment of stress perceptions among members of a homicide. *Justice Quarterly* 30(5): 811–836.

Deschamps, F., Paganon-Badinier, I., Marchand, A. and Merle, C. 2003. Sources and assessment of occupational stress in the police. *Journal of Occupational Health* 45: 358–364.

Franke, W. D., Collins, S. A. and Hinz, P. N. 1998. Cardiovascular disease morbidity in an Iowa law enforcement cohort, compared with the general Iowa population. *Journal of Occupational and Environmental Medicine* 40(5): 441–444.

Garbarino, S., Cuomo, G., Chiorri, C. and Magnavita, N. 2013. Association of work-related stress with mental health problems in a special police force unit. *BMJ open* 3(7).

Gershon, R. R., Barocas, B., Canton, A. N., Li, X. and Vlahov, D. 2009. Mental, physical, and behavioral outcomes associated with perceived work stress in police officers. *Criminal Justice and Behavior* 36: 275–289.

Gudjonsson, G. H. and Adlam, R. 1983. Personality patterns of British police officers. *Personality and Individual difference* 4(5): 507–512.

Hart, P. M. and Cotton, P. 2002. Conventional wisdom is often misleading: Police stress within and organizational health framework. In M.F. Dollard, A.H. Winefiled, and H.R. Winefiled (Eds), Occupational stress in the service professions. (pp. 130–138), London: Taylor & Francis.

Kohan, A., & Mazmanian, D. (2003). Police work, burnout, and pro-Organizational behavior a consideration of daily work experiences. *Criminal Justice and Behavior*, 30(5): 559–583.

Komarovskaya, I., Maguen, S., McCaslin, S. E., Metzler, T. J., Madan, A., Brown, A. D. and Marmar, C. R. 2011. The impact of killing and injuring others on mental health symptoms among police officers. *Journal of psychiatric research* 45(10):1332–1336.

Kop, N. and Euwema, M. C. 2001. Occupational stress and the use of force by Dutch police officers. *Criminal Justice and Behavior* 28(5): 631–652.

Kop, N., Euwema, M. and Schaufeli, W. 1999. Burnout, job stress and violent behaviour among Dutch police. *Work and Stress* 13: 326–340.

Liberman, A. M., Best, S. R., Metzler, T. J., Fagan, J. A., Weiss, D. S. and Marmar, C. R. 2002. Routine occupational stress and psychological distress in police. *Policing: An International Journal of Police Strategies and Management* 25: 421–439.

Malloy, T. E. and Mays, G. L. 1984. The police stress hypothesis: a critical evaluation. *Criminal Justice and Behavior* 11:197–224.

Martinussen, M., Richardsen, A. M. and Burke, R. J. 2007. Job demands, job resources and burnout among police officers. *Journal of Criminal Justice* 35: 239–249.

Morash, M., Haarr, R. and Kwak, D. 2006. Multilevel influence on police stress. *Journal of Contemporary Criminal Justice* 22: 26–43.

Stinchcomb, J. B. 2004. Searching for stress in all the wrong places: combatting chronic organizational stressors in policing. *Police Practice and Research* 5(3): 259–277.

Storch, J. and Panzarella, R. 1996. Police stress: state anxiety in relation to occupational and personal stressors. *Journal of Criminal Justice* 24: 99–107.

Suresh, R. S., Anantharaman, R. N., Angusamy, A. and Ganesan, J. 2013. Sources of job stress in police work in a developing country. *International Journal of Business and Management* 8(13): 102–110.

Violanti, J. M. 1995. The mystery within: Understanding police suicide. *FBI Law Enforcement Bulletin* 64(2): 19–23.

Violanti, J. M. and Aron, F. 1994. Ranking police stressors. *Psychological Reports* 75: 824–826.

Wang, Z., Inslicht, S. S., Metzler, T. J., Henn-Haase, C., McCaslin, S. E., Tong, H., ... & Marmar, C. R. 2010. A prospective study of predictors of depression symptoms in police. *Psychiatry Research* 175(3): 211–216.

Webb, S. D. and Smith, D. L. 1980. Police stress: a conceptual overview. *Journal of Criminal Justice* 8: 251–257.

Webster, J. H. 2013. Police officer perceptions of occupational stress: the state of the art. *Policing: An International Journal of Police Strategies and Management* 36(3): 636–652.

Zhao, J., He, N. and Lovrich, N. 2002. Predicting five dimensions of police officer stress: Looking more deeply into organizational settings for sources of police stress. *Police Quarterly* 5(1): 43–62.

STRESS AND WELLBEING AT WORK: AN UPDATE

Andrew P. Smith

*Centre for Occupational and Health Psychology,
School of Psychology, Cardiff University,
Cardiff, UK*

This paper describes recent developments in research on stress and wellbeing at work. The starting point is the case definition of stress at work which suggests that one must assess many aspects of a transactional stress process. The stress process can quickly be assessed by examining the combined effects of potential stressors. This usually shows linear relationships between the total negative factors score and reports of stress. A similar approach can be applied to wellbeing at work. Here the good job score reflects the difference between positive job characteristics/appraisals and negative job characteristics/appraisals. A multivariate approach is essential and this can be achieved by use of a range of brief questions.

Introduction

The aim of this paper is to review some of the research on stress and wellbeing at work that has been carried out in the last 10–15 years. The general approach adopted here is described in more detail in Smith (2011a, b). The starting point for this is a case definition of work-related stress (Cox et al. 2006).

A case definition of occupational stress

This research considered the feasibility and possible nature of a case definition of work-related stress that is suitable for application in a variety of stakeholder domains. A case definition is needed in occupational health research as the basis for surveillance, and for monitoring the effectiveness of interventions.

Cox et al. (2006) examined definitions already applied within epidemiological surveys recently conducted into work-related stress in the UK. Their second study involved identifying key stakeholders and collecting information on (i) the case definitions employed in their various fields and (ii) their views on the feasibility of developing a single case definition that could span all domains while remaining consistent with epidemiological case definitions.

The research suggested that it is possible to develop a case definition for application in surveys that is broadly compatible with thinking and practice in other domains studied. A case definition and associated assessment framework was arrived at by consensus and acknowledged across stakeholder groups as appropriate for application within the occupational health domain.

This case definition requires the person to report:

- High levels of stress
- Unreasonable job characteristics
- Mental health problems
- Work-related problems (e.g. high sick-leave)
- The above to be work related and not due to confounding factors

Cox et al. (2006) suggest that this will largely be used for research purposes and it fits in with recent models of occupational stress. One such model, the Demands-Resources-Individual Effects (DRIVE) model (Mark and Smith, 2008) is described in the next section.

The DRIVE model of stress

This model is shown in Figure 1. It has many of the features of earlier models of stress but puts a greater emphasis on individual characteristics and personal resources. The issue of balance between complexity and simplicity in stress and well-being assessment has been discussed by Mark & Smith (2008). In this paper the authors suggest that an ideal approach would allow for the model to account for circumstances, individual experiences, and subjective perceptions without too much complexity. Their proposed basic model included factors from the Demand-Control-Support model (DCS) model, the Effort-Reward Imbalance (ERI) model, coping behaviours, and attributional explanatory styles as well as outcomes including anxiety, depression, and job satisfaction. These variables were categorised as work demands, work resources (e.g. control, support), individual differences (e.g. coping style, attributional style), and outcomes, although the model is intended as a framework into which any relevant variables can be applied (Mark & Smith, 2008). This simple DRIVE model proposed direct effects on outcomes by each of the other variable groups, as well as a moderating effect of individual differences and resources on demands. A more complex version (the enhanced DRIVE model) was also developed to acknowledge a subjective element and included perceived stress as well as further interactive effects. Research using the DRIVE model has supported the direct effects of these variable groups on outcomes, although support was found for interactions (Mark & Smith, 2008). Stronger support of direct effects compared to interactions has also been found in research on other models such as the DCS model, where review has shown that the buffering effect of control and support had less evidence than the direct effects of these variables on outcomes.

This model has been tested in studies using university staff (Mark and Smith, 2011) and nurses (Mark and Smith, 2012). The results support inclusion of the concepts described in the model but showed that these factors were independent and that there were few interactions between the variables. This supports the "combined

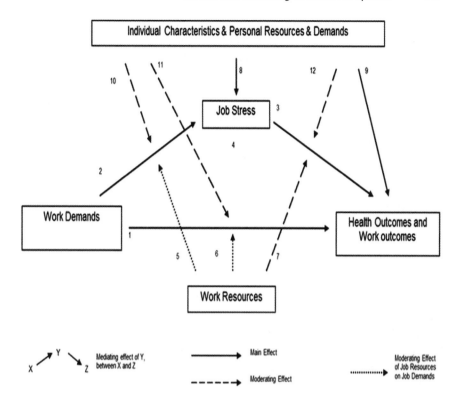

Figure 1.   The demands-resources-individual effects model (Mark & Smith, 2008).

effects" approach to the study of occupational stressors (Smith et al., 2004) which is described in the next section.

Combined effects of occupation hazards

There has been much previous research on a large number of workplace hazards, and for the most part the nature and effects of such factors have been considered in isolation. Such an approach is not likely to be representative of the real-life workplace where employees are often exposed to multiple hazards. For example, individuals are very unlikely to work in a noisy environment that does not also expose them to other stressors that have considerable potential to harm. There is limited information on the combined effects of these hazards on health and safety. Indeed, there have been no systematic literature reviews, no attempt to produce a coherent framework for studying these factors, and a dearth of studies using a variety of methods to investigate the topic.

The basic combined effects approach (see Smith, McNamara & Wellens, 2004, 2011) involves summing the number of negative job characteristics (or absence of

Table 1. Associations between the total negative factors score (split into quartiles) and being in the high stress category (defined as reporting being very or extremely stressed at work).

	OR	CI
1st Quartile	1.00	
2nd Quartile	1.60	1.32–1.93
3rd Quartile	2.08	1.72–2.53
4th Quartile	3.84	3.17–4.66

positive job characteristics) a person is exposed to. This "Total Negative Score" is then sub-divided into quartiles and logistic regressions used to examine associations between this score and outcomes. Table 1 shows the associations between the total negative score and high stress at work. The lowest quartile was set as the comparison group and the odds ratios show that the likelihood of being in the high stress group increases as one goes from quartile 2 to quartile 4.

The above results show a linear relationship between total negative job characteristics and perceived stress at work. Mental health outcomes and accidents at work can also be examined in this way. Other results showed that a measure of exposure to combinations of workplace factors (the Negative Occupational Factors Score) was associated with a number of health and safety outcomes, many of which were consistent across different industry sectors. Some of the associations reflected levels of perceived stress at work whereas others did not involve stress mediation. Negative occupational factors combined with non-work factors (e.g. demographic characteristics) to increase the risk of stress and minor injuries at work. Dissection of the negative occupational factors score identified outcomes influenced by job demand-control-support, effort-reward imbalance and physical hazards/working hours. Some outcomes were associated with all of these dimensions, others by pairs of dimensions and some were specific to a particular dimension. Stress at work was associated with all dimensions but a combination of high demands/high effort had the major effect. Analyses of longitudinal data confirmed effects observed in the cross-sectional analyses and gave a better indication of causality. The results from the Accident and Emergency Unit study showed that the negative occupational factors score also predicted accidents at work. Similarly, objective measures of performance and physiology were influenced by exposure to combinations of occupational factors. The combined effects approach has also been shown to be important in assessing specific problems in certain occupations (e.g. seafarers' fatigue) and in clarifying the effects of drug use on safety at work. In addition, it has strong implications for the development of stress management standards. Similarly, one can use the approach to examine wellbeing at work and address the question of what is a good job (or what factors are associated with greater wellbeing and/or the absence of negative outcomes) and research on this topic is described in the next section.

Well-being at work: What is a good job?

There is a huge amount of research on negative job characteristics, occupational stress and mental health problems. But positive and negative emotions are not just the opposite ends of a continuum, and the absence of negative emotion doesn't mean the presence of positive emotion. Recent approaches (e.g. Waddell & Burton, 2006) have suggested that "Work is good for you." However, detailed consideration of the literature suggests that it is the absence of work that is bad for you. Indeed, work per se is not necessarily going to be good – but good work is good for you (Smith & Wadsworth, 2011). This then leads to the question of what is a good job. This could be answered in many different ways (e.g. from an economic point of view). However, within the present context the question is what psychosocial characteristics associated with work are associated with positive outcomes. A literature review (Wadsworth et al., 2010a) showed that, compared to the negative effects of work, there is very little published evidence on the positive. Indeed, the literature on positive aspects has many problems, such as a lack of theory, lack of data to support views and weak methodology. Measures of wellbeing are mainly outcomes and do not reflect the 'wellbeing process', which is necessary to understand the topic.

Secondary analyses of large-scale surveys (Wadsworth et al., 2010b) compared the effects of the presence and absence of positive/negative job characteristics. For example, the analyses considered questions such as: 'Is the presence of social support good, the absence of social support bad, or are both true?' This was done by splitting the scores into tertiles (three equal parts), using the mid-value as the reference value, and examining whether equal and opposite changes occurred at opposite ends of the continuum. The results from these analyses showed that dose response did not occur for all types of association. This shows that one must examine both ends of the continuum – presence of positive features and absence of negative features – rather than inferring the effects of one from the other.

New survey data, including positive jobs characteristics, appraisals and outcomes were also collected (Smith et al., 2011). The major question addressed was what predicts positive outcomes? Again, a combined effects approach was used and the "good job score", which best predicted positive outcomes (e.g. good health; wellbeing) was the sum of the presence positive job characteristics and appraisals and the absence of negative characteristics and appraisals. An example of this can be seen in Table 2. This shows that those with the highest good job score were nearly 23 times more likely to be in the high positive health group than those in the lowest good job category.

These pieces of research showed that there is a need for a multi-dimensional model of wellbeing at work that measures a wide range of job characteristics, job attitudes, individual characteristics and outcomes. This has been addressed by developing surveys involving short measures of a large number of concepts, and an example of this approach has been the development of the Wellbeing Process Questionnaire (WPQ) which has been used to address many of the above issues (Williams & Smith, 2011).

**Table 2. Associations between the good job
score (shown as quartiles) and positive mental
health (a median slit into high/low groups).**

	Odds ratio
Low good job	1.00
Second quartile	2.89
Third quartile	5.24
High good job	22.83

Development of the Wellbeing Process Questionnaire (WPQ): This relates back to the case definition of occupational stress, the DRIVE model and measurement of wellbeing. Research on the WPQ showed that single items are often highly correlated with longer scales. This meant that it is possible to have a single question measuring perceived stress, single items measuring job characteristics, and single items measuring health outcomes. In addition, possible confounding factors (e.g. personality; life outside of work) can be measured by single items. The single questions provide examples of the concept being measured and responses are made using a scale of 1–10 which allows a greater potential range of responses. An example is shown below:

Job Demands: I feel that I do not have the time I need to get my work done (for example I am under constant time pressure, interrupted in my work, or overwhelmed by responsibility or work demands)

Response: on a 10-point scale from Disagree strongly to Agree strongly

An initial study with a sample of University staff showed significant correlations between single items and full scales (average correlation for work characteristics: 0.7; average for personality: 0.66). The predictive validity was examined by testing the JDCS and ERI models with full scales and single items. Very similar results were obtained (i.e. predictive validity of single items is comparable to full scales; at risk groups based on the models can be identified with single items).

This approach also allows removal of overlapping constructs. Using single items enables one to use a lot more concepts but these often overlap and one can determine which variables remain in the model after all have been entered into the regression. Using this technique the following constructs remained in the model:

- Job characteristics:
  - Negative: Demands; Effort; Over commitment.
  - Positive: Rewards; Control; Support; Consultation on change; Good supervisor relationship.
- Positive life circumstances: Uplifts; Flourishing; Social Support.
- Negative life circumstances: Hassles.
- Personality: Optimism; Self-esteem; Self-Efficacy; Emotional Stability.
- Negative Coping: Avoidance; Self-blame; Wishful thinking.

Using the above variables selective effects are observed. Only certain variables predict specific outcomes. For example, work characteristics are more important for job satisfaction and job stress, whereas personality is a better predictor of positive and negative affect.

Where next? A great deal of research is in progress using the approaches described here to address two main themes. First, research is investigating stress and wellbeing at work in different sectors (call centres; the police; offshore; healthcare professionals). Secondly, additional constructs are being examined to see how these fit into the model (ethnicity; personality; and religion). Thirdly, different outcomes are being investigated to determine whether the approach is appropriate for them (musculo-skeletal disorders; accidents and incidents). The research is also being extended internationally to determine which effects are general and which may be culture specific. Future research will also include using the approach to evaluate interventions that change working practices and offer occupation support (Smith & Cowap, 2011).

References

Cox, T., Griffiths, A. & Houdmont, J. (2006) *Defining a case of work-related stress.* Health and Safety Research Report 449. Sudbury: HSE Books.

Mark, G.M. & Smith, A.P. (2008) Stress models: A review and suggested new direction. In: *Occupational Health Psychology: European Perspectives on research, education and practice. Vol. 3.* EA-OHP series. Edited by J. Houdmont & S. Leka. Nottingham University Press. 111–144.

Mark, G. & Smith, A.P. (2011) Effects of occupational stress, job characteristics, coping and attributional style on the mental health and job satisfaction of university employees. *Anxiety, Stress and Coping,* **25**, 63–78. Doi: 10.1080/10615806.2010.548088

Mark, G. & Smith, A.P. (2012) Occupational stress, job characteristics, coping and mental health of nurses. *British Journal of Health Psychology,* **17**, 505–521. Doi: 10.1111/j.2044-8287.2011.02051.x

Smith, A.P. (2011a) *A holistic approach to stress and well-being. Occupational Health (At Work)* ISSN 1744-2265. 7(4), 34–35

Smith, A.P. (2011b) A holistic approach to stress and well-being. Part 2: Stress at work: models, practice and policy. *Occupational Health (At Work),* **8**(1), 33–35. ISSN 1744-2265.

Smith, A.P., McNamara, R. & Wellens, B.T. (2011) A holistic approach to stress and well-being. Part 3: Combined effects of job characteristics on stress and other outcomes. *Occupational Health (At Work),* **8**(2) 34–35. ISSN 1744-2265.

Smith, A.P. & Cowap, D. (2011) A holistic approach to stress and well-being. Part 4: Reducing stress and sick leave by providing occupational support. *Occupational Health (At Work),* **8**(3), 19–21. ISSN 1744-2265.

Smith, A.P. & Wadsworth, E. (2011) A holistic approach to stress and well-being. Part 5: what is a good job? *Occupational Health (At Work),* **8**(4), 25–27. ISSN 1744-2265.

Smith, A., McNamara, R. & Wellens, B. (2004) *Combined effects of Occupational Health Hazards*. HSE Contract Research Report 287. HSE Books. ISBN 0-7176-2923-6

Smith, A.P., Wadsworth, E.J.K., Chaplin, K., Allen, P.H. & Mark, G. (2011) *The relationship between work/well-being and improved health and well-being*. Report 11.1 IOSH. Leicester.

Waddell, G. & Burton, A.K. (2006) *Is work good for your health and well-being?* The Stationery Office: Norwich.

Wadsworth, E.J.K., Chaplin, K., Allen, P.H. & Smith, A.P. (2010a) What is a Good Job? Current Perspectives on Work and Improved Health and Well-being. *The Open Health & Safety Journal*, **2**, 9–15.

Wadsworth, E.J.K., Chaplin, K., & Smith, A.P. (2010b) The work environment, stress and well-being. *Occupational Medicine*. **60**, 635–639. Doi: 10.1093/occmed/kqq139

Williams, G.M. & Smith, A.P. (2012) A holistic approach to stress and well-being. Part 6: The Wellbeing Process Questionnaire (WPQ Short Form). *Occupational Health (At Work)*, **9**(1). 29–31. ISSN 1744-2265.

THE COMBINED EFFECTS OF OCCUPATIONAL STRESSORS ON HEALTH AND WELLBEING IN THE OFFSHORE OIL INDUSTRY

Andrew P. Smith & Rachel L. McNamara

Centre for Occupational and Health Psychology,
School of Psychology, Cardiff University,
Cardiff, UK

This paper describes results from a survey of stress, fatigue, and health of seafarers and installation personnel working in the UK sector of the offshore oil industry. Potential stressors and fatiguerelated variables were considered in terms of their combined effects on subjective outcome measures. A total negative factors score demonstrated a linear association with depression, cognitive failures, social functioning, lack of/poor quality sleep, fatigue, and the home-work interface. Effects were more pronounced amongst installation personnel than seafarers. The results suggest that exposure to a combination of stressors has a significantly greater negative effect on health than any of these risk factors in isolation.

Introduction

There has been global concern with the extent of offshore stress and fatigue and its potential environmental cost is widely evident across the industry. Regulators, ship and installation owners, trade unions and protection and indemnity clubs are all alert to the fact that a combination of minimal manning, sequences of rapid turnarounds, adverse weather conditions and high levels of traffic may find offshore personnel working long hours and with insufficient recuperative rest. In these circumstances stress, fatigue and reduced performance may lead to environmental damage, ill-health and reduced life-span among highly skilled workers who are in increasingly short supply. A long history of research into working hours and conditions in manufacturing as well as road transport and civil aviation industries has no parallel in the offshore industry. There are huge potential consequences of stress and fatigue offshore both in terms of both operations (accidents, collision risk, and poorer performance, economic cost and environmental damage) and the individual worker (injury, poor health and well-being). Not only has there been relatively little research on this topic but what there has been has been largely focused on specific jobs (e.g. watch keeping), specific sectors and specific outcomes (e.g. accidents). This reflects general trends in stress and fatigue research where the emphasis has often been on specific groups of workers and on safety rather than quality of working life (a crucial part of current definitions of occupational health).

Recent research has addressed these areas by studying seafarers (e.g. Allen et al., 2007, 2008; Kingdom & Smith, 2011, 2012; Smith et al., 2006; Wadsworth et al., 2006, 2008) and installation workers (Gann et al., 1990; Parkes, 1997, 1998, 2002, 2004, 2010, 2012). One of the conclusions from the research on seafarers' fatigue was that a combination of risk factors was involved. This view is consistent with results from large scale surveys of occupational stress (Smith et al., 2004). The results showed that a measure of exposure to combinations of workplace factors (the Negative Occupational Factors Score) was associated with a number of health and safety outcomes, many of which were consistent across different industry sectors. Some of the associations reflected levels of perceived stress at work whereas others did not involve stress mediation. Negative occupational factors combined with non-work factors (e.g. demographic characteristics) to increase the risk of stress and minor injuries at work. Dissection of the negative occupational factors score identified outcomes influenced by job demand-control-support, effort-reward imbalance and physical hazards/working hours. Some outcomes were associated with all of these dimensions, others by pairs of dimensions and some were specific to a particular dimension. Stress at work was associated with all dimensions but a combination of high demands/high effort had the major effect. Analyses of longitudinal data confirmed effects observed in the cross-sectional analyses and gave a better indication of causality. This approach has also been shown to be important in assessing specific problems in certain occupations and in clarifying the effects of drug use on safety at work. In addition, it has strong implications for the development of stress management standards which demonstrated linear effects between negative job characteristics (e.g. the physical environment; working hours; job demands; lack of control or support; and effort-reward imbalance). The aim of the present study was to use this combined effects approach with an offshore sample.

Material and methods

Survey content

The questionnaire was designed to encompass all aspects of life offshore. It was divided into the following three sections:

1. Offshore: included questions relating specifically to work patterns, and subjective measures of attitudes towards work.
2. On leave: included subjective measures of health and well-being, and health-related behaviours such as eating, drinking, smoking and exercise.
3. Life in general: included a number of standardised scales of well-being, such as the General Health Questionnaire (GHQ: Goldberg 1972), the Profile of Fatigue-Related Symptoms (PFRS: Ray 1991), the Cognitive Failures Questionnaire (CFQ: Broadbent et al. 1982) and the MOS Short Form Health Questionnaire (SF-36: Ware et al. 1993).

Procedure

The questionnaire was distributed to the home addresses of members of the seafaring officers' union (NUMAST – National Union of Marine, Aviation and Shipping Transport Officers) and the installation workers' union (MSF – Manufacturing, Science and Finance Union). Secondly, questionnaires were distributed to seafarers onboard offshore oil support vessels operating in the UK sector, by visiting researchers.

Participants

Questionnaires were returned by 555 seafarers (75% officers; mean age: 44 years SD 9.55) and 385 installation workers (20% managers; mean age 46 years SD $= 7.82$). The seafarers sample was most highly represented by seafarers working on supply vessels, support vessels, standby vessels, pipe layers and dive support vessels. The installation workers did a variety of jobs on both drilling rigs and platforms.

Results

Differences between seafarers and installation workers

Both seafarers and installation workers reported excessive hours of work (over 90% of the total sample worked more than 60 hours a week). The most common tour of duty for the seafarers was 4 weeks on – 4 weeks off, whereas the installation workers were more likely to work 2 weeks on – 2 weeks off. Seafarers were more likely to work fixed shifts whereas over half the installation workers had rotating shifts. Most of the installation workers had 12 hour shifts whereas the shift length of the seafarers was more variable (41.4% 12 hr; 22.4% 6on/6off; 18% 4on/8off). These differences in working hours, and the different jobs carried out by the two groups made it reasonable to expect differences in the outcome measures and these are now described.

Installation workers were significantly worse for the following outcomes (see Table 1):

- GHQ (F $[1, 711] = 25.28$, p < 0.0001)
- SF-36 social functioning (F $[1, 709] = 10.46$, p < 0.001)
- PRFS fatigue (F $[1, 709] = 11.94$, p < 0.001)
- Physical fatigue (F $[1, 720] = 6.42$, p < 0.01)
- Mental fatigue (F $[1, 720] = 13.02$, p < 0.0001)
- Job stress (F $[1, 708] = 4.42$, p < 0.04)
- Life stress (F $[1, 712] = 8.25$, p < 0.004)

Combined effects

Median splits of potential stressors were [where $1 =$ low exposure, $2 =$ high exposure], and summed to create a 'total negative indicators score'. These comprised

Table 1.   **Significant differences between seafarers and installation workers (means and SDs; high scores = more mental health problems [GHQ], fatigue, stress but better social functioning).**

	Seafarers	Installation workers
GHQ	1.48 (2.44)	2.32 (3.03)
Social functioning	84.85 (20.49)	22.22 (1.14)
PFRS fatigue	25.36 (12.42)	28.02 (13.31)
Physical fatigue	2.34 (0.71)	2.50 (0.66)
Mental fatigue	2.56 (0.74)	2.73 (0.70)
Job stress	2.47 (0.97)	2.60 (0.99)
Life stress	1.66 (0.69)	1.78 (0.75)

(a) variables relating to working hours: 'number of hours worked per week', night work, shift work, and unsociable hours; (b) the physical environment: breathing fumes/harmful substances, touching/handling harmful substances, ringing in the ears, background noise and vibration; (c) Effort-Reward imbalance: 'I have constant time pressure due to a heavy workload', 'I have many interruptions and disturbances in my job', 'I have a lot of responsibility in my job', 'I am often under pressure to work overtime', 'I have experienced or expect to experience an undesirable change in my work', 'my job promotion prospects are poor', 'my job security is poor' and 'I am treated unfairly at work'. A quartile split of this composite variable was then entered into a series of analyses of co-variance (ANCOVA) co-varying for: age, education and socio economic status (SES) and stratifying for occupational group (i.e. Seafarers or installation personnel).

Significant effects of the composite negative score were found on virtually all subjective measures of health and well-being, including mental health (GHQ Score $F [1, 739] = 35.38$, $p < .0001$), cognitive failures (CFQ $F [1, 723] = 29.62$, $p < .0001$), fatigue (PFRS fatigue $F [1, 732] = 43.37$, $p < .0001$), physical functioning (SF-36 physical functioning $F [1, 735] = 5.55$, $p < .0001$), social functioning (SF-36 social functioning $F [1, 732] = 36.32$, $p < .0001$), job stress ($F [1, 732] = 53.38$, $p < .0001$), life stress ($F [1, 737] = 7.17$, $p < .0001$), lack of sleep ($F [1, 735] = 25.66$, $p < .0001$), poor quality sleep ($F [1, 739] = 34.31$, $p < .0001$), physical fatigue ($F [1, 745] = 31.32$, $p < .0001$), mental fatigue ($F [1, 744] = 38.93$, $p < .0001$), and aspects of the home-work interface, including: 'problems at work make you irritable at home' ($F [1, 728] = 9.55$, $p < .0001$) and 'job takes up too much energy' ($F [1, 731] = 34.65$, $p < .0001$). Means and standard deviations for these outcome measures are shown in Table 2.

These results suggest that physical and psychosocial hazards in the offshore environment combine additively to produce a linear effect on a wide range of health and well-being outcome measures.

Table 2. Combined effects of negative factors (quartiles) and outcomes.

Outcome	1st quartile	2nd	3rd	4th
GHQ	0.59 (1.28)	1.25 (2.09)	2.22 (2.75)	3.17 (3.44)
CFQ	32.50 (11.85)	36.62 (11.79)	40.40 (13.20)	44.13 (13.65)
SF-36 Social Functioning	92.41 (14.14)	88.24 (17.67)	80.10 (22.16)	72.62 (22.68)
Poor quality sleep	1.97 (0.75)	2.36 (0.83)	2.55 (0.86)	2.80 (0.88)
Job stress	1.95 (0.78)	2.40 (0.94)	2.64 (0.90)	3.08 (0.95)
Physical fatigue	2.08 (0.62)	2.32 (0.65)	2.50 (0.66)	2.72 (0.68)
Mental fatigue	2.22 (0.66)	2.55 (0.67)	2.77 (0.64)	2.93 (0.71)
Problems at work = Irritable at home	1.32 (0.50)	1.53 (0.60)	1.76 (0.60)	1.67 (0.65)

Discussion

These results clearly demonstrate that exposure to a combination of workplace stressors has a significantly greater negative impact on subjective measures of health and well-being than any one hazard in isolation. Furthermore, installation workers in this study appear even worse off in terms of well-being than their seafaring counterparts. This may be explained in part by the differences in shift systems between the two groups: installation workers in this study tended to work fast rotating as opposed to fixed shifts, which have previously been demonstrated to be the most detrimental shift pattern in terms of health and performance. This idea requires further clarification however: future research in the area might therefore wish to investigate this issue.

There are a number of problems inherent in the type of methodology used in this study. It is not possible to determine causal relationships from a cross sectional survey. Although the results suggest that working in an offshore environment is detrimental to health, the possibility that poor health may lead to a more negative perception of working patterns cannot be ruled out. Individual differences such as negative affectivity may create reporting biases amongst those who seem to be most affected. These difficulties could be overcome in future by employing longitudinal or intervention studies, although an approach of this nature might prove difficult to implement from a practical point of view. Co-varying for negative affectivity may provide a more suitable alternative.

The current research highlights the potential for fatigue in an offshore environment. Although it was not clear in this instance what the consequences of this might be in terms of injury and accident causation, future research should seek to examine this link as the environmental, financial and personal costs of such a relationship are potentially devastating. It is already clear that a revision of working practices within the industry would greatly improve the well-being of the workforce. Further research is needed to compare offshore samples with those working onshore. About

20% of onshore workers report high levels of stress and fatigue. There is considerable variation across occupations and it is not clear which onshore jobs provide the best comparison for the samples studied in this project.

Acknowledgement

This research was funded by the Maritime and Coastguard Agency, the Health and Safety Executive and the Seafarers International Research Centre.

References

Allen, P. H., Wadsworth, E. J. and Smith, A. P. (2007) The prevention and management of seafarers' fatigue: A review. *International Maritime Health* **58**: 167–177.

Allen, P. H, Wadsworth, E. J. and Smith, A. P. (2008) Seafarers' fatigue: a review of the recent literature. *International Maritime Health* **59**: 1/4, 81–92.

Broadbent, D. E., Cooper, P. F., Fitzgerald, P. and Parkes, K. R. (1982) The cognitive failures questionnaire (CFQ) and its correlates. *British Journal of Clinical Psychology* **21**: 1–16.

Collinson, D. L. (1988) "Shifting lives": Work-home pressures in the North Sea oil industry. *Canadian Review of Sociology and Anthropology* **35**: 301–324.

Gann, M., Corpe, U. and Wilson, I. (1990) The application of a short anxiety and depression questionnaire to oil industry staff. *Journal of the Society of Occupational Medicine* **40**: 138–142.

Goldberg, D. (1972) *The detection of a psychiatric illness by questionnaire.* Oxford University Press, London.

Kingdom, S. E. and Smith, A. P. (2011) Psychosocial risk factors for work related stress in Her Majesty's Coastguard (HMCG). *International Maritime Health* **62(3)**: ISSN 1641-9251.

Kingdom, S .E. and Smith, A. P. (2012) Combined effects of Work-Related Stress in Her Majesty's Coastguard (HMCG). *International Maritime Health* **63(1)**: 63–70. ISSN 1641–9251.

Parkes, K. R. (1997) *Psychosocial aspects of work and health in the North Sea oil and gas industry, part 1: review of the literature.* Health and Safety Executive, HSE Books. Publication No OTH 96 523, Sudbury, Suffolk, 1997.

Parkes, K. R. (1998) Psychosocial aspects of stress, health and safety on North Sea installations. *Scandinavian Journal of Environment and Health* **24**: 321–333.

Parkes, K. R. (2002) *Psychosocial aspects of work and health in the North Sea oil and gas industry: Summaries of reports published 1996–2001,* HSE RR 002, HSE Books. ISBN 0 7176 2156 1. http://www.hse.gov.uk/ research/rrhtm/rr002.htm

Parkes, K. R., Farmer, E. and Carnell, S. (2004) *Psychosocial aspects of work and health in the North Sea oil and gas industry: A survey of FPSO installations, and comparison with platforms and drilling rigs.* Research Report 202. Health and Safety Executive. http://www.hse.gov.uk/research/rrhtm/rr202.htm

Parkes, K. (2010) *Offshore working time in relation to performance, health and safety: A review of current practice and evidence*. Research Report RR772. Health and Safety Executive. http://www.hse.gov.uk/ research/rrhtm/rr772.htm

Parkes, K. R. (2012) Shift schedules on North Sea oil/gas installations: A systematic review of their impact on performance, safety and health. *Safety Science* **50**: 1636–1651.

Ray, C. (1991) Chronic fatigue syndrome and depression: Conceptual and methodological difficulties. *Psychosomatic Medicine* **21**: 1–9.

Siegrist, J. (1996) Adverse health effects of high-effort – low reward conditions at work. *Journal of Occupational Health Psychology* **1**: 27–43.

Smith, A. P., Johal, S. S., Wadsworth, E., Davey-Smith, G. and Peters, T. (2000) *The scale of occupational stress: The Bristol stress and health at work study*. HSE Books Report 265/2000.

Smith, A., Allen, P. and Wadsworth, E. (2006) *Seafarer fatigue: the Cardiff Research Programme*. MCA: Southampton. http://www.dft.gov.uk/mca/research_report_464.pdf

Wadsworth, E., Allen, P., Wellens, B., McNamara, R. and Smith, A. (2006) Patterns of fatigue among seafarers during a tour of duty. *American Journal of Industrial Medicine* **49**: 836–844.

Wadsworth, E. J., Allen, P. H., McNamara, R. L., Wellens, B.T. and Smith, A. P. (2008) Fatigue and health in a seafaring population. *Occupational Medicine*. Doi: 10.1093/occmed/kqn008. **58**: 198–204.

Ware, J. E., Snow, K. K., Kosinski, M. and Grandek, B. (1993) *SF-36 health survey manual and interpretation guide*. Boston: MA The Health Institute, New England Medical Centre.

TRANSPORT

COMMUNICATION ON THE BRIDGE OF A SHIP

Paul Allen & Andrew P. Smith

*Centre for Occupational and Health Psychology,
School of Psychology, Cardiff University,
Cardiff, UK*

This paper reports research from the CASCADe (Model-based Co-operative and Adaptive Ship based Context Aware Design) project. The aim of the research was to generate ideas for practical changes on the ship's bridge that would improve efficiency and safety. A multi-method approach was used, starting with a literature review and followed by onboard visits, focus groups of seafarers, and a survey of serving ships' officers. The idea of a new communication tool for the bridge emerged, one which could be used to log information, aid in hand-overs, and aid in sharing information and situational awareness around the bridge team.

Introduction

The purpose of this study was to produce hypotheses on how the bridge, as a cooperative system, processes information. Featuring at an early stage in the CASCADe project, a decision was to use this particular work package for two key purposes:

- To bring in an end-user perspective.
- To help focus the scope of the project.

By conducting surveys, focus groups and observations of seafarers, the aim was to build up a conceptualisation of the bridge from the end-user's perspective, with a particular focus on identifying those areas where there are currently problems and therefore there is scope for improvement. The theoretical framework for the project was the 'Combined Effects approach', as used in the Cardiff Seafarers' Fatigue study (Smith, Allen and Wadsworth 2006). This approach posits that dangerous situations are built up from multiple factors. Whilst the addition of one or more dangerous factors (e.g. fatigue, bad weather) can have an enormous impact on the likelihood of a serious outcome (e.g. collision or grounding), the reverse of this equation is that by addressing one or two problems, significant benefits can be felt. The present research was focused upon identifying areas where such improvements might be focused.

Methods

Literature review

As a starting point, five papers/reports of direct relevance to the area were reviewed.

1. MAIB Bridge Watchkeeping Safety Study (2004)

The UK's Marine Accident Investigation Branch (MAIB) reviewed 66 collisions, groundings and contacts investigated by the organisation between 1993 and 2004, involving vessels over 500 gt. The study highlighted three major areas of concern, as follows:

1. Groundings and fatigue: A third of all the groundings involved a fatigued officer alone on the bridge at night.
2. Collisions and lookout: Two thirds of all the vessels involved in collisions were not keeping a proper lookout.
3. Safe manning and the role of the master: A third of all the accidents that occurred at night involved a sole watch-keeper on the bridge.

This report therefore highlighted the significant role of fatigue and proper lookouts in terms of understanding accidents at sea.

2. IMO Results of a worldwide e-navigation user needs survey (2009)

This 2009 paper was not based on accident reports but on the results from a worldwide survey of seafarers. The focus was upon establishing the e-navigation needs of seafarers, with questions also addressing those areas where seafarers considered there to be problems. In total 353 participants completed the survey, with relevant results as follows:

1. Communication: 62% had either high or moderate concerns about communication between ships in terms of language skills.
2. Reporting: Participants commented on the burden of reporting to shore authorities and the duplication of information sent.
3. Human-Machine interfaces (HMI): The main areas for improvement in HMI's were: user friendliness, standardisation, integration, ergonomic issues, alarm management, reliability and training.
4. S-Mode: 80% of participants were in favour/rather in favour of the idea of an S-mode concept – where equipment can be returned to a standardised mode at any time.

3. Perceptions of technology at sea amongst British seafaring officers, Allen (2009)

A survey was conducted of British seafaring officers to assess perceptions of technology at sea as part of the European FP6 funded 'Flagship' project. A sample of 819 seafarers completed the questionnaire survey, with older seafarers and those with lower computer literacy showing greater resistance to new technology. Training was highlighted as a key area of concern for seafarers, with a perception that equipment is often introduced without proper instruction on how to use it.

4. Human and organisational factors in maritime accidents: analysis of collisions at sea using the HFACS, Chauvin et al. (2013)

This 2013 paper approached the analysis of accident reports in a systematic way using the HFACS (Human Factors Analysis and Classification System) framework. Chauvin et al. looked in depth at 27 accidents between 1998 and 2012 involving a collision between two merchant vessels or between a merchant vessel and a fishing vessel. The conclusions were as follows:

1. Most collisions are due to decision errors.
2. Important factors found were: poor visibility, misuse of instruments, loss of situational awareness or deficits of attention, deficits in inter-ship communications or Bridge Resource Management (BRM).
3. At the leadership level, the authors found problems with the planning of operations, and non-compliance with the Safety Management System (SMS).
4. The study showed the importance of Bridge Resource Management for navigation situations with a pilot onboard in restricted waters.

Bridge Resource Management is of particular relevance as it concerns how information – and resources – are shared amongst the crew. This is at the heart of a cooperative system. Recommendations regarding proper Bridge Resource Management (or Bridge Teamwork Management as it can be understood) were brought in with the 1995 revisions to the Standards of Training, Certification and Watchkeeping (STCW). Critically, the Manila amendments, brought into force in 2012, make BRM training compulsory. Leadership and teamwork awareness has therefore been given new prominence in the industry. Chauvin et al. found that failures in BRM, manifesting in poor coordination, communication and shared situational awareness, are involved in more than 1 in 3 of the collisions investigated.

5. Behind the headlines? An analysis of accident investigation reports, Tang et al. (2013)

A paper presented at the Seafarers International Research Centre (SIRC) biennial symposium (July 2013) also described analysis of accident reports. The authors analysed 319 accident investigation reports from the UK (148), Australia (110), New Zealand (43) and the National Transportation Safety Board in the US (18), covering a 10 year period from 2002 to 2011. Of the accidents reviewed, 31% were collisions and contacts, 19% were groundings and 10% were fires and explosions. From the analysis, the authors found that 6 factors were either an immediate or contributory cause in 20% or more of accidents:

- Inadequate risk management
- Third party deficiency
- Poor judgement/operation
- Inadequate training/experience

- Failure in communication/co-ordination
- Weather/other environmental factors

In its description of the analysis process, the paper also serves as a reminder of the complexity of the task in terms of defining causes. Many factors are related, so that, for example, poor judgement/operation may be related to fatigue, training or poor communication. Trying to delineate one cause can therefore become a somewhat artificial exercise. The positive side of this argument, as put forward by the combined effects approach, is that an improvement in one area can have a positive impact on several other areas. This conceptualisation will underpin the solutions that are put forward in the CASCADe project: by addressing one or two specific areas of concern on the bridge, the impact on safety (and efficiency) will be significant.

Onboard ferry observations

The first piece of direct research involved a fact-finding visit onboard a ferry in March 2012. This visit had three aims:

- To familiarise partners not working in the maritime domain with the way a bridge operates.
- To start the process of thinking about where the CASCADe project might focus its efforts for maximum benefit.
- To speak to active seafarers – in their environment – to understand how they perceive their roles and the supporting function of technology.

The ferry visit proved important in terms of helping to define the scope of the work in CASCADe. Critically, it was identified that a full re-design of the bridge was likely to be a poor use of resources in the project as bridges have evolved over time for a specific purpose and are largely effective in this task of navigation. There are clearly examples of poorly designed bridges, but the new Integrated Navigation System (INS) standards, adopted by the International Maritime Organisation (IMO), and brought into force at the beginning of 2011, will go a significant way towards addressing such fundamental problems – particularly around integration. It was therefore felt that CASCADe should be more forward thinking, looking for innovative solutions that complement and enhance the current technology, rather than trying to re-invent it. Discussion following the ship visit, including at the second whole consortium meeting in Cyprus in April 2013, focused on the need to produce technological solutions that would be of practical help in the real work environment. It was therefore decided that early design ideas should be developed in order to help stimulate discussion with end-users (i.e. seafarers) around where problems currently lie. This approach complimented parallel work on the CASCADe project focused on modelling the way in which seafarers move and operate on the bridge using a virtual simulation platform (see Denker et al., 2014).

Design ideas

Four broad design ideas were developed as a starting point:

- A drawing tool that allows seafarers to annotate over the top of key pieces of equipment e.g. electronic navigational charts
- An operation adaptive display that automatically changes the display of instrumentation depending on the mode of operation.
- A captain's portable display relaying key information from the bridge to bring the captain up to speed in an emergency situation.
- A handover tool to present key information during handover.

Three means of gathering user feedback were employed:

- A questionnaire survey of seafarers to provide a broad idea of seafarer opinions
- Focus groups to provide more in-depth perspectives.
- Direct observation of seafarers completing simulation exercises.

A survey

A short questionnaire was developed and distributed online to navigational officers. The survey was designed with mostly open questions as a means of gathering some initial impressions concerning the areas that are of most concern to seafarers. The survey was advertised by Nautilus International, an associate partner on the project, both online and in their monthly publication, the Nautilus Telegraph. The survey was completed by 31 seafarers, the majority of whom were British. A representative sample was not required, as this was designed as an exploratory piece of work to gather initial impressions.

Key issues identified included:

- Many seafarers were concerned about over-reliance on technology by new seafarers. Respondents frequently mentioned the problem of new seafarers looking at displays rather than out of the window.
- Poor communication was a common theme. In addition to problems of basic communication skills, cultural and language problems were often highlighted.
- When under pilotage, many respondents identified the danger of crew members becoming complacent and becoming over-reliant on the pilot.

Survey respondents were also directly asked about areas where they believed new technology might be of benefit, and to describe any new design ideas they might have. Interestingly, two seafarers talked about large touch-screen charts, not too dissimilar from the drawing tool that had been independently mocked up in the project. Another idea, suggested by two seafarers, was the idea of some form of heads-up display so that seafarers could keep an eye on instruments whilst also keeping a visual lookout of the window.

Focus group

The questionnaire survey was complimented by two visits to maritime colleges. The first was to a UK maritime college to conduct focus groups and learn more about the issues highlighted in the survey. The second, in Germany, was focused upon observing seafarers during simulator exercises to learn about the way seafarers interact, and to see how simulation exercises are conducted.

The visit to a UK maritime college involved conducting three 1 hour focus groups. These were chosen to present three distinctly different perspectives. The three groups were as follows:

- Trainers on navigational courses (n = 3)
- Cadets (n = 12)
- Experienced seafarers (n = 19)

In the same way as the survey, the focus group participants were asked about specific modes of operation and areas of concern, but were also presented with mock-ups of the early design ideas to help develop the themes in a practical way. Many themes identified in the survey resurfaced in the focus groups, including problems with communication, problems with English language skills for non-native speakers, crew complacency when a pilot comes onboard, and over-reliance on technology by new seafarers. Other problems highlighted included the way in which a pilot communicates with a bridge team. A number of the focus group participants had experienced situations where a pilot only communicated with the captain, with information not shared amongst the rest of the bridge team. The course trainers also highlighted that general training in communication and leadership is currently lacking, with an almost exclusive focus on technical rather than 'soft skills' on simulator-based courses. A surprise for the research team was the extent to which the use of checklists was supported by both the cadet and expert seafarer focus groups (although the trainers were considerably less positive). Participants explained that checklists can be useful, but only if used in the correct way, and if they are regularly updated to make them as relevant and useful as possible (i.e. they are kept 'live').

Response to design ideas

In terms of the early design ideas presented, the drawing tool was the most positively received by the respondents. Seafarers liked the idea of adding extra notes and annotations, but only if these could easily be removed, or were only shown on an auxiliary rather than primary display. Large elements of the operation adaptive display and portable captain's display were felt to be familiar to the seafarers, with the added elements of automatic mode switching and portability felt to be more of a potential danger than benefit. The handover tool was also seen to have some benefits, with a cross-over with the drawing tool discussed. Seafarers liked the idea of adding notes and annotations during a watch that could then be used as a means of passing over key information at hand-over. A tie-in with electronic logbooks was also discussed. In terms of the actual means of display, a participant in the cadet group suggested the idea of a large interactive whiteboard at the back of the bridge which could be used as a means of sharing situational awareness.

Simulator exercises

Simulator exercises were observed during the visit to the UK maritime college, primarily to help in formulating ideas for later evaluation. Observations were also carried out in Germany at another maritime college, with a focus again upon learning about the way in which simulation exercises are carried out, but also to learn more about the way seafarers interact with each other, and interact with bridge technology. A key observation from the German maritime college concerned the way in which seafarers spend different amounts of time monitoring different equipment during different modes of operation. During berthing, for example, the seafarers were observed to mostly look out of the window, whilst during departure the radar was used considerably more. Any solution will therefore have to be adaptive to the flexible nature of seafarer working.

Conclusions

From these initial studies a picture emerged concerning where it might be most beneficial for CASCADe to focus. This can be seen as a 'seafarers' eye-view' of the bridge; an identification of areas for improvement based on the direct experience of seafarers. In a broad sense, the most promising area for potential innovation appeared to be around communication. The idea of some form of communication tool for the bridge emerged, one which could be used to both log information, aid in hand-over, and aid in sharing information and situational awareness around the bridge team. This tool could also aid in bringing a pilot 'into the loop', as he/she could use the tool as a means of communicating intentions amongst the bridge team. The seafarers approached were receptive to the idea of using touch as the means of interaction, and having space to both annotate bridge screens (such as electronic navigational charts) and write down notes and messages on some form of 'scribble pad'. Seafarer suggestions included the idea of presenting information on a large interactive white board at the back of the bridge, or a heads-up display at the front. These results will feed into the next phase of the CASCADe project where a prototype tool to enhance cooperation and communication on the bridge will be produced.

References

Allen, P. 2009. Perceptions of technology at sea amongst British seafaring officers. *Ergonomics* 52(10): 1206–1214.

Chauvin, C., Lardjane, S., Morel, G., Clostermann, J. P. and Langard, B. 2013. Human and organisational factors in maritime accidents: Analysis of collisions at sea using the HFACS. *Accident Analysis and Prevention*, 59, 26–37.

Denker, C., Sobiech, C., Mextorf, H., Randall, G., Allen, P., Mikkelsen, T., Adami, E. and Javaux, D. Modelling human-machine cooperation for human-centered ship bridge design. *Proceedings of the 5th Transport Research Arena Conference*. April 2014.

IMO Sub-committee on safety of navigation, 55th Session, Agenda item 11 2009. Development of an E-Navigation strategy implementation plan: Results of a worldwide e-navigation user needs survey.

MAIB, MAIB Bridge Watchkeeping Safety Study 2004. MAIB. URL: http://www.maib.gov.uk/cms_resources/Bridge_watchkeeping_safety_study.pdf

Smith, A., Allen, P. and Wadsworth, E. 2006. Seafarer Fatigue: The Cardiff Research Programme. MCA/HSE report.

Tang, L., Acejo, I., Ellis, N., Turgo, N. and Sampson, H. 2013. Behind the headlines? An analysis of accident investigation reports. Seafarers International Research Centre Symposium 2013. URL: http://www.sirc.cf.ac.uk/Uploads/Symposium/Symposium%20Proceedings%202013.pdf

WHERE IS THE PLATFORM? WRONG SIDE DOOR RELEASE AT TRAIN STATIONS

Dan Basacik & Huw Gibson

RSSB

There is potential for harm to railway passengers if train doors are operated where there is no platform to disembark onto. In order to assist with risk mitigation, this study investigated the human errors that led to doors being released on the wrong side of a train. Methods used to conduct the study included a review of incident investigation reports, train driver observation and human reliability assessment. The causes of door release errors were mostly slips and were influenced by factors such as train cab and platform equipment design, along with distractions and fatigue. The study found that train drivers are performing the task extremely reliably when error rates derived from incidents are compared with those predicted by human reliability assessment.

Introduction

On modern trains, passenger doors must be operated by the guard or driver before passengers can board or alight. Incidents can occur if the doors are operated where there is no platform for passengers to disembark onto. This may be because the train is not in a station, because the doors on the wrong side of the train are opened at a station or because doors over the end of the platform are operated.

The immediate cause of these incidents is usually a human error in activating the door controls. Human errors manifest themselves in a limited number of ways (Reason, 1990). By understanding the types of error that cause or could cause an incident, the rate at which it is likely to occur and the factors that affect this, measures that reduce its impact or frequency can be identified (Kirwan, 2005).

The Incident Factor Classification System database (IFCS – Gibson et al., 2013) is being developed as part of the cross-industry Safety Management Information System (SMIS) and uses established classifications to describe the human error types and underlying factors in railway incidents. It applies two distinct classifications depending on the type of factors identified during investigations:

- Human error (categories related to slips, lapses, mistakes and violations).
- 10 categories of systems failure and performance shaping factors.

Human reliability assessment provides a methodology for predicting the likelihood of errors based on characteristics of the task being performed. Examples of techniques include the Human Error Assessment and Reduction Technique (HEART – Williams, 1986), the Technique for Human Error Rate Prediction (THERP – Swain and Guttman, 1983) and more recently Railway Action Reliability Assessment (RARA – Gibson, 2012). The latter is based on HEART but is tailored towards train driver tasks and errors. The predictive power of these techniques is most obvious when no human error data are available in relation to a proposed design or redesign. However, even when human error data are available, for example from incident or near miss reporting systems, theoretical error rate predictions from error quantification methods can be useful. Comparing error rates experienced within the system to the rate predicted by human error quantification methods helps to set the experienced error rate in context, and understand the extent to which human reliability improvements can be expected by addressing performance shaping factors. For example, if a repetitive task is being performed extremely reliably in comparison to the level predicted by tools such as HEART, a step change in task reliability may more likely be achieved by allocation of the task to a machine, than by, for example, small tweaks to an already suitable task environment.

London Overground Rail Operations Limited (LOROL) operate an intensive passenger train service on the urban and suburban railway network around London. LOROL have experienced door release incidents, though to date there have been no incidents where passengers have fallen from the train. Other operators have also had door release incidents, but without detailed analysis it is unclear how different operators' incident rates compare to one another. In 2012, LOROL requested that RSSB provide a human factors review of train door release, the overall aim being to support their considerations for the management of this risk. This paper describes one part of the overall project. It focuses on some of the work done to identify human errors and performance shaping factors leading to door operation errors at stations and to compare the error rate derived from incidents to the rate predicted by RARA.

Method

A review of incident reports was carried out. The aim was to identify the human errors which led to LOROL's door release incidents, and any causal or contributory factors. Incident investigation reports are compiled by incident investigators based on evidence including driver interview, witness testimony, CCTV, on-train data recorders and other available and relevant data sources. In line with the scope of the project, the analysis only included incidents which took place on the new Class 378 rolling stock on which doors are released by the driver. As the introduction of these trains was completed in October 2010, the analysis was carried out on data from then until March 2013, when the study began. Thirty-nine reports were available for review (i.e. investigations had been completed and the report signed off) from a total of 42 incidents recorded in SMIS. The reports were reviewed and analysed using the IFCS, to categorise human errors, systems failures and performance shaping factors.

Driver observations were carried out on LOROL passenger services, during two weekdays. The aim of this activity was to develop a sound understanding of how the task of stopping the train at a station occurs in practice and context, and to identify factors that could potentially influence the likelihood of error.

The data collected during the observations were used to carry out a structured task analysis of approaching the station, stopping the train and releasing the doors (e.g. Shepherd and Stammers, 2005).

The human error quantification method RARA (Gibson, 2012), was followed to estimate the theoretical likelihood of human error when performing the task. The theoretical rate of human error was then compared to the error likelihood based on incident data. This was derived through an analysis of irregular working and door incident data from the cross-industry Safety Management Information System (SMIS) database, which were cleansed. An error rate was calculated by dividing the number of incidents by the number of times a LOROL train was timetabled to call at a station (and therefore release its doors).

Results

The door release task

On most passenger trains it is possible for the doors to be released or opened by the driver or guard. If they are released, the doors are unlocked but remain closed until passengers use the controls provided to open a door. If doors are opened by the driver or guard, the doors are not only unlocked, but all of the doors on the train open without any intervention from passengers.

LOROL's procedure for stopping at a station and releasing the doors is:

- Driver stops the train at the correct stop board.
- Driver uses left hand to select step 3 braking using the combined power-brake controller.
- Driver uses left hand to set direction switch to neutral.
- Driver uses left hand to operate buttons on the left hand side of the cab desk to release doors on the left hand side of the train, or their right hand to operate buttons on the right hand side of the cab desk to release doors on the right hand side of the train.

LOROL advises their drivers to take 1 to 2 seconds of thinking time between stopping the train and releasing the doors. They instruct their drivers to release the doors but not open them, to add another layer of protection against doors becoming open on the wrong side. For doors to open drivers would have to release the doors on the wrong side and passengers would then have to open the doors where there is no platform, or the driver would have to incorrectly use the 'open' button in addition to selecting the buttons for the wrong side of the train. To further protect

Table 1.   Types of human error identified in the review of incident reports.

Error Type	Description	Number identified
Perception slip	Errors in visual detection and searching, and listening errors.	1
Action slip	Errors in actions or speech are not being performed as planned (i.e. action not as intended).	14
Memory lapse	Failures of short or long-term memory.	2
Decision error	Errors in acts of judgement, decisions or strategies. They typically rely on knowledge and information being correctly recalled but wrongly applied.	14
Situational violation	This is when people break the rules because of particular pressures or circumstances arising from a specific job.	2
Routine violation	This is when breaking the rule or not following the procedure has become the normal way of working.	0
Not classified	Used when there was not enough information in the report to classify the error.	39

against the latter scenario, on LOROLs' Class 378 trains, the 'door open' buttons in the driver's cab are covered with flaps.

Incident report review

The Incident Factor Classification System uses the categories in Table 1 to distinguish between different types of human failure. Across the 39 investigation reports reviewed, a total of 72 human failures were identified. These were classified as shown in Table 1.

Errors rather than violations dominated the figures; there was no evidence that drivers were intentionally releasing the doors on the non-platform side at stations. The violations that were identified took place during incident reporting.

Unsurprisingly, action slips were the most common type of error in the incident reports. The incident reports identified that the driver in question knew which side the platform was on but released the doors on the wrong side in error. They mainly occurred when the driver became accustomed to opening the doors on one side of the train but then came into a station where the platform was on the other side. Decision errors were present in the data, and tended to exist where drivers had adopted faulty strategies to remain alert, or used unofficial cues on the non-platform side rather than the official (but perhaps harder to see) stop markers on the platform. These decision errors were not errors in the task of releasing the doors, but combined with other types of error (action slip or memory lapse) to result in the doors being released where there was no platform.

Memory lapses, involving drivers forgetting which station they were at, which side the platform was on, or which step in the task sequences they needed to carry out,

Table 2.   Factors contributing to door release errors.

Factor	Number identified
Equipment	29
Personal	20
Knowledge, skills and experience	10
Practices and processes	3
Workload	2
Information	1

were uncommon. Perception slips, for example those involving drivers mis-seeing which side the platform was on, were also rare.

Unfortunately nearly half of the errors/violations in the incident reports remained unclassified as there was not enough information in the investigation reports.

Underlying factors

Across the thirty nine incidents, a total of 65 factors (between one and four factors per incident, and approximately one and a half, on average) were identified that influenced drivers' performance of the door release task and increased the likelihood of error. The factors were categorised according to the 10 Incident Factor Classification, and are shown in Table 2.

Equipment design and layout was most frequently identified in the reports, occurring a total of twenty-nine times. Twelve of these related to rolling stock and seventeen to infrastructure. Some examples of equipment problems were:

• The platform being on the opposite side of the train compared to a run of recent station stops.
• In-cab CCTV screens (which are used when preparing the train for departure) being located near the right hand side door release buttons and therefore prompting the driver to use those controls on rare occasions when the doors are being re-released after being closed.
• Poor visibility from the cab window making it difficult to see stop markers on island platforms, leading to the use of unofficial markers on the non-platform side, which then focuses the driver's attention on the wrong side.
• Platform signals being on the opposite side to the platform, again attracting the driver's attention to the wrong side.

Factors to do with the person's physical or mental state (i.e. 'Personal' factors) were identified a total of twenty times across the incidents, and were present in seventeen out of the thirty-nine incidents. These factors included distraction and fatigue, for example, drivers thinking about an upcoming holiday towards the end of their shift, and drivers becoming tired following an early start. 'Knowledge, skills and experience' factors were identified ten times and were present in nine out of the thirty-nine incidents, mostly where the drivers were identified as being trainees, but on one occasion in relation to inadequate training. The remaining factors were each identified three times or less.

Human error quantification

Incident data were analysed to understand whether the rate at which door release incidents occur is in line with the expected rate of error based on the type of task. In the period between October 2010 and March 2013 there were 42 reported door release incidents which took place at stations and involved LOROL services. In the time period covered by the data, doors were released 12,599,599 times (based on timetable data and discussions with LOROL) on LOROL passenger services running on the three lines being studied. This yields a rate of 3.33×10^{-6}, which means an incident occurs just over three times in every million station stops.

According to RARA, a 'completely familiar, well-designed, highly practised task which is routine' would be expected to have a human error probability of 4×10^{-4} (lower bound 8×10^{-5}, upper bound 7×10^{-3}). It can be seen that drivers at LOROL are much more reliable in carrying out door release at stations than would be expected for a human task of this nature, based on RARA. The human error rates in RARA are based on data in HEART.

Discussion and conclusion

The task of activating the door release buttons is a simple sequence of actions that is highly practised. The stopping frequency of LOROL services reinforces the practice effect. The analysis supports the hypothesis that the task of releasing the doors of a train is largely automatic, and completed as part of the station stop routine, involving use of the brake control, direction switch and door controls.

Action slips are associated with highly practised and largely automatic tasks involving movement, and occur when the action is not as the person intended. The investigation report review suggests that drivers know which side of the train the platform is on and in a vast majority of cases the platform is clearly visible. However, errors can occur when the driver's 'stream of intended actions is captured by the similar, well-practised behaviour pattern' (Wickens, 1992); in other words, the driver gets used to opening the doors on one side of the train but then comes into a platform on the other side. Understanding the type of error is useful because it allows appropriate mitigations to be considered. Slip and lapse type errors, dominant within the data analysed here, are part of human variability are not straightforward to address (e.g. with training), instead requiring system design changes such as reminders, error detection systems or interlocks.

Many of the errors identified in incident reports were not described in sufficient detail to classify them, and this remains a learning point for incident investigators. If all of the errors that could not be classified turned out to be situational violations, for example, this would significantly alter the approach needed to tackle wrong side door release.

The incident analysis considered underlying factors and identified issues to do with equipment, driver physical or mental state and driver experience as being key contributors to door release incidents. Identification and classification of these factors facilitates wider system learning, and suggests that the following are important considerations in mitigating wrong side door release incidents:

- A wide field of view from the cab to support uninterrupted observation of stop boards, whether platforms are on the left or right.
- Placement of the in-cab CCTV door monitor away from door controls.
- The CCTV image appearing on the same side as the platform.
- Consistency in terms of which side the platform is on.

When compared with the error rate predicted by RARA, LOROL drivers are performing the door release task at the limits of human reliability. The rate of about one in three hundred thousand is better performance than RARA and HEART predict for any task, and is therefore likely to be of interest to authors of human reliability assessment tools and practitioners who use them. It may be that similar data from a broader range of train driving tasks could be used to update tools such as RARA.

Because the incident data suggest that drivers are performing at such high levels of reliability it is important to question the validity of the incident data. Under-reporting of door release incidents would mean that the error rate calculated using incident data is an underestimate. LOROL believe that they have a culture that encourages reporting of these incidents. Generally, if their drivers report door release incidents they are not subjected to disciplinary action; however, if they do not report an incident which the company becomes aware of through different channels, disciplinary action is pursued. There are examples of incidents that have been reported by drivers, which the company would not necessarily have been aware of by other means (e.g. the door close buttons were activated so quickly after the door release buttons that the on-train data recorder did not detect the doors as having been released). Additionally, incident data is gathered via a number of channels including members of public who report incidents via customer helplines and social media, and staff travelling as passengers on trains. LOROL believe that this comprehensive scrutiny of door release errors, combined with their policies, gives confidence that the number of incidents reported is an accurate reflection of the number of incidents that occur.

For LOROL it is good news that drivers are performing the door release task very reliably, and this is supported by elements such as good interface design and strategies adopted by drivers to reduce the likelihood of such incidents. Nevertheless, exposure to the task is high, with over one hundred thousand station stops per week, so the potential for harm to passengers does exist.

Given the high level of reliability with which drivers perform the door release task, and the fact that most of the performance shaping factors were to do with the positioning of platforms or design of train cabs (neither of which are easy to alter), a step change in performance seems unlikely without allocating the door release task, or at least the choice of which doors to release, to a machine which works reliably. One design solution is correct side door enabling equipment. This equipment uses track based tags containing information about which side of the train the platform is on and is read by equipment on board the train. This system only allows the doors on the correct side of the train to be released. Indeed following completion of this project, LOROL has negotiated funding to fit their trains and stations with correct side door enabling equipment.

Acknowledgement

The authors of this paper would like to thank LOROL for supporting this project.

References

Gibson, H. 2012. Railway Action Reliability Assessment user manual: A technique for quantification of human error in the rail industry. Retrieved 2nd September, from: http://www.sparkrail.org.

Gibson, W.H., Smith, S., Lowe, E., Mills, A.M., Morse, G., and Carpenter, S. 2013. Incident Factor Classification System. In N. Dadashi, A. Scott, J.R. Wilson and A. Mills (eds.) *Rail Human Factors,* (CRC Press, Leiden).

Kirwan, B. 2005. Human reliability assessment. In: J.R. Wilson and N. Corlett *Evaluation of Human Work*, 3rd Edition. (Taylor and Francis, Boca Raton).

Reason, J. 1990. *Human Error*. (Cambridge University Press, Cambridge).

Swain, A.D. and Guttman, H.E. 1983. *A handbook of human reliability analysis with emphasis on nuclear power plant applications. NUREG/CR1278.* Washington D.C. USNRC.

Shepherd, A. and Stammers, R.B. 2005. In: J.R. Wilson and N. Corlett *Evaluation of Human Work*, 3rd Edition. (Taylor and Francis, Boca Raton).

Wickens, C.D. 1992. *Engineering Psychology and Human Performance*. 2nd Edition. (New York: Harper Collins).

Williams, J.C. 1986. HEART – a proposed method for assessing and reducing human error. In *Proceedings of the 9th advances in reliability technology symposium*. University of Bradford, UK.

IMPLEMENTATION OF REMOTE CONDITION MONITORING SYSTEM FOR PREDICTIVE MAINTENANCE: AN ORGANISATIONAL CHALLENGE

Luminita Ciocoiu, Ella-Mae Hubbard & Carys E. Siemieniuch

Engineering Systems of Systems Research Group, School of Electronic, Electrical and Systems Engineering, Loughborough University

The "Health and Prognostic Assessment of Railway Assets for Predictive Maintenance" project is developing a Remote Condition Monitoring (RCM) system to manage asset degradation to enable predictive maintenance. Despite the benefits of the RCM systems, many of the programmes that seek to introduce them fail. Previous research shows that, beside technological challenges, there are organisational factors that contribute to the success of these programmes; the paper presents a three step approach taken to meet these challenges and some initial findings of the research.

Introduction

RCM systems are revolutionary e-systems that make use of contemporary advances in information technology solutions (Jonsson et al., 2010) to enable a shift in the maintenance of industrial equipment from mainly corrective (after failure) and preventive maintenance (maintenance performed on fixed schedule regardless of the condition of the asset) to predictive maintenance (maintenance in advance of failure) (Koochaki and Bouwhuis, 2008). Despite the obvious benefits that these systems can bring to the maintenance regime and although there is an increase in reported success stories in literature (Koochaki and Bouwhuis, 2008) many of the programmes that seek to introduce a maintenance regime based on asset condition monitoring, fail to achieve their objective (Mobley, 2002; Mitchell, 2007; Koochaki and Bouwhuis, 2008).

While it is widely accepted that there are technical challenges associated with the introduction of new condition-based maintenance systems (Jonsson et al., 2010), what is perhaps less understood is the human and organisational factors (Koochaki and Bouwhuis, 2008; Jonsson et al., 2010) involved in the successful implementation of such programmes and the changes that these systems demand in the organisation's maintenance processes. The successful implementation of condition-based maintenance programmes and introduction of RCM systems requires changes and updates to existing processes within the organisation implementing the system.

Project background

The "Health and Prognostic Assessment of Railway Assets for Predictive Maintenance" project wants to produce and introduce a Remote Condition Monitoring (RCM) system that will provide reliable and dependable health assessment data for London Underground (LU) with the scope of managing the asset's degradation so that maintenance interventions can be planed and undertaken at the optimum time, in advance of failure (Loughborough University Projects, 2014). The asset, which is used as a case study in the current project, is the escalator. The LU escalators are different from the ones usually found in shops and offices; they are longer, wider and more robust in comparison, such that they can cope with high demand, running approx. 20 hours a day, 364 days a year and carrying up to 13,000 people an hour (Campbell, 2002). LU has more than 400 escalators with around 95% of them being operational at any given time (Campbell, 2002).

Maintaining these machines is also challenging due to the tight space constraints in which the escalators are located and the underground environment (tight spaces, operating hours, access routes) in which escalators operate. Maintenance is usually allocated to the few hours when the stations are closed; however, Transport for London (TfL) recently announced the introduction of a 24-hour service at weekends to start in September 2015, which will impact on these available hours. There are standards for good operations and maintenance (such as BS EN 13015:2001+A1:2008) in place as well as inspection and maintenance methods and schedules; however, as the escalators age and due to increasing operational demands, unplanned outages are increasing (Campbell, 2002) and this is one of the areas where the introduction of the CRM system will assist i.e. maintenance in advance of failure through provision of prognostic data.

Organisational factors

As mentioned previously organisational factors play an important role in the successful implementation of condition-based and remote condition-based maintenance systems. Koochaki and Bouwhuis (2008) in their case study carried out at an industrial plant found that knowledge sharing processes (retrieval and transfer), lack of shared ontology, lack of clarity between the roles and responsibilities between the teams and absence of a clear and shared organisational strategy are key organisational factors that contribute to a decrease in the effectiveness of a condition based monitoring system and hence failure of condition-based maintenance programmes.

More recently, Jonsson et al. (2010) looking at the opportunities and challenges related to the design and implementation of e-based systems for remote diagnostic of industrial equipment found that, for the RCM systems to be successful, they must not only be built based on cutting-edge technology but also they have to be based on comprehensive organisational models of the maintenance processes, such that the RCM system and the organisation(s) will achieve their predefined goals. In addition their findings show that the way knowledge is shared between organisations (customer, maintenance service providers, manufacturer) or domains within the organisations is a key factor in implementing RCM systems and that the introduction of these types of systems requires re-organisation of the business functions and maintenance routines.

Research approach

In order to address the organisational challenges related to the building and imple-mentation of the RCM system into LU maintenance regime, a three-step approach has been devised. This is presented in Figure 1. First a map of the maintenance processes within the organisation, as they should be according to the organisation's formal documentation is produced. In parallel, a model mapping the organisation maintenance processes as they are in practice is being created based on information gathered through semi-structured interviews and on-site observations. Secondly a gap analysis of the two models will be performed to build a comprehensive picture of the existing maintenance processes and highlight possible discrepancies between the formal and in-practice processes.

The last step in the research approach is to construct a comprehensive model of the future 'ideal' or required maintenance processes as they will be ("to be") following

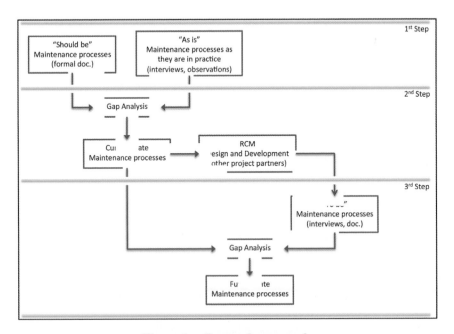

Figure 1.   Research approach.

the implementation of the CRM system. The model will be built based on empirical data gathered from the partner organisation building the CRM system using semi-structured interviews and analysis of the system's documentation. The result of these phases in the research are then going to be compared with the model of the extant maintenance processes.

It is expected that the results of this comparison analysis will highlight the changes that the organisation will need to undertake to accommodate the introduction of the new system. The analysis of these organisational models will highlight the impact that the new system will have on the extant organisational structure and support LU in building a shared and integrated organisational strategy across the business, prior to the introduction of the new RCM system.

Finally to ensure the validity of the study, several strategies are going to be employed, such as triangulation of data (data collected through interviews and observations), member's check (ongoing dialogue with the project partners regarding the interpretation of data), participatory modes of research (LU are going to be involved in all stages of the research) as well as providing detailed description of procedures, methods and techniques employed in the study (Creswell, 2009).

Preliminary findings

The first phase of the research has been completed, a series of work instructions/standards examined and the organisational maintenance processes mapped out. The analysis revealed that, although models of maintenance processes can be mapped based on these documents, a comprehensive model of the overall maintenance process cannot be built based only on the documents analysed. This suggests that there is a fractioned maintenance process within the organisation or that the maintenance regime does not entirely comply with the formal documentation. Format issues/inconsistencies within the documents have also been highlighted during the analysis. Further analysis is required to determine the causes of the issues seen and determine what impact such issues could have on the organisation, both currently and when future changes are implemented.

Discussions and conclusions

The key human factors issues associated with the introduction of a new maintenance regime are: need for clarity between teams regarding roles and responsibilities; need for shared ontology across business domains; need for clear knowledge sharing processes and need for clear and shared organisational strategy (Koochaki and Bouwhuis, 2008; Jonsson et al., 2010). Preliminary findings showed that there might be a fractioned maintenance process from an organisational point of view or a lack of rigorous formal processes in place to guide the organisation maintenance activities, which might indicate a lack of shared and/or integrated organisational strategy. Ongoing work includes a series of observations and validation interviews with Subject Matter Experts (SMEs) to determine if there is any missing information regarding the formal processes and to gather information regarding the maintenance processes as they are in practice, so that a comprehensive map of the existing organisation's processes can be built. Future work will address the issues identified by mapping out the interactions involved in the implementation of the new RCM system, investigating the impact of the new system on existing business structures/work processes and produce guidelines to aid LU, make the necessary organisational changes for a successful implementation of the RCM system.

Acknowledgments

The authors would like to thank London Underground and other partners in the project: Telent, Humaware and University of Nottingham, as well as the project funding bodies Innovate UK and the Rail Safety and Standards Board (RSSB). The views expressed in this paper are those of the authors and do not necessarily represent those of the other project partners or funding bodies.

References

BS EN 13015:2001+A1 (2008) "Maintenance for lifts and escalators – Rules for maintenance instructions." Retrieved 24th June 2014 from: http://shop.bsigroup.com/ProductDetail/?pid=000000000030176472.

Campbell, P. (2006) "Why does it take so long to amend an escalator?" Retrieved 21st July 2014, from: http://www.lrb.co.uk/v24/n05/peter-campbell/why-does-it-take-so-long-to-mend-an-escalator.

Creswell, J. W. (2009). *Research design: Qualitative, quantitative, and mixed methods approaches 3rd Edn.* California, USA: SAGE Publications.

Jonsson, K., Holmstrom, J. & Leven, P. (2010) "Organisational dimensions of e-maintenance: a multi-contextual perspective." *International Journal of System Assurance Engineering and Management* 1(3): 210–218.

Koochaki, J. & Bouwhuis, I. M. (2008) The Role of Knowledge Sharing and Transactive Memory System on Condition Based Maintenance Policy. *IEEE International Conference on Industrial Engineering and Engineering Management,* Singapore.

Loughborough University Projects. (2014) "Health and prognostic assessment of railway assets for predictive maintenance." Retrieve 24th September 2014, from: http://www.lboro.ac.uk/departments/eese/research/systems/system-of-systems/projects/seese-hpa.html.

Mitchell, J. (2007) "From vibration measurements to condition based maintenance." *Sound and Vibrations* 41(1): 62–75.

Mobley, R. (2002) *An Introduction to Predictive Maintenance* (2nd Edition), Butterworth-Heinemann.

IDENTIFYING A SET OF GESTURES FOR IN-CAR TOUCH SCREENS

Ayse Leyla Eren[1], Gary Burnett[1], Simon Thompson[2], Catherine Harvey[1] & Lee Skrypchuk[2]

[1] *University of Nottingham Human Factors Research Group, Nottingham*
[2] *Jaguar Land Rover HMI, Coventry*

Touchscreens are commonplace in vehicles, but the distraction burden can be high. This paper presents a driving study aiming to identify and assess a set of task-independent gestures to be used as user-defined shortcuts on in-car touch screens. Ten common gestures were identified by users in a preliminary study and then assessed in a driving simulator study – where 12 participants drove two routes whilst undertaking each of the gestures on an in-vehicle touchscreen. The effects of visual feedback from the screen whilst drawing gestures were observed. Measures were taken of eye movements, drawing accuracy and subjective ratings. A final set of four gestures *(tick, roof, squiggle and triangle)* was recommended for interaction with in-vehicle touchscreens.

Introduction

In the last thirty years there has been a significant increase in the functionality of In-Vehicle Information Systems (IVIS). The functions provided by IVIS range from navigation, entertainment, comfort, media and communication such as hands-free calling and access to the internet whilst driving. These systems have a number of advantages, such as providing various functions in one small space (Burnett et al, 2013) and the ability to update the system easily if necessary. However, they also present a number of potential disadvantages, in particular driver distraction (Green 1999; Tu et al, 2012). There is a limited time available for a driver to allocate attention to secondary tasks when driving and secondary tasks that require extended interaction times, combined with a high mental workload, will increase the overall demands on the driver (Lansdown et al, 2004). Following the increase in in-car technologies, automotive companies are seeking new ways of decreasing the demand of secondary tasks for the safety of drivers and other road users. A major aim is to minimise driver workload, in particular to reduce the visual attention required by built-in IVIS.

The study reported in this paper aimed to identify a set of surface gestures which could be used as shortcuts for frequently performed tasks on touch screens in cars. Therefore, instead of browsing through hierarchical menus, drivers would be able to use these gestures to perform certain tasks quicker and with less visual demand. The implementation of these gestures would allow the user to execute any command (e.g. Navigate to Home, Move sound to rear of vehicle) that he/she assigned to a specific gesture from the predefined gesture set. Consequently, the gestures would be task-independent (not inherently tied to a task) and user-defined (set up within system preferences). As a related design issue, the effects of visual feedback from the touch screen whilst drawing gestures were also considered. It was hypothesised that visual feedback could potentially improve accuracy of drawing but increase visual demand.

Method

Participants

Twelve (3 females, 9 males) drivers recruited from the University of Nottingham (students and staff) took part in the study (mean age = 28 years: s.d. 7.6 years). All participants were right-handed and had been driving for at least two years with a UK license. They all owned at least one touch screen device and had experience using them for at least 1 to 3 years.

Design

In a within-subjects design, there were two independent variables in this study. The first was the type of gesture, of which there were 10: squiggle, roof, triangle, square, house, tick, star, spiral, infinity, and diamond (see Figure 1). These gestures were identified in a preliminary user-centered study in which 25 participants were asked to produce 10 shapes/symbols each using finger painting and were also given the numbers 0–9 and letters A-Z to ensure that their shapes/symbols were not the same as these. The order in which the gestures were drawn in the driving study was randomised for each participant. The second independent variable was the existence of visual feedback received from the touch screen. This had two levels: "feedback" during which participants were able to see what they were drawing on the touch screen and "non-feedback" during which they were not able to see anything on the screen. Dependent variables were the number/duration of glances made towards the touchscreen, the perceived difficulty and demands associated with the task and the accuracy of gestures.

The participants were asked to fill out a difficulty/easiness rating scale for each gesture immediately after they finished drawing it each time. The 5-point scale ranged from 1 (very easy) to 5 (very difficult) with 3 (neutral) being the middle point. Participants were also asked to fill out a NASA-TLX questionnaire to assess the effects of the existence (or absence) of visual feedback on perceived workload after each condition. Participants drove two short, identical routes in the simulator; one route for each of the feedback conditions. The order of presentation of the two visual feedback conditions was counterbalanced. For each condition participants

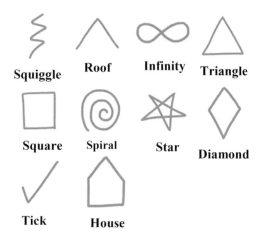

Figure 1.   10 gestures collected during the preliminary study.

were asked to draw each gesture three times, drawing 30 gestures in total per condition. The list order of the 30 gestures was also randomised for each participant. Participants were reminded that the task of driving was the primary task in this study and the task of drawing gestures was the secondary task.

All the gestures drawn by the participants were recorded and used to measure accuracy. A panel of judges of 7 Human Factors experts – 3 from the University of Nottingham and 4 from a vehicle manufacturer – assessed the 720 gestures collected during the driving study to decide whether or not each gesture accurately represented the original gesture they were based on. The assessment was carried out individually in order to avoid the judges affecting each other's perceptions of the gestures.

Apparatus and stimuli

A fixed-based, medium-fidelity driving simulator located in the Human Factors Research Group at the University of Nottingham was used for the study. The simulator consists of the front half of a right-hand drive Honda Civic car positioned in front of a curved screen providing approximately 270° viewing angle. Three overhead projectors are used to project the driving scenario onto the curved screen. Rear and side mirror projections are also provided. The simulated driving scenario and driving experience were created using STISIM Drive (version 2) software.

The driving scenario consisted of a UK motorway with trees and fields to both sides. The road itself was a straight road with no bends. The participants were required to follow a yellow car on the inside lane of the motorway and were asked to keep to the speed limit and follow at a safe distance. There were no lane changes involved in the scenario.

Participants drew their gestures by hand directly on an Apple iPad2, which was positioned in the center console of the car to the left of the driver horizontally. For the "feedback" condition the screen had a white background with the brightness of the screen set up to maximum brightness and the trace of the finger was represented with a dark purple, thick visible line.

The SensoMotoric Instruments (SMI) eye tracking glasses were used to track participants' eye movements. The eye tracking glasses were connected to a Lenovo laptop which had the eye tracking recording software iView by SMI with a USB cable for calibration and data recording. The dependent measures collected using the glasses included; total eyes-off-road time, number of glances made towards the touch screen, mean glance duration and individual glance durations.

Procedure

Before starting the study participants were given fifteen minutes for a test drive to get used to the driving simulator controls and the driving scenario. They were also presented with the ten gestures (see Figure 1) and were given the time to practice drawing each gesture up to three times on the touch screen whilst stationary.

Once participants had familiarised themselves with all the controls and gestures, they were asked to drive the simulator for approximately ten minutes under two conditions (feedback or non-feedback). In each condition participants were asked to draw the gestures verbally instructed by the experimenter using the gesture names. They were asked to complete the gestures as quickly and accurately as possible. After each gesture and at the end of the study participants were asked to fill out the necessary forms. The whole study took approximately one hour to complete.

Results and analysis

Accuracy of drawings

360 gestures were assessed in total for each condition. The number of correct gestures identified by the panel was calculated and represented as the accuracy of drawings across both feedback and non-feedback conditions – see Figure 2. As seen in the graph the most accurate gesture was *tick* and the least accurate gesture was *diamond.* It was hypothesised that the existence of feedback would result in more accurate gestures – however, this effect was not observed.

Difficulty and NASA-TLX workload ratings

According to participants, the most difficult gesture to draw whilst driving was *star* and the easiest gesture to draw was *tick*. A two-way repeated-measures ANOVA showed that the type of gesture had a significant effect on difficulty ratings $F(9,108) = 27.442$, $p < 0.05$ as expected. The existence of

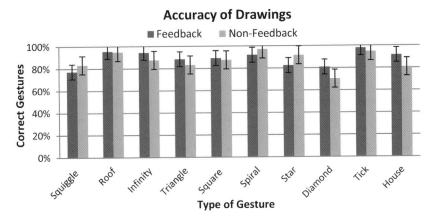

Figure 2.   Average percentage of correctly drawn gestures.

feedback F(1,12) = 0.006, p > 0.05 and the combination of feedback and type of gesture F(9,108) = 0.766, p > 0.05 did not have a significant effect on perceived difficulty – that is, participants did not think having visual feedback made it easier to draw the gestures.

The results of the NASA-TLX questionnaire ratings showed that there was no significant difference between the feedback and non-feedback conditions for each of the NASA-TLX components.

Eye tracking data

Due to technical problems with some of the eye trace videos, it was only possible to analyse the eye tracking data for 9 of the 12 participants. The results showed a significant difference in visual demand measures between the feedback and non-feedback conditions. The existence of feedback had a significant effect on the number of glances to the touch screen for all gestures (t = 3.02; p < 0.05) apart from *tick* and *roof* (t = 0.15 and t = 1.34; p > 0.05). More glances were made on average towards the touchscreen when feedback was provided, compared to when no feedback was given – see Figure 3.

There was also a significant difference between the total gaze duration values of the two conditions for all gestures, (t = 3.03; p < 0.05) apart from *tick* and *roof* (t = 0.15 and t = 1.35; p > 0.05). Glances made towards the touchscreen were significantly longer when feedback was provided, compared to when no feedback was given – see Figure 4.

Discussion

The increase in workload with the introduction of a secondary task while driving can require the division of attention between tasks leading to distraction (Alpern

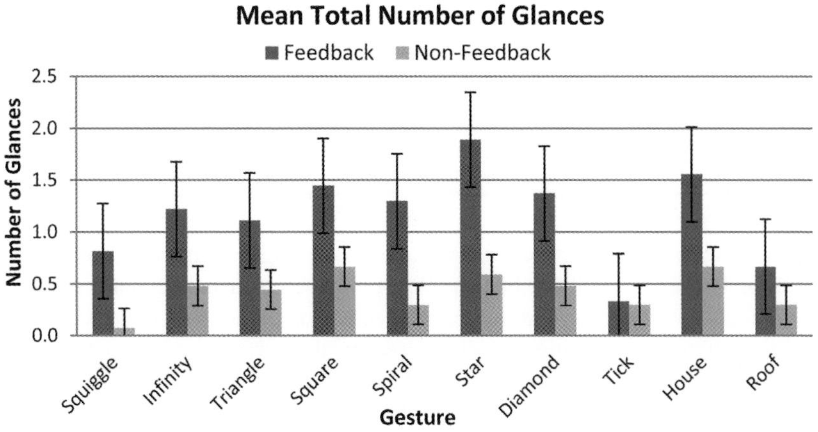

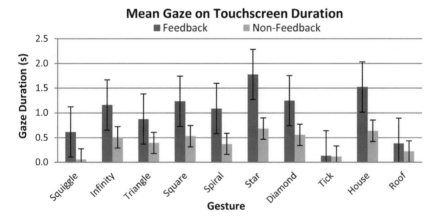

Figure 3.   Total number of glances on the touch screen for each gesture across two conditions for 9 participants.

Figure 4.   Gaze duration for each gesture across the two conditions for 9 participants.

and Minardo 2003). The data collected in the driving study provided an insight into the effects of using the gestures and the provision of visual feedback on workload and visual behaviour.

Contrary to assumptions, the accuracy of drawings was not significantly different in the feedback and non-feedback conditions. It was expected that the drawings obtained from the feedback condition would be more accurate as participants were able to see the result of the gesture they were producing while they were performing the task. However due to the dual task nature of driving they might not have had the resources to capitalise on visual feedback in this context.

Prior to the analysis of the results it was assumed that the most difficult gesture to draw would be *star* – as it required five directional changes and required the finger to return to the starting position – and the easiest gesture to draw would be *tick* – as there were only two directional changes and the finger did not need to return to the starting position. Unlike the accuracy results, the results of the difficulty ratings supported these predictions, demonstrating *star* as the most difficult and *tick* as the easiest gesture to draw.

In contrast with the accuracy results and subjective ratings, eye tracking data showed that the existence of feedback had an effect on objective visual demand. The eyes-off-road time was less in the non-feedback condition, which was expected. In this condition there was nothing directly to see on the screen, hence no reason for participants to take their eyes off the road. As the feedback condition had a higher total number of glances and higher gaze duration, it could be stated that providing visual feedback for gestures on the screen could cause distraction. In order to avoid visual distraction, other methods of feedback should be investigated such as audio\haptic modes. Specifically for tasks where there is no clear and immediate task-related feedback.

As expected, the gestures that were perceived to be the most difficult to carry out had the highest gaze duration and number of glances. These results demonstrate that the more difficult the gesture is, the more demanding visually it is to draw whilst driving. The two gestures that had the highest visual demand and difficulty ratings were *star* and *house*. Hence it is recommended that these gestures should not be used on touch screens as they will cause distraction for the driver. The two gestures that were the easiest to draw and had the lowest gaze duration were *tick* and *roof*. These two gestures were also the most accurately drawn gestures.

In terms of a recommended set of gestures, the ten gestures can be grouped under three categories. The gestures *tick* and *roof* are the ones that are most strongly recommended, as they performed well across all criteria. Conversely, the gestures *star* and *house* should definitely not be in the final set. The other six gestures could be chosen dependent on what is required from the gestures. For example, if the gestures are expected to be drawn accurately then *diamond* should not be included in the final set as it was the least accurately executed gesture in the set.

Conclusions and future work

The developments in the in-vehicle technologies industry, which allow drivers to perform more tasks while driving, are also potentially increasing workload for drivers. For common tasks that are performed regularly with touchscreens, a set of gestures can be used to make it easier for the drivers to execute these tasks. As no common gestures were found for typical tasks in previous studies there was a need to identify an initial set of gestures that could be used as short-cuts in the interface design and assign tasks to them later. This study only focused on identifying the gestures and testing the ease of input of the set. Based on the data as a whole, *tick, roof, squiggle and triangle* can be recommended to be used on in-car touch screens as a final gesture set as these gestures were the most accurate, least difficult and demanding gestures in the set.

The existence of feedback from the touch screen was also an important concept to take into account. Visual feedback was chosen to be assessed in the driving study. The results of the studies showed that there was no significant effect of feedback on perceived workload, difficulty of gestures, accuracy of gestures and various other measures such as performance, temporal demand and frustration. However, the results of the eye tracking data showed that the feedback condition required significantly more eyes-off-the-road time compared to the non-feedback condition. Therefore, it is recommended that gestures should not include any visual feedback from the touch screen to minimise visual distraction effects. Conversely, more research could be carried out in looking into other types of feedback for this gesture set.

References

Alpern, M. and Minardo, K. 2003. Developing a Car Gesture Interface for Use as a Secondary Task, in *CHI 2003: New Horizons*. Human-Computer Interaction Institute (HCII), Carnegie Mellon University.

Burnett, G., Crundall, L., Large, D., Lawson, G. and Skrypchuk, L. 2013. A Study of Unidirectional Gestures on In-Vehicle Touch Screens. *Conference on Automotive User Interfaces and Interactive Vehicular Applications (AutomotiveUI)*. Eindhoven, The Netherlands, October 28–30, 2013.

Green, P. 1999. Visual and Task Demands of Driver Information Systems. The University of Michigan Transportation Research Institute (UMTRI).

Lansdown, T.C., Brook-Carter, N. and Kersloot, T. 2004. Distraction from multiple in-vehicle secondary tasks: vehicle performance and mental workload implications. *Ergonomics*, **47**, 1, pp. 91–104.

Tu, H., Ren, X. and Z, S.. 2012. A Comparative Evaluation of Finger and Pen Stroke Gestures. *Conference on Human-Computer Interaction, Session: Space: The Interaction Frontier.* Austin, Texas, USA, May 5–10, 2012.

A SYSTEMS-BASED APPROACH TO UNDERSTANDING SLIPS, TRIPS, AND FALLS AMONG OLDER RAIL PASSENGERS

Victoria Kendrick[1], Patrick Waterson[1], Brendan Ryan[2],
Thomas Jun[1] & Roger Haslam[1]

[1]*Loughborough University*
[2]*Nottingham University, UK*

A system based approach was used to identify factors that contribute to slips, trips, and falls in older rail passengers, involving stakeholder interviews; station observations; passenger interviews; and a survey of station managers. Factors were categorised into organisational influences, station environment, and individual influences that contribute to older passenger slips, trips, and falls. Our findings identified priority areas for future rail safety interventions, aiming to enhance the safety and satisfaction of older rail passengers, and reduce slips, trips, and falls.

Introduction

On average 30% of adults over the age of 65 years fall each year, with 1 in 5 of those falls requiring medical attention (Gillespie et al, 2012). With an ageing population across parts of the world (e.g. United Kingdom, Australia, and the United States of America) (Hollnagel, 2008; Gillespie et al., 2012), increased proportions of passengers are using public transport (i.e. railways) into older age, with implications for the design and use of the railway (Currie and Delbosc, 2010; DTI, 2000; Palacin, 2011; RSSB, 2008, 2011).

In common with other areas of rail research (e.g., reliability improvement) a 'whole systems' approach to passenger STF is useful. This considers how sub-systems within the railway interact with one another and how interdependencies between these function (e.g., interdependencies between changing passenger characteristics, station design and capacity forecasts). The aim of the approach is to move towards the 'joined up' services and operations across the rail network. The adoption of a 'whole systems' approach also explicitly acknowledges that successful risk management involves a set of dynamic and complex interdependencies between sub-systems, people, materials, tools, machines, software, facilities and procedures.

Method

A systems based approach was used to identify factors that contribute to slips, trips, and falls (STF) in older rail passengers. The research consisted of stakeholder interviews (n = 44) with experts from within rail and across other industries (i.e. healthcare, aviation), station observations (n = 11); older passenger interviews (n = 18, aged 67–94 years, 8 females: 10 males); and a survey of station managers (n = 66). Participants were recruited on a structured convenience basis, sampling participants from the chosen sample groups most likely to be able to provide useful insights into the problem under investigation. Data were analysed iteratively using hybrid thematic analysis, with data driven codes developed, and emergent overarching themes identified (Bryman, 2004).

Results

Factors that impact slips, rips, and falls (STF) in older rail passengers were categorised into organisational influences, influences within the station environment, and individual influences (based on previous research by Haslam and Stubbs, 2006; Bearfield et al., 2013; and Rasmussen and Svedung, 2000). Issues were then classified as high, medium and low priorities for targeting older rail passenger STF in the future (2014–2050). High priorities included:

- Organisational influences: Assistance services
- Station environment: Impacts of weather, Sensory distractions, Escalator safety, and Crowding
- Individual influences: Mental health

Organisational influences – Assistance services

Public announcements, 'help points' available at the entrance to the station and the platforms, vehicle available to assist passengers to the platform. Overall the service appears to be very good; however it is not always well advertised due to inappropriate use by some passengers as a 'porter service'. Differences in the approach to older passengers emerged across GB stations, with some station managers indicating a 'proactive' approach by staff to identify vulnerable users and ask if assistance is required. As shown by Stakeholder interviewee 23 – Station Manager:

> *"Our station staff target vulnerable passengers and ask if they require assistance" (Stakeholder interviewee 23 – Station Manager)*

However, other station policies regarded this approach as discriminatory.

Station environment – Impacts of weather

Weather including snow, ice, rain, wet and slippery flooring (doorways), leaking roof, and uncovered platforms all impact STF within stations across Great Britain. For example Stakeholder interviewee 3 – Station Manager suggested:

> *"The roof is being repaired, however it still leaks at this end of the station, so when it rains we roll out a large mat and that soaks up the water and stops*

passengers slipping on the platform. However the mat doesn't sit flat at the ends and so it sort of creates a trip hazard at the same time" (Stakeholder interviewee 3 – Station Manager)

Station environment – Sensory distractions

Sensory distractions within the station include the use of headphones (hearing), and other electronic device (music, games, social media), as well as physical distraction (posters, advertisements, people) displayed and observed by passengers around the station.

Station environment – Escalator safety

Signage encouraging passengers to us alternative lifts if carrying luggage (on floor, posters between escalators, barriers in front of escalators, holograms to warn of dangers). Older passenger interviews identified a misbelief of the dangers of escalators, with passengers Passenger interviewee 1 suggesting:

"I know that the escalators are safer, but I would prefer to struggle up the stairs with my bag" (Passenger interviewee 1, Older passenger)

Seminars are currently offered by one of the train operating companies in Great Britain, aiming to educate older passengers of the dangers of escalators. However the success of the intervention is unclear, and would require further observation.

Station environment – Crowding

Population increase, 12% of the population will be over the age of 65 years by 2050, impacting fire safety and evacuation within train stations, as well as pedestrian flow, and passenger comfort and experience (Gillespie et al., 2012). During station observations it was observed that:

"...to reduce the crowding on platforms pedestrians are forced to walk longer distances to reach the platform, to disperse the crowd – not always most convenient for older passengers" (Field notes – Station observation)

Individual influences – Mental health

Dementia (Alzheimer's disease) and depression for example are likely to rise with the ageing population, likely to require increased awareness from staff and passengers, including staff training to increase awareness (Gillespie et al, 2012).

Conclusion

Our research used the Haslam and Stubbs (2006) model of STF as a basis, along with the Bearfield et al. (2013) Bayesian model of STF at the train platform interface specifically, to enhance understanding surrounding STF of older passenger at train stations across Great Britain. Our findings provide insight into priority areas to target future rail safety interventions from 2014–2050, in order to improve the experience (safety, comfort, satisfaction and performance) of older passengers travelling by train, with the overall aim of reducing STF.

References

Bearfield, G., Holloway, A., Marsh, W. 2013. Change and safety: decision-making from data. *Proceedings of the Institution of Mechanical Engineers Part F-journal of Rail and Rapid Transit*, 227(6), 704–714.

Bryman, A. 2004. *Social Research Methods* (3rd ed.). Oxford: Oxford University Press.

Currie, G., Delbosc, A. 2010. *Exploring Public Transport usage trends in an aging population, Social Research in Transport*, Monash Research for an Ageing Society Forum.

DTI 2000. Ageing Population Panel Report, Design for Living, Department for Trade and Industry, UK.

Gillespie, L.D., Robertson, M.C., Gillespie, W.J., Sherrington. C., Gates, S., Clemson, L.M., Lamb, S.E. 2012. Interventions for preventing falls in older people living in the community (Review), *The Cochrane Library*, 9.

Haslam, R., Stubbs, D. (Eds.). 2006. *Understanding and Preventing Falls*. London: Taylor & Francis.

Hollnagel, E. 2008. The changing nature of risks. *Ergonomics Australia*, 22(1–2), pp. 33–46

Palacin, R. 2011. Ageing and mobility: challenges for a sustainable future, *World Congress on Railway Research*, Lille, France.

Rasmussen, J. 1997. Risk management in a dynamic society: A modelling problem. *Safety Science*, 27, 183–213.

RSSB, 2008. The effects of an ageing population on rail travel, S065, RSSB, UK.

RSSB, 2011. The implications of an ageing population for the railway, T661, RSSB, UK.

THE ANALYSIS OF PILOTS FIXATION DISTRIBUTION FOR PERFORMING AIR-TO-AIR AND AIR-TO-SURFACE TASKS

Wen-Chin Li[1], Chung-San Yu[2], Graham Braithwaite[1] & Matthew Greaves[1]

[1]*Safety and Accident Investigation Centre, Cranfield University, Bedfordshire, UK*
[2]*Department of Industrial Engineering and Engineering Management, National Tsing Hua University, Hsinchu, R.O.C.*

Thirty-seven F-16 pilots participated in this research who were between 372 and 3,200 total flying hours (M = 1280, SD = 769). Eye movement data were collected by a head-mounted eye-tracker combined with an F-16 fighter simulator. The scenarios included air-to-air and air-to-surface tasks. The results showed pilots distributed a similar percentage of fixation and fixation duration in performing both air-to-air and air-to-surface tasks. However, there were significantly different distributions in use of the HUD and looking outside of cockpit respectively. The findings of this research are valuable for developing training syllabi to improve flight safety. In addition, the integration of eye-tracking and flight simulator could facilitate pilot training efficiency.

Introduction

Visual scan pattern distributed across the interior of an aircraft flight deck and the exterior of the cockpit is a sensory register process for investigating pilots' operational behaviours connected with what areas they scanned, what stimulus they perceived, and what actions they projected. Eye movements are closely linked with visual attention, which can reveal pilots' attention allocation on the areas of interested (AOIs) and can be analysed to explore how much effort and shifting attention occurred during performing visual tasks (Kowler, 2008). An eye tracker is a relevant tool for understanding pilots' perceptual errors which are in turn related to 75% of human factors incidents in flight operations (Jones and Endsley, 1996).

A previous study indicated that a human's fixation points are not strongly related to salient objects, but rather the meaningful places for the task that is being undertaken (Henderson, 2003). Fixation duration came from deliberate consideration and induces more fixation points for acquiring more detail information (Mecklenbeck, Kühberger, and Ranyard, 2011). However, it is necessary to monitor those eye movement data combined with a holistic task flow for a precise interpretation based upon frequency of fixation and fixation duration data recoded from an operator's behaviour patterns (Kilingaru, Tweedale, Thatcher and Jain, 2013). Pilots not only have to distribute attention for seeking and decoding the information based on visual scans, but also perceive and filter those salient cues by attention distribution during flight operations. Therefore, it was suggested that visual attention is a precursor to initiate the cognitive process (Lavine, Sibert, Gokturk, and Dickens, 2002).

Previous research found that the fixation trajectory could be a key component to observe pilot's SA performance (Ratwanti, McCurry and Trafton, 2010) and fixation duration on the relevant AOIs also could be an indicator for situational awareness (SA) performance (Moore and Gugerty, 2010). In addition, understanding the process of SA can also directly impact the development of training programs and interventions to prevent control flight into terrain (CFIT) due to loss of SA (Dirso and Sethumadhavan, 2008). The objectives of this research were (1) to investigate pilots' percentage of fixation among AOIs during air-to-air and air-to-surface aiming tasks; (2) to investigate pilots' fixation duration among AOIs during air-to-air and air-to-surface aiming tasks; (3) to compare the difference of visual scan patterns between air-to-air and air-to-surface aiming tasks; (4) to develop guidelines by applying eye movement information in training design.

Methodology

Participants

Thirty seven qualified mission-ready F-16 pilots participated in this research. All of the participants were volunteers and well informed that they have right to cease the experiment and withdraw information they provided without any reason. The treatment of all participants complied with the ethical standards required by the Research Ethics Regulations of Cranfield University.

Apparatus
- Flight Simulator
    The flight simulator used in the experiment is a formal F-16 trainer. It is a high-fidelity and fixed-base type consisting of identical cockpit displays to those in the actual aircraft.
- Eye Tracking Device
    Pilots' eye movement data were collected using a mobile head-mounted eye-tracker (ASL Series 4000) which is designed by Applied Science Laboratory. It is portable and light (76 g) so participants can move their head without any limitations. The sampling frequency of this type of eye-tracker is 30 Hz. The definition of a fixation point in the present study was three gaze points occurred within an area of 10 by 10 pixels with a dwell time (the time spent per glance at an AOI) being more than 200 msec (Salvucci and Goldberg, 2000). Five AOIs were set up to observe participants' eye movement during tactical operations. Those AOIs were selected based on the requirements of the standard operating procedures (SOP) for tactical missions. They are AOI-1: Head-up Display (HUD); AOI-2: Integrated Control Panel (ICP); AOI-3: Right Multiple Function Display

(RMFD); AOI-4: Left Multiple Function Display (LMFD); and AOI-5: Outside of cockpit (OC).

Scenarios

- Experiment I: Air-to-Air Task (Aiming Dynamic Target)

  The scenario of pursuing a dynamic target is an air-to-air (AA) manoeuver. The altitude of the interceptor (participant) in the combat patrol area was 20,000 feet with a cruise speed of 300 knots indicated airspeed (KIAS) and heading 050° under the weather conditions of 7-mile visibility and scattered clouds. A foe unexpectedly appears at the same altitude as the target moving from left to right with heading of 090° and air speed of 300 KIAS. The participants have to search the airspace to establish eye contact with the target, and intercept it by executing tactical manoeuvers. The target would then change its heading, altitude, speed and attitude in an attempt to escape from the interceptor's pursuit.

- Experiment II: Air-to-Surface Task (Aiming Stationary Target)

  The scenario of pursuing a stationary target is an air-to-surface (AS) mission. Participants were flying in a patrol area at an attitude of 20,000 feet, heading of 050° with a cruise speed of 300 KIAS under visual flight rules (VFR). While pilots were dispatched to attack one surface stationary target, they not only needed to execute tasks precisely by operating the aircraft, but also to follow the navigation system, entering appropriate codes by using various flight deck interfaces. When approaching the target, participants needed to roll-out, level the aircraft, aim at the target, lock-on and pick-off the weapons. Finally participants have to pull up with a 5 ～ 5.5 G-force to break-away from the range.

- Measurement of Situation Awareness Performance

  To avoid affecting participants' attention and SA, the current study adopted the measurement of embedded task to evaluate pilot's SA performance (Endsley, 1995). The generator would be failed unexpectedly with the warning signal appearing on the HUD. The malfunction light for the generator on the Warning Light Panel (WLP) would be illuminated simultaneously. It remained on for 5 seconds during the flight. This unexpected event was controlled by an instructor pilot (IP) sitting at the instructor console of the simulator. When the participant called out "generator failed" and put off the master caution light, this was recorded as "good SA performance" or conversely, "poor SA performance".

Procedures

All participants undertook the following; (1) complete the demographical data form including rank, job title, age, education level, qualifications, type hours and total flight hours; (2) a short briefing explaining the purposes of the study and the introductions of AA and AS scenarios without mentioning any potential aircraft equipment failure or target information; (3) being seated in the F-16 simulator and then the eye-tracker was put on for calibration by using three points distributed over the cockpit display panels and outer screen; (4) conducted flights which incorporated the AA and AS tasks. Simultaneously, the instructor pilot in the simulator

Table 1.   Demographical Variables of Participants (n = 37).

Variables	Groups	Frequencies
Age	25–30	13 (35.1%)
	31–35	11 (29.7%)
	36–40	7 (18.9%)
	41–45	6 (16.2%)
Rank	Lieutenant	1 (2.7%)
	Captain	16 (43.2%)
	Major	9 (24.3%)
	Lieutenant Colonel	10 (27%)
	Colonel Above	1 (2.7%)
Qualification	Combat ready	13 (35.1%)
	Two fighter team leader	4 (10.8%)
	Four fighter team leader	9 (24.3%)
	Daytime back seat instructor	2 (5.4%)
	Training instructor	9 (24.3%)
Total Flight Hours	500 and less	3 (8.1%)
	501–1000	13 (35.1%)
	1001–1500	11 (29.7%)
	1501–2000	4 (10.8%)
	2001 and above	6 (16.2%)

console evaluated participants' SA performance. It took around 55 minutes for each participant to complete the trials.

Results

The ages of participants were between 26 and 45 years old (M = 33, SD = 5); total flying hours between 372 and 3,200 hours (M = 1280, SD = 769). Subjects' demographical information including age, rank, qualification and total flight hour are shown as table 1. The subjects' fixation distributions among five areas of interested (AOIs) during air-to-air and air-to-surface are shown as table 2.

The 'percentage of fixation' is proportional data, and it is necessary to perform an arcsine transformation for further statistical analysis (Howell, 2013). Therefore, the data of pilots' percentage of fixation on five AOIs during the tasks of air-to-air and air-to-surface were transformed into arcsine values before conducting analysis of variance (table 2). There is a significant difference at the percentage of fixation on the HUD between air-to-air and air-to-surface, $F (1, 72) = 11.47$, $p < .01$, $\eta2\rho = .242$. Further comparisons by post-hoc Bonferroni adjusted tests showed that pilots' percentage of fixation on HUD during the air-to-surface task had a significantly higher than during the air-to-air task. There is a significant difference on the LMFD, $F (1, 72) = 13.203$, $p < .01$, $\eta2\rho = .268$ and further comparisons by post-hoc Bonferroni adjusted tests showed that pilots' percentage of fixation on the air-to-air task was significantly higher than during the air-to-surface task. Similarly,

Table 2.   Subjects' Percentage of Fixation and Fixation Duration among AOIs during Air-to-Air and Air-to-Surface.

| Variables | AOIs | Task Types | | | |
| | | Air-to-Air | | Air-to-Surface | |
		M	SD	M	SD
Percentage of fixation (arcsin)	HUD	45.31	13.01	54.15	11.19
	ICP	5.35	6.31	4.41	4.59
	RMFD	2.15	4.23	1.26	2.69
	LMFD	3.03	4.36	0.37	1.56
	OC	42.55	12.77	34.78	10.38
Average fixation duration (msec)	HUD	595	234	542	110
	ICP	191	207	180	213
	RMFD	91	197	63	141
	LMFD	142	202	14	61
	OC	399	102	395	81

AOI-1: HUD (Head-up Display); AOI-2: ICP (Integrated Control Panel); AOI-3: RMFD (Right Multiple Function Display); AOI-4: LMFD (Left Multiple Function Display); AOI-5: OC (Outside Cockpit)

significance was observed at the percentage of fixation on the outside cockpit, $F (1, 72) = 9.187$, $P < .01$, $\eta2\rho = .203$. Further comparisons by post-hoc Bonferroni adjusted tests showed pilots' percentage of fixation during the air-to-air task was significantly higher than during the air-to-surface task. There were no significant differences between the air-to-air and air-to-surface tasks at ICP ($F = 0.67$, $p > .05$, $\eta2\rho = .018$) and also at RMFD ($F = 1.21$, $p > .05$, $\eta2\rho = .033$).

There is a significant difference in average fixation duration on the LMFD between air-to-air and air-to-surface tasks, $F (1, 72) = 15.331$, $p < .001$, $\eta2\rho = .299$. Further comparisons using post-hoc Bonferroni adjusted tests showed that pilots' fixation duration during the air-to-air task was significantly higher than during the air-to-surface task on the LMFD. However, there are no significant differences in fixation duration between the air-to-air and air-to-surface tasks in terms of the HUD ($F = 2.019$, $p > .05$, $\eta2\rho = .053$), ICP ($F = 0.049$, $p > .05$, $\eta2\rho = .001$), RMFD ($F = 0.429$, $p > .05$, $\eta2\rho = .012$) and outside cockpit ($F = 0.051$, $p > .05$, $\eta2\rho = .001$).

Discussion

Pilots' eye movements are closely linked to attention distributed across the interior and exterior of the cockpit in order to achieve the requirements of tactical operations whilst optimizing situational awareness, decision-making and flight performance. According to the results in table 2, regardless of whether it was an air-to-air or air-to-surface task, the highest percentage of fixation of pilots was located on the

HUD (AOI-1) and outside cockpit (AOI-5). The results showed that information provided by the HUD and outside cockpit are the main sources of information for completing the tasks successfully. The fixation distribution is relatively even on the HUD (45.31%) and outside cockpit (42.55%) for the air-to-air task, however, there is a significant difference in the air-to-surface task of 54.15% on the HUD and 34.78% outside cockpit respectively. The percentage of fixation distribution on the HUD and outside cockpit has apparent differences between the air-to-air and air-to-surface tasks, although the sum of fixation distribution during these two tasks is very similar which are 87.86% (air-to-air) and 88.93% (air-to-surface) respectively. In addition, LMFD (AOI-4) is the unique AOI with significant difference in fixation duration. Pilots distributed 142 ms on LMFD during air-to-air mission and only 14 ms during the air-to-surface task (table 2).

The percentage of fixation indicates the attention distribution across AOIs for pilots performing a task (Ratwanti, McCurry and Trafton, 2010). Fixation is meaningful and is closely linked to where pilots' main attention is employed. Table 2 reveals that pilots distributed a relatively even attention on the HUD and outside cockpit in the air-to-air, but not in the air-to-surface task. The phenomenon indicates that pilots pursuing the dynamic target not only relied on the relevant information present on the HUD, but also need to distribute similar attention on the target's manoeuvers to integrate the interior digital signal and exterior eye-contact for successful lock-on to the target. On the other hand, pilots pursuing the stationary target mainly relied on the information present on the HUD (54.15%); pilots focus on aiming at the target by using information present on the HUD as soon as the target is in range. The difference between pilots performing air-to-air and air-to-surface tasks is due to the context of either aiming a dynamic or stationary target, which matches the findings of Sarter, Mumaw and Wickens (2007) that pilots did not employ a standardized scanning pattern but monitored the situation of the surrounding environment based on expectations associated with specific flight contexts.

The length of fixation duration on the AOIs allows aviation professionals to diagnose the cognitive process of pilots' attention distribution. Longer fixations are generally believed to be an indication of a pilot's difficulty interpreting information from a display (Kotval and Goldberg, 1998). Fixation duration indicates the importance of information for achieving the mission (Henderson, 2003). Furthermore, an increase of fixation duration also reveals the complexity of content that pilots scanned (Rayner, 1998). Table 2 shows pilots had much longer fixation duration on LMFD during the air-to-air task (142 msec) than during the air-to-surface (14 mesc) task. The result indicates pilots have to perceive radar information of the dynamic target's trajectory and predict the target's position in the near future in order to intercept the moving target effectively. In contrast, pilots did not rely on the information provided by LMFD during operating air-to-surface task for aiming the stationary target, as pilots' priority information for performing the air-to-surface task is the rapid changing altitude, speed and vertical speed whilst the target on the surface is in sight. It explains the low percentage of fixation (0.37%) and short fixation duration (14 mesc) on the LMFD recorded in the air-to-surface task.

Pilots pursue either the moving target during air-to-air or the stationary target during air-to-surface showed a rapid and constant decision-making based on constantly estimation, modification and prediction of the relative dynamic trajectory positions between pilots themselves and the target. Those pilots' operational behaviors were the results of cognitive processes which are closely linked with the fixation points and the length of fixation duration among AOIs. With the findings of this study, different eye movement patterns across the interior and exterior of cockpit could be identified and applied for developing training syllabi. Also, eye-tracking device can be integrated with flight simulator to facilitate pilot's smooth pursuit training, as the eye movement information is the main support of pursuit control (Spering and Montagnini, 2011).

Conclusions

In summary, the present study found pilots' percentage of fixation and fixation duration was significantly different on the HUD and outside cockpit while performing air-to-air and air-to-surface task. It shows that pilots did apply different approaches of attention distribution for pursuing the moving target and stationary target. Therefore, it is applicable to apply the findings of this research for developing air-to-air and air-to-surface training syllabi for improving flight safety. In addition, the integration of eye-tracking technology and traditional flight simulator has potential benefits to facilitate pilot's training efficiency.

References

Dirso, F. T., and Sethumadhavan, A. 2008. "Situation awareness: Understanding dynamic environments." *Human Factors* 53: 447–463.

Endsley, M. R. 1995. "Toward a theory of situation awareness in dynamic systems." *Human Factors* 37: 32–64.

Henderson, J. M. 2003. "Human gaze control during real-world scene perception." *TRENDS in Cognitive Sciences* 7(11): 498–504.

Howell, D. C. (2013). "*Statistical methods for psychology*", Belmont: Wadsworth, pp 346–352.

Jones, D. G., and Endsley, M. R. 1996. "Sources of situation awareness error in aviation." *Aviation, Space and Environmental Medicine* 67: 507–512.

Kilingaru, K., Tweedale, J. W., Thatcher, S., and Jain, L. C. 2013. "Monitoring pilot situation awareness." *Journal of Intelligent and Fuzzy Systems* 24: 457–466.

Kotval, X. P., and Goldberg, J. H. 1998. *Eye movements and interface components grouping: An evaluation method.* In Proceedings of the 42nd Annual Meeting of the Human Factors and Ergonomics Society. Santa Monica: Human Factors and Ergonomics Society, 486–490.

Kowler, E. 2008. *Attention and eye movements.* In R. Krauzlis (Ed.), Encyclopedia of neuroscience (Elsevier Ltd, Amsterdam), 605–616.

Lavine, R. A., Sibert, L., Gokturk, M., and Dickens, B. 2002. "Eye-tracking measures and human performance in a vigilance task." *Aviation, Space and Environmental Medicine* 73: 367–372.

Mecklenbeck, M.S., Kühberger, A., and Ranyard, R. 2011. "The role of process data in the development and testing of process models of judgment and decision making." *Judgment and Decision Making* 6 (8): 733–739.

Moore, K., and Gugerty, L. 2010. *Development of a novel measure of situation awareness: The case for eye movement analysis.* In Proceedings of the 54th Annual Meeting of the Human Factors and Ergonomics Society. Santa Monica, CA: Human Factors and Ergonomics Society.

Ratwanti, R. M., McCurry, J. M., and Trafton, J. G. 2010. *Single operator, multiple robots: An eye movement based theoretic model of operator situation awareness.* In Proceedings of the Fifth ACM/ IEEE International Conference on Human-Robot Interaction. Nara, Japan.

Rayner, K. 1998. "Eye movements in reading and information processing: 20 years of research." *Psychological Bulletin* 124(3): 372–422.

Salvucci, D. D. and Goldberg, J. H. 2000. *Identifying fixations and saccades in eye-tracking protocols.* Eye Tracking Research and Applications Symposium. Palm Beach Gardens, CA, USA.

Sarter, N. B., Mumaw, R. J., and Wickens, C. D. 2007. "Pilots' monitoring strategies and performance on automated flight decks: An empirical study combing behavioral and eye-tracking data." *Human Factors* 49 (3): 347–357.

Spering, M., and Montagnini, A. 2011. "Do we track what we see? Common versus independent processing for motion perception and smooth pursuit eye movements: A review." *Vision Research* 51: 836–852.

RAIL INDUSTRY REQUIREMENTS AROUND NON-TECHNICAL SKILLS

Ruth Madigan[1], David Golightly[1] & Richard Madders[2]

[1]*Human Factors Research Group, University of Nottingham, Nottingham, UK*
[2]*Arcadia Alive Ltd., Stafford, UK*

The aim of this study is to identify UK Rail Industry requirements for the incorporation of Non-Technical Skills (NTS) into current Competency Management Systems. A series of interviews and workshops with rail employees were conducted to identify these requirements. Results highlight the importance of training people in the support roles around an individual, in order to maximise their ability to provide individualised support and to understand the link between technical and non-technical competency. The issue of releasing those in safety-critical roles for training also emerged as a problem when trying to implement NTS training. Overall, the results suggest an individualised and adaptable approach will be required to incorporate NTS in the rail industry.

Introduction

Non-Technical Skills can be defined as *"the cognitive and social skills that complement a worker's technical skills"* (Flin, O'Connor, & Crichton, 2008). Researchers have for some time recognized Non-Technical Skills (NTS) and Human Factors Skills as having high importance in safety critical roles across industries such as aviation, petrochemical, and medicine (e.g. Diehl, 1991; O'Connor, Flin, & Fletcher, 2002; Patankar & Taylor, 2008).

Crew Resource Management (CRM) training has been used to improve NTS and performance in aviation teams since the 1980s. Reviews of CRM training effectiveness across aviation and other industries have shown that it is generally well received, and results in a positive change in attitudes and behaviours (O'Connor et al., 2002). Studies in military aviation indicate that CRM training decreased the accident rate for US Navy A-6 Intruder crew members by 81% (Diehl, 1991) and research has also shown a Return-of-Investment (ROI) of approximately 23% from the implementation of CRM for airline maintenance workers (Patankar and Taylor, 2008). Feedback from pilot NTS training courses in the rail industry has been positive, with drivers finding the course useful and showing significant improvements across NTS over time (Bonsall-Clarke & Pugh, 2013). NTS are important for optimising safety and performance during routine work conditions as well as for managing critical situations or emergencies (Civil Aviation Advisory Publication, 2011).

There is no standardized methodology for developing CRM training but core skills modules often include teamwork, leadership, situation awareness, decision making, communication, and personal limitations. Skills are identified from accident analyses, incident reports, trainers experience and simulator research studies (Flin, O'Connor, & Mearns, 2002). Training is generally delivered over two or three days, and tends to be classroom-based including lectures, practical exercises, role play, cases studies, and films of accident re-enactments (O'Connor & Flin, 2003). Research in the medical industry has shown that that the impact of this type of training varies with participant specialty (i.e. surgeons, anesthesiologists, nurses etc.) (Suva et al., 2012), and rail industry research suggests that employee engagement is dependent on the relevance of course content to a specific Train Operating Company or depot (Russell, Bailey, & Moore, 2013). It is therefore very important that any training regimes are designed around the specific requirements of a trainee's job.

Project T869 by the RSSB has provided a thorough analysis of the NTS required to help UK train drivers to anticipate, identify and mitigate threats and error (RSSB, 2012a). From this analysis a total of 7 NTS have been identified as having vital importance to the driver role, along with 26 underlying behavioural markers. The 7 NTS are situational awareness, conscientiousness, communication, decision making and action, cooperation and working with others, workload management, and self-management. The RSSB recommend that NTS should be integrated across the business to all relevant staff at key stages of the 'life cycle' (recruitment, initial training, ongoing training and support, measurement of competence and incident investigation). Training courses should not be delivered in isolation, and the principles of the course must be reinforced throughout the organisation (Bonsall-Clarke & Pugh, 2013). It is also recommended that this not be a purely top-down exercise – participants must drive the discussion, and feedback suggestions for how organisations can help them in their role. Finally, the RSSB does not recommend that NTS are assessed in isolation. As is the case in the aviation industry, a pass/fail decision should be based on technical competence, and NTS should be used to help inform that decision (RSSB, 2012b).

The 2012 Office of Rail Regulation (ORR) Railway Guidance Document recommends that all rail companies consider whether there may be elements of NTS they should incorporate into their Competency Management Systems (CMS). They emphasize the need to ensure that NTS training is integrated into the wider CMS, rather than included as a stand-alone "add-on", along with the importance of committing to on-going development and reinforcement of staff NTS, rather than a one-off fix.

The development of such an integrated approach raises a number of issues for rail companies around how to develop the skill-sets within the company to manage and assess NTS, and how to ensure that employees understanding of these skills are increased given the constraints around having workers released for training on a continuous basis. The current study provides a first step in addressing this issue, through the identification of industry requirements for the incorporation of NTS into CMS for the driver and conductor roles. The specific research questions this study aims to address are as follows:

- What is the current understanding of NTS across different roles within the rail industry?
- What are the current gaps in rail companies' performance, particularly in relation to NTS, both at individual and organisational levels?
- What requirements do Rail Industry members have for the incorporation NTS training into current competency management frameworks?

Methodology

Data collection consisted of workshops and interviews with industry representatives. The workshops were based within two meetings of an industry advisory board for University of Nottingham and Arcadia Alive's Knowledge Transfer Partnership (KTP) project. These meetings included representatives from nine Train Operating Companies (TOCs), Network Rail, NSARE, ASLEF and RMT. The aim of these meetings was to identify the main concerns rail companies have around the incorporation of NTS into CMS.

The interviews were conducted with 22 participants from various roles across the industry. Participants included 2 Operations Standards Managers, 1 Depot Manager, 1 Routing Instructor, 6 Driver Managers, 3 Conductor Managers, 6 Drivers, and 3 Conductors. The interviews were semi-structured and lasted between 14 minutes and 80 minutes. Questions revolved around three main areas of interest:

- Participants' understanding of NTS.
- Identification of the current gaps in companies' performance, particularly in relation to NTS, both at individual and organisational levels.
- Requirements for NTS training.

Results

Industry requirements from workshops

The two workshops identified a number of industry concerns regarding the incorporation of NTS into current CMS including:

- The importance of training people in the support roles around an individual, to ensure that they fully understand NTS prior to assessing or training.
- Ensuring NTS training is targeted at specific roles, as the type of training required will vary across roles.
- Uncertainty as to the best manner in which to integrate NTS and technical skills within current competency frameworks.

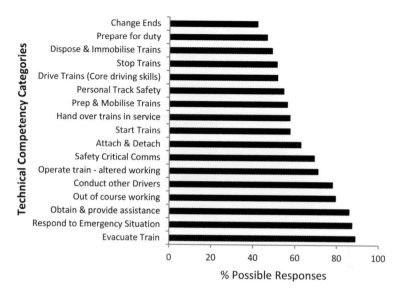

Figure 1. Average NTS contribution.

- Uncertainty as to how NTS policies would fit in with company's current procedures around issues such as fatigue management and sickness management.
- Concerns around ownership and responsibility – the risk that NTS is perceived as another point of compliance that front-line staff (e.g. drivers) need to adhere to, rather than a set of skills that will enhance their job.

Participants of the second Focus Group (N = 7) were also asked to evaluate which of the RSSB's 7 NTS were required for each of 34 competency statements (measuring 17 categories) that are used in a current driver technical skills assessments.

Figure 1 shows participant's evaluation of the importance of NTS to each of the competency categories.

Figure 2 provides a synopsis of the perceived impact of each NTS category on the technical competency statements.

Rail employee perceptions of NTS

The interviews with rail employees identified a number of NTS and Human Factors skills which were believed to be vital to the drivers' role. These included concentration, preparation, professionalism, communication, the ability to adapt to the driver lifestyle (i.e. sleeping and eating properly, adapting social activities), awareness, and ability to work alone. A number of interviewees mentioned the importance of a professional attitude and careful preparation for work as being key indicators of a "good" driver. The majority of drivers believed that NTS were important for their role, and felt that technical and non-technical competencies were linked.

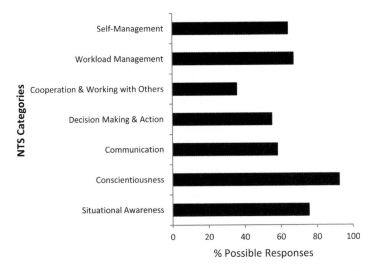

Figure 2.   Average technical competency contribution to NTS.

For managers, a number of concerns about the implementation of NTS arose, including:

- The lack of systematic training for managers to ensure that they fully understand NTS and how to support them
- How to identify which NTS were being used within a given task, particularly as the current NTS evaluation for a lot of companies is in a separate section to the technical evaluation.
- The importance of providing examples that drivers/guards can relate to when trying to understand the contribution of NTS.
- Difficulties in tailoring assessments or feedback to an individual's needs.
- Difficulties in releasing drivers for training because of the shortage of people to cover their routes.

Discussion

This study aimed to identify rail industry requirements for the incorporation of NTS into current CMS. Through the focus groups and interviews with industry representatives a number of requirements have emerged.

Current understanding of NTS

The interviews and focus groups showed that the majority of participants have a basic understanding of what NTS are and believe that NTS play an important part in how safety critical roles are performed. The behavioural markers associated with Conscientiousness (e.g. preparation, professionalism) and Situational Awareness (e.g. concentration, awareness) consistently emerged as being of vital importance to the train driving role. However, it would appear that the behavioural markers associated with decision making and action, communication, and cooperation are not considered as vital. Thus, it would appear that the team-based focus of CRM training in the aviation industry (Flin et al., 2002) is not appropriate for the train driver role. Workshop participants rated NTS to be of more importance during out of course working and emergency situations, than in everyday working. This suggests that people are more aware of the need to maintain NTS in out of course situations, but may not be as conscious of the dangers in more routine conditions.

Current gaps in company performance relating to NTS

Both the RSSB (2012b) and the ORR (2012) have emphasized the importance of integrating NTS and Technical skills evaluation. However, the workshops and interviews show that there is uncertainty in the industry as to the best manner in which to do this, and current NTS evaluations are often placed in a separate section of CMS to the technical evaluation. If NTS are to be used to inform decisions around technical competence (RSSB, 2012b), companies need to understand the underlying NTS which can impact their current performance criteria. The results also highlight the need to provide adequate training for those in management roles prior to rolling out any NTS initiatives, as currently managers lack confidence in assessing NTS and providing feedback. Finally, it would appear that there is some confusion about the difference between NTS and other personal lifestyle factors such as fatigue, health and wellbeing etc. Interview participants included a number of these factors in their definitions of NTS, and it emerged from the workshops that there is currently uncertainty as to how NTS policies should fit in with company's current procedures around issues such as fatigue management and sickness management.

Industry requirements for the incorporation of NTS

Across both the interviews and the workshops the importance of training people in the support roles around an individual emerged. This suggests that rail companies are aware of the importance of reinforcing NTS throughout the organisation (RSSB, 2012b). However, from the interviews it emerged that many managers struggle to explain NTS and feel it would be beneficial to be able to provide more examples and case studies for front line staff to relate to.

There are some concerns about the risk that NTS is perceived as another point of compliance that front-line staff (e.g. drivers) need to adhere to, rather than a set of skills that will enhance their job. Previous evaluation studies have found an improved attitude towards NTS after attending training (Bonsall-Clarke & Pugh, 2013; O'Connor et al., 2002) so it may be that this perception can be adjusted once training is rolled out. If course principles are reinforced throughout the organisation, providing a supportive culture for employees (RSSB, 2012b) this will aid the process. It also emerged from the interviews that the strong focus on compliance in current CMS means that managers find it difficult to tailor feedback to suit particular individual's requirements. As previous research has shown that the impact of NTS training will vary according to an individual's position (Suva et al., 2012), it is important to ensure that individual's receive training that is appropriate to their needs at a given point in their career.

A final issue which regularly emerged was the difficulty of releasing drivers for training because of the shortage of people to cover their routes. This makes it very difficult for companies to enable all those in safety critical roles to attend a two/three day training course plus refresher days. This suggests that NTS training in the rail industry may need to take a different format than is typical of the courses in the aviation and medical industries.

Conclusion

The aim of this study was to identify rail industry requirements for the incorporation of NTS into current CMS. The results indicate that there is currently a basic understanding within the rail industry of what NTS entail. However, there is uncertainty as to how these skills can best be integrated with technical skills evaluation, and where lifestyle issues regarding fatigue, health etc. should fit in. In addition, managers are currently struggling to provide individualized feedback and training for front-line staff, particularly as it is very difficult to release staff for training. Overall, the results highlight the importance of taking a systems approach to the implementation of NTS, ensuring that people at all levels of the organization gain an understanding of the skills required to support NTS. The difficulties faced in releasing those in safety critical roles for training suggests that the type of training programmes which have been implemented in other safety critical industries such as aviation and medicine may not be appropriate for the rail industry, and a more tailored and flexible approach may be required to ensure maximum effectiveness.

References

Bonsall-Clarke, K., and Pugh, S. 2013, Non-technical skills for rail: development, piloting, evaluation, and implementation of courses for front-line staff and managers. In N. Dadashi, A. Scott, J.R. Wilson, and A. Mills. (eds.) *Rail Human Factors: Supporting Reliability, Safety and Cost Reduction* (Taylor and Francis, London), 519–528.

Civil Aviation Safety Authority 2011, Non-Technical Skills trainnig and assessment for regular public transport operations. *Civil Aviation Advisory Publication* (CAAP SMS-3). Australia: Australian Government.

Diehl, A. 1991, *Does cockpit management training reduce aircrew error?* 22nd International Seminar of the International Society of Air Safety Investigators, Canberra, Australia.

Flin, R., O'Connor, P., and Crichton, M. 2008, *Safety at the Sharp End. A Guide to Non-technical Skills.* (Ashgate Publishing, Aldershot, UK).

Flin, R., O'Connor, P., and Mearns, K. 2002. Crew resource management: Improving team work in high reliability industries. *Team Performance Management*, 8, 68–78.

O'Connor, P., Flin, R., and Fletcher, G. 2002. Methods used to evaluate the effectiveness of CRM training. *Journal of Human Factors and Aerospace Safety*, 2(3): 217–234

O'Connor, P., and Flin, R. 2003, Crew Resource Management training of offshore oil production teams, *Safety Science*, 41: 591–609.

ORR 2012, "Non-Technical Skills (NTS) for rail staff, and RSSB's NTS guidance." 2014, from: http://orr.gov.uk/__data/assets/pdf_file/0017/2285/rgd-2012-03-web.pdf.

Patankar, M.S., and Taylor, J.C. 2008, MRM training, evaluation, and safety management. *The International Journal of Aviation Psychology*, 18: 61–71.

RSSB 2012a, "Non-technical skills for rail: A list of skills and behavioural markers for drivers, with guidance notes." Retrieved 1st May, 2014, from: http://www.rssb.co.uk/pages/research-catalogue/t869.aspx

RSSB 2012b, "Non-technical skills required in train driver role: Developing an integrated approach to NTS training and investment." Retrieved 1st May, 2014, from: http://www.rssb.co.uk/pages/research-catalogue/t869.aspx

Russell, A.J., Bailey, S., and Moore, A.C. 2013, The introduction of non-technical skills into a train driver competence management system. In N. Dadashi, A. Scott, J.R. Wilson, and A. Mills. (eds.) *Rail Human Factors: Supporting Reliability, Safety and Cost Reduction* (Taylor and Francis, London), 529–538.

Suva, D., Haller, G., Lübbeke, A., and Hoffmeyer, P. 2012, Differential impact of a Crew Resource Management Program According to Professional Speciality. *American Journal of Medical Quality*, 27(4): 313–320.

LAB TRIALS ON TRIAL: PREFERRED POSTURES IN AN AUTOMOTIVE TEST RIG (BUCK) AND THE REAL VEHICLE

Neil J. Mansfield & Brendan Hazlett

Loughborough Design School, Loughborough University, Loughborough, UK

During research and during development of vehicles, it is common-place for tests of concepts to be performed in a test rig (often referred to as a 'buck'). Test volunteers might be asked to set a comfortable driving position in the buck. It is assumed that the position set up in the buck is representative of the production vehicle. This paper tests this assumption, by comparing the postures that were selected by 20 drivers in the laboratory using a buck with dimensions, pedals, steering wheel and seat identical to a production vehicle (Nissan NV200) with those that were selected in the real vehicle. It is shown that differences in postural angles between a buck and real vehicle were generally small but varied according to location.

Introduction

Excellent occupant packaging requires the *'design of a vehicle around a specified range of drivers and passengers'* and *'design a vehicle to meet his or her needs and capabilities'* (Herriotts and Johnson [2012]). It addresses the need to fit vehicles to drivers and their passengers; allowing them comfort and access to all necessary features and allowing adjustment in order to encompass the diverse range of shapes and sizes that humans can be.

Due to the nature of the task of driving, and the associated costs with building prototypes, much experimental investigation into occupant packaging is conducted by using simulated conditions such as driving simulators or driving rigs (bucks). While this is a well-proven method, it is possible that drivers may behave differently in an environment that they perceive to be different from the 'real thing'. This could bring into question the transferability of subsequent designs and conclusions.

The automotive industry makes great efforts to measure the posture of occupants in vehicles as accurately as possible during vehicle research and development. However, the measurement of postures is notoriously difficult due to the requirement for occupants to be clothed, wearing shoes, and occupying an envelope that is designed to give the impression of roominess, whilst being as compact as possible. Pheasant and Haslegrave (2006) referred to the concept of the 'anthropometric inch' (25 mm) which is an expected level of accuracy in measurement in most cases, and that it is 'virtually impossible' to measure to a precision of better than 5 mm. Despite these efforts to achieve highly accurate measures, if the posture itself is not transferrable, then implied precision in measurement is unnecessary.

This paper explores the repeatability of postures selected in a vehicle, and the similarity between those and postures selected in a buck built to replicate the same vehicle.

Methods

A buck was built for laboratory trials. This buck replicated a Nissan NV200 and used original NV200 pedals, steering wheel and seat. The layout was identical to the production vehicle to an accuracy of 1 mm, and checked with CAD and vehicle measurements (Figures 1a and 1b).

Ten males and ten females (age 18–24) participated in the experiment and were required to attend on two occasions a week apart. On one occasion they sat in the buck in the lab; on the other in the production vehicle. Order of presentation was balanced.

For each trial, the seat was initially set to have a forward leaning backrest, and the seat at the mid-point on the seat slide. Participants were required to set the seat in their ideal driving position by adjusting the backrest and seat slide, ensuring that they could operate and view primary controls and displays. They were required to hold the steering wheel at the 'ten to two' position and rest their right foot on the accelerator pedal.

Measurements of joint angles were made using a goniometer with telescopic extensions at four locations: ankle, knee, hip, elbow (Figure 2). In order to allow this to happen, before entering the vehicle subjects had bright circular stickers placed on anatomical landmarks at the shoulder (acromion), elbow (lateral humeral

Figure 1a.   Male subject in NV200 ('car').

epicondyle), wrist (ulnar styloid), hip (trochanterion), knee (lateral femoral epicondyle) and ankle (lateral malleolus) and also at the most anterior point of the shoe. Measurements were made with an undepressed and with a depressed accelerator pedal. Measurements were made three times for each 'test'. Backrest angle and fore-aft position on the seat slide were measured.

The process was repeated a total of three times, with the seat being returned to its mid-point and forward leaning backrest each time. Thus each postural variable was measured 9 times for each condition.

Prior to testing with human subjects, measurements were made of similar angles for an H-point manikin to determine the experimenter repeatability. Results showed that measurements were repeatable to <1°. The experiment was approved by Loughborough University Ethical Advisory Committee.

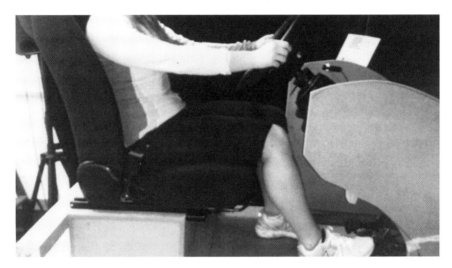

Figure 1b.   Female subject in driving rig.

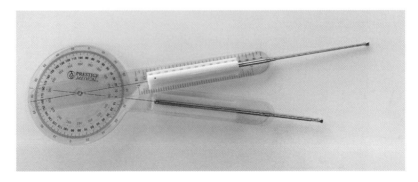

Figure 2.   Goniometer with telescopic extensions.

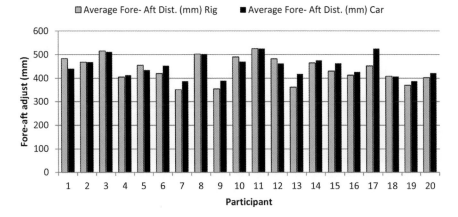

Figure 3.   Comparison of fore-aft adjust for the rig and car conditions (each data point represents an average of nine measurements).

Results

Fore-aft adjustment

Overall, the mean selected fore-aft position for the rig condition was 438 mm, in comparison for 449 mm in the car. This suggests that participants on average selected a position 11 mm further away from the pedals in the 'car' condition. This was true for 11 out of the 20 participants, with a further three subjects selecting exactly the same mean position for both conditions (Figure 3).

A two-tailed paired t-test was returned a p-value of 0.083, meaning that the difference the position the drivers selected between conditions is not statistically significant at the 95% confidence level.

Postural angle variation

The initial ankle angle showed small differences between the rig and the car. Twelve participants had a smaller ankle angle for the car (i.e. more acute) than the rig. Four had identical angles and four greater (Figure 4). The mean angles were 92° for the rig and 88° for the car. The differences between initial ankle angles were significant ($p < 0.05$, t-test, 2-tailed).

Further investigation showed that there was a slight difference in the initial pedal force between the rig and the car and therefore there might have been a slight depression of the pedal for the rig that was not felt by the participants, thus changing the angle of the foot, and explaining the difference between conditions.

The variability in individual settings of the joint angles was investigated. Across all lower limb angles the coefficient of variation (s.d./mean) was about 2%. This means that there was some angular variation each time the seat was set up, even though the set ups occurred immediately after each other.

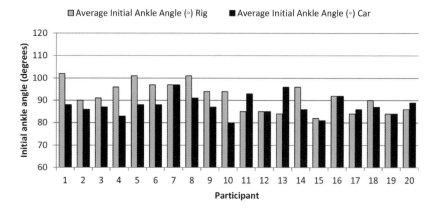

Figure 4.   Comparison of ankle angle for the rig and car conditions (each data point represents an average of nine measurements).

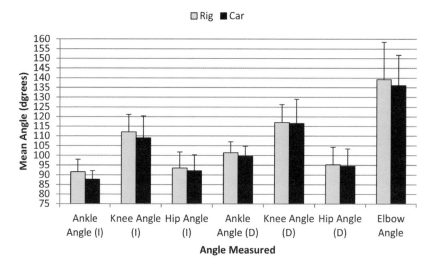

Figure 5.   Comparison of joint angles for the rig and car conditions (each data point represents an average across all 20 participants).

Other postural angles showed no significant differences between the measurements made in the rig and in the car ($p > 0.05$, t-test, 2-tailed). Even through the ankle angle showed differences for the undepressed pedal, when it was depressed the angles were 100 and 101 degrees for the car and rig respectively (Figure 5). Similarly, knee angles showed a greater difference when the pedal was undepressed compared to depressed. A similar pattern was observed for the hip angle.

The largest inter-subject variation was observed for the elbow angle. This was expected as the elbow angle is known to have a large range. It is also a joint angle that is easy to vary during driving by grasping different points on the steering wheel.

Discussion

This study sought to determine whether there are consistent differences between selected driving postures in a lab buck and in a production vehicle. Only one dimension was shown to be statistically significantly different between the two conditions, the initial ankle angle. However, this difference disappeared when the accelerator pedal was operated. Most driving simulations in the laboratory require the accelerator pedal to be operated and so this difference may not be of practical importance for many investigators.

There was a slight difference in the fore-aft adjustment of the seat between the rig and the car, although it did not reach statistical significance. The fore-aft adjust was achieved through a standard seat guide with fixed increments, which could have forced postures to quantised positions.

Results show that it is unlikely that a driver will select exactly the same driving position each time they enter a vehicle. Despite this, they do suggest that drivers will consistently select seat positions that are within a small envelope of variation. Similarly, the lower limb driving posture, in terms of joint angles at the hip, knee and ankle, can be expected to vary by just over 2%. In conclusion, driver posture and the seat position they select can be considered to be closely repeatable, yet not to the extent that it will be an exact replica on each occasion.

The large variability in the elbow is likely to be due to participants being able to self-select how they held the steering wheel. This was a deliberate strategy aimed at replicating procedures used in many lab-test trials.

Apart from initial ankle angle, there was no significant difference found for any criteria between the rig and the car measurements. This study suggests that whilst using a driving rig will not necessarily produce a driving posture identical every time to that seen every time in a vehicle, it is close enough to be suitable for replication of real driving postures.

Whilst the results can be generalised beyond the automotive industry into other sectors where there is a requirement to simulate postures in a test-setting, there is no indication provided here of where the limitations could occur and therefore care should be taken over application of these results with alternative products.

References

Herriotts, P. and Johnson, P. 2012, Are you sitting comfortably? A guide to occupant packaging in automotive design. In Gkikas, N. *Automotive Ergonomics* (CRC Press, Boca Raton, USA)

Pheasant, S. and Haslegrave C. 2006, *Bodyspace: Anthropometrics, Ergonomics and Design at Work.* 3rd edition. (CRC Press, Boca Raton, USA).

A REVIEW OF COMPLIANCE WITH PERMANENT, TEMPORARY AND EMERGENCY SPEED RESTRICTIONS

A. Monk, P. Murphy, M. Dacre, R. Johnson, A. Mills & S. Cassidy

RSSB, London, UK

Evidence suggests that some trains are exceeding permanent, temporary and emergency speed restrictions. A literature review, interviews, surveys and workshops with frontline staff and senior managers were conducted to identify the scale of the problem, understand the reasons why a driver may exceed the designated speed and identify potential controls to prevent or recover from underlying causes of over speeding. The data indicates that there are a relatively small number of speeding incidents per year but the scale of the problem cannot be accurately determined because of the limited data available. A number of reasons why a driver may exceed a designated speed have been identified. Issues apply to drivers, designers, track workers, signallers and driver managers.

Introduction

Evidence suggests that some trains are exceeding permanent (PSRs), temporary (TSRs) and emergency (ESRs) speed restrictions. The scale of the problem is unknown but recent checks in association with Adjacent Line Open (ALO) working have identified that the occurrences might be higher than previously thought. The reasons why drivers may be exceeding the required speed limits are unclear. Without a clearer understanding of the problem it is difficult to develop and implement effective mitigations to prevent occurrences of over speeds.

In light of this, the 4 main objectives for the project were:

1. Identify (as far as possible) the scale of the problem of drivers exceed the speed limit in PSRs, TSRs and ESRs using the limited data available.
2. Consider the whole process, which includes the design and setting up of a speed restriction and understand how all roles/activities can influence driver behaviour in response to speed restrictions.
3. Gain an understanding of the reasons why drivers exceed the speed limits in PSRs, TSRs and ESRs.
4. Identify potential controls to prevent or recover from the underlying causes of over speeding and evaluate these mitigations.

Methodology

There were phases of the project as follows:

1. Literature review (Murphy, 2014)
2. Frontline staff interviews (Monk, 2014)
3. Driver surveys (Monk, 2014)
4. Workshops with frontline staff and senior managers (Johnson, 2014)

The aim of the literature review was to use existing data sources to try to establish if there is an issue of speeding and if so, the scale of the issue. Existing literature and data from other countries were used to start identifying 'what could go wrong' in the design, laying out of speed restriction signage and equipment, driving through of speed restrictions and monitoring of speeding incidents. The findings from the literature review shaped the questions for the frontline staff interviews.

The frontline staff interviews were held with 33 drivers, 16 driver managers, 6 designers of speed restrictions, 9 signallers and 18 track workers. This gave an understanding of the process of how speed restrictions are designed, laid out, driven through and monitored. A driver survey was then conducted to validate the interview findings and explore reasons why speeds may be exceeded with a larger/more representative sample of drivers. There were 566 respondents to the survey; 493 from Train Operating Companies (TOCs) and 60 Freight Operating Companies (FOCs)/On-Track Machine (OTM) drivers.

The workshops with frontline staff and senior managers were held to identify potential mitigations and evaluate these in terms of their priority, strengths/weaknesses and feasibility to implement. Two workshops were held with frontline staff and 2 workshops with senior managers.

Results

Risk from speeding

Speeding-related derailments have the potential to result in large numbers of casualties. For example, the accident in July 2013 at Santiago de Compostela, Spain resulted in 79 fatalities. However, such accidents are rare. It is more than 30 years since the last fatal derailment caused by speeding in Britain but the potential for such an accident remains. Speeding also has the potential to result in accidents other than derailments, such as a train striking track workers. More recent examples of incidents of drivers exceeding the designated speed can be found in the Rail Accident Investigation Branch (RAIB) investigations at Ty Mawr (August 2007) and Bletchley (February 2012).

The Safety Risk Model (SRM) v8.1 [1] estimates the safety risk from derailments caused by train speeding to be 0.020 fatalities and weighted injuries (FWI) per year [1]. Most of this arises from passenger train derailments because, although they are less frequent than non-passenger train derailments, the expected consequences are more severe. SRMv8.1 estimates that speeding accounts for around 1% of all derailment risk.

The SRM calculates risk by combining estimates of frequency and consequence. Frequency estimates for speeding-related derailments are based on the past occurrence of such events adjusted to account for the introduction of Train Protection Warning System (TPWS) at high-risk PSRs in the early part of the last decade. They are based on a small number of derailments (6 since 1990, including only 1 involving a passenger train) and so are subject to a fairly high level of statistical uncertainty. The SRM derailment consequence models take into account derailment speeds and the potential sequence of events that could subsequently occur (for example, carriages overturning, striking a structure, secondary collision or fire). The consequence estimates are derived by combining the likelihood of each accident scenario with the estimated number of fatalities and injuries that would typically result. There is also uncertainty in the consequence estimates. For example, the same derailment consequence model is used for different causes and may not fully reflect factors specific to speeding-related derailments.

Safety risk is also not the only 'cost' associated with speeding. For example, the derailed train at Bletchley in 2012 caused significant damage to the locomotive, track and overhead electrification equipment and major disruption to the West Coast Main Line. More commonly, TPWS OSS (Train Protection Warning System Over Speed Sensor) trips can cause delays to train services.

Frequency of speeding incidents

The Safety Management Information System (SMIS) is the rail industry's national database for recording safety-related events that occur in Great Britain (GB). A search was conducted for recorded incidents of over speeds between 2006 to October 2013. There were on average 39 recorded incidents per year.

During the frontline staff interviews, track workers said that they would make a report if they thought a train was exceeding the designated speed. However, unless the train was travelling excessively fast it would be difficult to determine if there had been an over speed.

Frontline staff interviews with driver managers revealed that they mainly use On Train Data Recorder (OTDR) downloads to monitor speeding incidents. Driver managers can see PSRs on downloads but not TSRs/ESRs (these need to be mapped on to the downloads). Seven out of 16 driver managers found this difficult and

---

[1] FWI is a combined measure of risk or harm in which injuries are weighted according to their relative severity. For example, a major injury is deemed to be statistically equivalent to one-tenth of a fatality. Appendix A of RSSB (2014) defines the different injury degrees and their weightings.

time consuming. This suggests that the exact locations of the commencement of TSRs/ESRs are not easy to map on to OTDR downloads which makes the monitoring of speeding more difficult. An experienced driver who is not on a competency development plan (CDP) will be 'downloaded' once per year. Therefore the OTDR download is a snapshot of one journey rather than a constant monitor of driver compliance with speed restrictions. Over-speeds are internally monitored by keeping a record on a driver's file, or in an incident log/register/database. Driver managers in the sample were asked to detail which speeding events were entered in to SMIS. There were a variety of different answers. For example, two driver managers said that all incidents of speeding over 3 mph were entered in SMIS and one driver manager said that they did not enter speeding incidents in to SMIS.

In addition to SMIS, data exists from a number of radar gun studies. At Saltley depot in August 2011, in the 'worst' instance, 5.8% of trains checked exceeded the TSR speed. There was a difference in the percentage of trains exceeding the speed according to route with one location having more incidents than the other. A more extensive speed monitoring programme where data was collected from 2 sites over 2 years found that 1.6% of trains checked exceeded the designated speed for TSRs. These radar gun studies give a useful indication of the frequency that drivers exceed the designated speed but are not large enough to be generalised across the rail network. Therefore, the driver survey was used to collect data on drivers' perceptions of the frequency of over speeds. Respondents were asked 'How often do you think drivers exceed the designated speed upon entering and/or driving through a speed restriction for a PSR/TSR/ESR'. The majority of respondents thought that drivers exceeded the designated speed 'rarely' or 'sometimes'. Firm conclusions cannot be made about the size of the issue of non-compliance with speed restrictions.

Reasons why a driver may exceed the designated speed

A summary of the reasons (from the frontline staff interviews and driver surveys) why a driver may exceed the designated speed is shown in Table 1.

Respondents in the driver survey were given a list of possible reasons for why a driver may exceed the designated speed. Respondents were required to select five reasons from the list and rank those selections from one to five for:

a) A high risk speeding incident (a high risk speeding incident was described as speeding at 11+mph over the designated speed and/or exceeding the designated speed at a location where a train might derail).
b) A lower risk speeding incident (a lower risk speeding incident was described as speeding at 3–10 mph over the designated speed at a location where it is very unlikely to result in an accident or damage the infrastructure)

The most frequently cited reasons by drivers for a high risk speeding incident and a low risk speeding incident are shown in Table 2.

Table 1.   Why a driver may exceed the designated speed.

Designing of speed restrictions	• Differences in methodology between how restrictions are designed in and out of office hours leads to inconsistent braking distances • Documentation is not up-to-date
Laying out of speed restriction signs	• There is no formal training for laying out speed restriction signage/ equipment • Correct signage is not always available • Documentation can contain errors • Sign can fall over due to vandalism, sabotage and human error • Time pressures to erect signage and equipment • Signs can be positioned incorrectly (on the wrong line, in the wrong order or behind obstructions) • Lack of awareness of Rule Book changes
Signallers	• Quality of communication with track workers • Providing ESR information too far in advance of a speed restriction
Drivers	• Documentation about speed restrictions are not fit for purpose • Drivers rely on the signs being positioned correctly • Signage can be dirty, incorrectly placed, poorly maintained or not visible in all lighting conditions • Little reporting of issues (poor reporting culture) • Erratic braking distances • In-cab warnings insufficient
Driver errors	• Attention • Situational awareness • Fatigue • Memory • Confusion • Expectation • Workload
Monitoring of speeding incidents	• Difficulty in mapping of ESRs/TSRs on to OTDR downloads • Frequency of OTDR downloads • Consistency of what is entered in to SMIS

Table 2.   The most frequently cited reasons by drivers for a high and low risk speeding incident.

High risk speeding incident (n = 366)	Low risk speeding incident (n = 368)
Lack of knowledge about where the speed restriction is	1) Lack of knowledge about where the speed restriction is
Fatigue	2) Memory/forgetting there is a speed restriction
Memory/forgetting there is a speed restriction	3) Fatigue
Signage missing or incorrectly placed Attention	4) Attention 5) Signage missing or incorrectly placed

Misjudgements Visibilty and lighting Dagger speed restrictions
Length of speed restriction
Temporary magnets Train clear
Consecutive in short distances Information
Traction capabilities

Braking distances

Boredom Discipline Repetitious work Repeater signs Sign size Non compliance
Reporting Rules Contradictory information Driving styles Vandalism
Management bullying Attitudes Complacency Attention Irrelevant information Ignorance
Conflicting speed restrictions Workload Fatigue Lack of information Repetition over routes
restrictive signals Monitoring Not achievable Braking conditions Memory Planning
Misjudgement Dirty Missing signs Not caring Gradient Other Mile posts
Maintenance Knowledge Expectation Positioning and layout Inconsistent signs
Driving style Sign faults Up to date notices
Faulty signs Late running Frustration Locomotive faults
Rostering Concentration Cab environment AWS Conformity
Stupidity Stations length of time speeds in place Reactions Familiarisation
Signs outdated boards Sectional appendix Incomplete information

Distractions

Shift work
Repeater boards Location information Route knowledge Situational awareness
WON Communication Visibility and lighting Height of signs
Not thinking of consequences Driver error Understanding why they're in place Train formation
Changes to restrictions Yards and sidings

Sign design Personal problems
Repeater

Figure 1.   Other reasons why a driver may exceed the designated speed.

Respondents were also asked whether they could think of any other reasons why a driver may exceed the designated speed. The results are shown in a work cloud in Figure 1 where the most frequent answers appear in a larger font size than answers less frequently given.

Conclusion

It has not been possible to establish the size of the issue of non-compliance with permanent, temporary and emergency speed restrictions. SMIS indicates that there are on average 39 incidents of over speeding each year. However, the frontline staff interviews revealed differences within and between companies in the incidents that are entered in to the database. Driver managers can also find it difficult to map exact locations of TSRs/ESRs onto OTDR downloads, which in turn means they may not be able to accurately tell if a driver has entered a speed restriction too fast. OTDR data is downloaded once per year for an experienced driver that is not on a competence development plan (CDP). The download is therefore only a snapshot of one moment in time. Track workers on the ground do report drivers exceeding the designated speed. However, only excessive speeds may be reported because it is difficult to tell if a train is only exceeding the designated speed by a small amount. For these reasons, it is highly likely that incidents of speeding are under-reported and the data in SMIS is not telling the entire story.

Speeding has the potential to result in a high-consequence derailment, such as the accident at Santiago del Compostela, Spain. It could also cause other types of accident, such as a collision with track workers. Figures from SRMv8.1 suggest that the risk from over speeding is low. In addition to safety consequences, there are other potential costs relating to speeding-related accidents and incidents include damage to trains or infrastructure and delays to services.

Small scale radar gun studies have shown varying percentages of drivers speeding (1.6%–5.8%) but have not been conducted with a large enough sample for a sufficient amount of time to be representative of the rest of the railway network. There did seem to be differences in the number of incidents according to location in the studies. The driver survey data indicate that PSRs/TSRs/ESRs are exceeded rarely or sometimes, but it is a measure of perceptions rather than incidents. Overall, it appears that a relatively small number of drivers are exceeding the designated speed, but it is very likely that over speeds are an area of under reporting which should be considered when interpreting these results.

The frontline staff interviews and driver survey have given a greater understanding of the process of designing, laying out, driving through and monitoring of speeding incidents. They have also identified a number of different reasons why a driver may exceed a designated speed. These reasons are not exclusive to drivers but also apply to designers of speed restrictions, track workers that lay out signs, signallers and driver managers.

References

Johnson, R. and Dacre, M. 2014, T1044: A review of compliance with permanent, temporary and emergency speed restrictions: Interim report on the workshops with frontline staff and senior managers, RSSB, London.

Monk, A., Mills, A. and Jones, A. 2014, T1044: A review of compliance with permanent, temporary and emergency speed restrictions: Interim report of interview data, RSSB, London.

Monk A. and Mills, A. 2014, T1044: A review of compliance with permanent, temporary and emergency speed restriction: Interim report of survey data, RSSB, London.

Murphy, P. 2014, A review of compliance with permanent, temporary and emergency speed restrictions: A literature review, RSSB, London.

RAIB, 2008 Retrieved 30th July 2014 from http://www.raib.gov.uk/cms_resources.cfm?file=/081030_R222008_Ty_Mawr.pdf.

RAIB, 2012, RAIB's Initial Report Into Bletchley Derailment. Retrieved 30th July 2014 via http://www.rail.co.uk/rail-news/2012/raibs-initial-report-into-bletchley-derailment/.

RSSB, 2014, Guidance on the use of cost-benefit analysis when determining whether a measure is necessary to ensure safety so far as is reasonably practicable, RSSB, London.

ARE OUR STREETS SAFE ENOUGH FOR FEMALE USERS? HOW EVERYDAY HARASSMENT AFFECTS MOBILITY

Jane Osmond & Andree Woodcock

Coventry School of Art and Design, Coventry University, UK

This paper reports on the harassment women in a UK city face as they go about their everyday lives. It is argued that such experiences are not only traumatic, but have a long term effect on women's sense of worth and on their mobility patterns. Looking at the transport system as a whole as required by new mobility paradigms and whole journey experiences, this may be seen as a perpetuating, deep-rooted system failure leading to inequality and reductions in inclusivity. After having summarised the results, the participants' suggestions are put forward as solutions to address this issue.

Introduction

Woodcock, Lenard and Walsh (2003) conducted a study on the safety and security of women drivers and their passengers, part of which looked at the harassment of women drivers. One of the responses contained a 'thank you' for enabling the women's voices to be heard. This paper is a companion piece, examining women's experiences of harassment as travellers in the public realm.

Street harassment represents perhaps the most common and frequent type of sexual harassment encountered by women. However, little research has been conducted into it, and it is often dismissed as a trivial and natural fact of life that women must tolerate. Bowman (1993) defines its characteristics as 1) the targets of harassment are female; 2) the harassers are male; 3) the harassers are unacquainted with their targets; 4) the encounter is face to face; 5) the forum is a public one; 6) the content of the speech, if any, is not intended as public discourse. The remarks are aimed at an individual, though they may be loud enough to be overheard and they are objectively degrading, objectifying, humiliating and frequently threatening in nature. Davis (1994) defined harassment as 'spirit murder', with a drip feed of harassment affecting women's life and liberty.

Street harassment is a way of silencing women. It inhibits dialogue and promotes sexual oppression. Although men may argue that it is harmless flirtation, Langelan (1993) pointed out that *'it is not only ineffective, but consistently counterproductive; women react with disgust, not desire, with fear, not fascination'* and *'women never really ignore harassment ... they must deal with all the emotional repercussions of victimization; fear, humiliation, feelings of powerlessness, rage.'*

Thus, street harassment is an invasion of women's privacy, an intrusion into personal space which has a negative impact on women's self esteem.

Benard and Schlaffer (1984) commented on the way in which sexual harassment restricted women's mobility, with public spaces being the male prerogative, and women denied access to the streets, and enjoyment of public resources. Bowman (1993) termed this as a '*ghettoization to the private sphere of hearth and home*'.

Adopting a feminist, phenomenological approach Turkheimer (1997) argued that men are 'blind' to the pain women suffer. Many do not 'see' sexual harassment unless it is pointed out to them and they do not understand the effects of such harassment. The only way in which action can be taken on this is if information is shared, named and articulated.

This study is timely: if we are to achieve sustainability targets, the needs of different users have to be considered. There is an obvious (Maffi et al., 2014) gender gap in transport and mobility, e.g. women travel differently from men in relation to modes of transport, distance travelled, the daily number of trips and patterns and purposes of travel. They use public transport more and are more supportive of sustainability agendas. However there is still a lack of knowledge regarding gender and mobility and the experiences of women as travellers. This gap needs to be addressed if inclusive mobility is to become a reality.

Methodology

The results are based on 193 online surveys and 16 telephone interviews with women in and around the city in 2013. The respondents were recruited via a variety of local online networks. The prevalent age of respondents was 17–29 years. Just over 90% of the respondents lived, worked or were attending educational institutions in the city (Osmond, 2013).

Results

Overall, just over 60% of the sample had experienced some form of harassment over the last 12 months. This broke down as 40% having received unwanted sexual comments, wolf whistling (32.8%), and groping (12.3%), and 20% some other form of harassment. Over half of the respondents ignored the incidents, 14% challenged it directly, and in just over 3% of the cases someone intervened.

The following results are presented as a reflection of the whole journey experience of women (i.e. from origin to destination). As such only the results pertinent to travel and use of public spaces have been included (i.e. work and school place harassment has not been included, neither have those cases which could not be specifically attributed to the public realm or transport). Bearing this in mind 14.2% of the incidents were on public transport and 52% on the street.

Asked how safe they felt on public transport or in the public realm, only 6% stated that they felt very safe, with 51.2% indicating that they felt fairly safe, 35% not very safe and 6.8% not safe at all. The following section looks at the incidents in more detail and the recommendations suggested by the respondents.

Examples of incidents in the last 12 months

Over 75 separate incidents were reported, most of which occurred in the street, e.g. near bars and nightclubs, on deserted roads or coming home from work, while women were walking, jogging, with children. The perpetrators were single men, men in groups, on bikes, in cars or on public transport. The incidents related to:

1. Unwanted physical contact (sexual and physical assaults, hair stroking of children, bottom slapping, being pelted with bottles/eggs, forced into cars).
2. Nonverbal behaviours including exposure of genitals, and sexual gestures, being followed (on foot and in car) and having movements blocked or copied, being 'accompanied', kerb crawling, being leered and beeped at.
3. Verbal behaviour included being sworn at by pedestrians, cyclists and motorists, wolf whistles, being subjected to crude, sexual and patronising comments, propositions and being forced to engage in unwanted (sexual) conversation.

All of these form part of the everyday experience of being a female traveller. They are neither novel nor astounding. What is astounding is that such behaviour is tolerated and almost to be expected regardless of the damage it does:

'I was walking to a friend's house and a man walked passed me, waited until I was a few steps away and whistled and stated making crude comments at me'

'... on a bus I was made to feel intimidated by two males sitting behind me wolf whistling, calling me sexy and asking me to talk to them – 'at least now we have something sexy to look at' was one comment. After ignoring them I became a 'stuck up slag' and when I got off the bus they were discussing the way my jeans made by bum look'

'Earlier in the year I had to change my route to work as a man was making me uncomfortable by staring at me as I stepped up on to the bus each morning, and I was informed by another passenger that he had ducked down to look up my skirt'

'... walked past a group of middle aged men, one muttered cunt under his breath. The others laughed.'

Walking to the bus stop feeling low – someone shouted 'cheer up love it will never happen' then 'give us a smile' then 'fuck off I was only trying to be friendly'. It is not my duty to look cheerful and smile for men'

Location of incidents

The location and time of incidents varied (outside bars, walking in the street, cycling, waiting outside work, in city arcades, in taxis) with one woman commenting *'hundreds of incidents, too many to articulate, this is the reality of day to day life'*. The duration of the incidents might be short (one off comment), persistent – does not stop until the woman is able to escape (forced to listen to lewd comments, or conversation), repeated (e.g. on commuter journeys) and prolonged (e.g. being followed home over 20 miles).

Actions of respondents

The actions of respondents revealed their lack of empowerment and victimisation. For example, on being subjected to day-time harassment outside a bar, one respondent commented that she did not go into the bar because she was frightened he would follow her and '*ultimately, my main fear is that perhaps, even with bouncers on the door at the time, I wouldn't have felt capable and sure enough of my position in the situation to ask them.*'

'*... the sad truth is that your only option is to ignore it, put up with it and internalise the self loathing that doing this brings with it*'

Clearly the nature of the harassment and the reaction of others to the situations makes the women feel guilty. They are too pretty, dressed wrongly, or are in a place where such behaviour is expected. If they react aggressively, they become the aggressor or are seen as unable to take a joke. Mostly, women chose to ignore it because they were afraid of an escalation, e.g. '*I just tried not to react in any way to avoid further engagement, in the hope they would stop*' and '*I had a combination of shock and of fear of what would happen if I challenged it. Spent the whole of the next day coming up with responses in my head.*'

Many women chose to carry on walking, putting distance between themselves and their assailants, an act which might instead led to more danger, '*ran away – ended up having to run across the ring road due to there being nowhere else to go – risking safety even more.*' One woman said she '*walked away fast and hoped the children wouldn't notice the bad language*' and '*I ignore it, because I worry that if I challenge it the situation may escalate, especially if I have children with me, or I fear for their safety*'.

Those who witness harassment rarely confront or apologise for the behaviour of the perpetrator. They sometimes attribute blame on the victim, ignore it or exploit the situation, '*one time after being harassed I sat down at a bus stop to cry and another man came along, asked if I was ok, and then stole my phone*'.

Perceptions of safety

From the results of the survey it would seem that women travelling alone or with children have experienced harassment at levels which significantly affect their mobility. '*It has almost become a part of life that us as women have to accept and put up with it as it is not tackled*'. They feel unsafe when they are alone, especially at dusk or night time, near groups of men, in public spaces and car parks, in taxis, in deserted precincts, in underpasses and poorly lit areas.

The effects on mobility are marked in comments such as '*I think I'm constantly waiting for someone to follow me, shout at me, engage me in conversation; 'I drive whenever I can as I hate being on my own in the streets*'; '*I actually feel safer when running because I am on the move and clearly engaging in an activity (which I probably naively feel protects me from more attention)*'; I just hate walking anywhere on my own ... am constantly waiting for someone to follow me and sexually demean *me in some way.'* and *'I am always on guard whenever I am out and about on my own'*.

The impact on mobility patterns is clear. *'I would not travel on a bus after 6 pm'*; *'commonplace for me to bolster my safety by not going to lonely places, by using public transport at night rather than return to a car park'; 'walking down the centre of my own street at night, and not going out alone at night'* and *'always aware of my surroundings and not to put myself in a vulnerable position'*.

Although often overlooked, the quality of the public realm was considered important. Most frequently mentioned was lighting, dark subways and car parks. Other items mentioned included the physical fabric of neighbourhoods – litter, cars parked randomly, poorly maintained roads and pavements and poorly maintained properties.

Recommendations

Over half of the respondents mentioned better lighting in either dark areas, in specific places, near bus stops or car parks. Increasing police or community warden presence and visibility was mentioned by over a quarter of respondents, especially at night, with additional comments around the need for sexual harassment to be treated more seriously, and to increase police powers moving to a zero tolerance of antisocial behaviour. This was accompanied by calls for better treatment of victims, better conviction rates and raising the awareness of support and increasing the number of support centres.

With regards to transport, respondents were able to specify locations where more lighting, patrols and traffic restrictions were needed and specific bus routes which had regular incidents of 'domestic violence, teenage violence, marijuana smoking'. Many comments pointed to a need for cultural change and education:

'Until society's attitudes change I don't think there's anything that can be done to make us feel safe. I'm getting really tired of being told as a woman it's not safe for me to walk alone especially at night: why not tell men it's not okay to treat us the way they do? It angers me that society has the ability to make women feel like victims just because of our gender'.

Discussion

Thomson (1994) discusses the need for transport operators to highlight harassment, as women form significant users of public transport. Administering a system to report harassment and publicizing anti-harassment regulations may create more public awareness and education. A similar approach was adopted in Medellin Colombia, to promote thoughtful travel behaviour on the new Metrocable (Atkinson et al., 2013). Additionally, the US captive audience legislation (where people are not able to remove themselves from a situation) could be used to uphold women's rights and freedom to travel in safety.

At the time of writing, there is little evidence that UK transport operators are specifically addressing the issues identified in this paper. However, there have been police-led moves to address unwanted sexual behaviour in two particular cities. In London's Project Guardian (British Transport Police 2014) the police are working closely with Transport for London to help reduce unwanted sexual behaviour on public transport. In the West Midlands Project Empower (West Midlands Police 2014) is training public transport staff to spot any incidents and support passenger reporting. This is backed up with an on-board and in-station marketing campaign

Although any attempt to address unwanted sexual behaviour on public transport is to be welcomed, the focus seems to be on policing behaviour, rather than designing public transport systems with passenger safety in mind. It is therefore recommended that public transport operators undertake a review of their existing services and include as a minimum the following issues:

1. Survey existing passengers to establish how safe they feel on the public transport network
2. Design stations and stopping places to include better lighting and staffing
3. Consider the placement of stopping places and also the routes to and from them
4. Train all public transport staff in how to deal with and report any incidents
5. Undertake a public marketing campaign displaying clearly what the unwanted behaviour looks like and reporting routes

Conclusions

Harassment in streets and on/around public transport is almost so commonplace as to be unnoticeable. Woodcock (2012) has previously argued that the whole journey experience is 'other' than the sum of its parts, and that the experience, from planning to arrival at destination should be the focus of attention, as depicted in the H-S model (2012b). While it has been contended that the experiences of the longest journey have the greatest effect on perceived quality (Susilo and Cats, 2014), evidence presented here may support a contrary view – that there is a threshold level at which experiences may be so unpleasant, even on short parts of a journey, that they not only bias perception but change mobility patterns. Therefore a microlevel analysis of the whole journey experience is needed. Such an analysis is again in line with the H-S model, which in its external layers looks at what influence external and cultural factors (in this case the objectification of women) have on person/transport interaction as exhibited through social interactions in the personal sector of the model.

This paper has clearly presented evidence of the everyday harassment women have to contend with when they are going to and from work, the shops or leisure activities. The harassment takes place in the public realm, on transport, whether women are walking, exercising or with their children. Both young and older women are targeted, with additional abuse levelled at older women who should be 'grateful' for the attention. Abuse can range from wolf whistling to serious sexual assault and molestation in public places, where women should be safe.

The sense of outrage the authors feel is immense. Transport policy abounds with schemes to encourage greater use of active and public forms of transport; technology providers are quick to develop apps to report crime, to track journeys, yet little investment is placed where it is needed – in providing heightened police presence, better lighting, more regular services, more serious regard towards the nature and effects of this harassment and in education. Harassment and perceived safety affects women's mobility patterns at planning, mode, time and route choice levels. Women incur additional direct/indirect costs associated with unsafe streets and poor system design in which they are truly third class citizens. We are left asking two questions 1) why would any women support a system where she is left feeling wounded and degraded; 2) why, with all the investment and hype surrounding smart, safe and inclusive transport is so little attention being placed on tackling inherent system inequalities.

It may be concluded that the commonplace harassment of women on streets is yet another symptom of the gender inequality in transport provision. The complexity of women's trips (trip chaining) has and continues to be ignored in many national surveys (e.g. National Household Survey in UK), and those conducted by transport authorities and operators. Despite the lack of research and support for the multimodal journeys that women take (e.g. when balancing multiple jobs, social and childcare duties), and the harassment that women face, it is still the expectation that women will play a vital and leading role in promoting and using sustainable and active forms of travel (Maffi et al., op cit). Without steps to guarantee their safety and security this hope will not be realised.

References

Benard, C. and Schlaffer, E. 1984. The Man in the Street: Why He Harasses. In A.M. Jaggar and P. S. Rothenberg (eds.) (2nd ed) *Feminist frameworks; Alternative theoretical accounts of the relations between women and men*

British Transport Police (2014) Project Guardian. Available online http://www.btp.police.uk/advice_and_information/how_we_tackle_crime/project_guardian.aspx. Accessed 14 November 2014

Bowman, C. 1993. Street harassment and the informal ghettoization of women, *Harvard L. Review*, 106, 3, 517–581

Davis, D. 1994. The harm that has no name: street harassment, embodiment and African American Women, *UCLA Women's Law Journal*, 4, 2, 135–178

Langelan, M.J. 1993. *Back off: how to confront and stop sexual harassment and harassers*, New York: Simon & Schuster

Osmond, J. 2013. An everyday occurrence: women and public sexual harassment. *Coventry Women's Voices*. http://bit.ly/1ySQHBA

Maffi, S., Malgieri, P. and Di Bartolo, C. 2015. CIVITAS Policy Note: *Gender equality and mobility; mind the gap,* CIVITAS

Roa-Atkinson, A., Atkinson, P., Osmond, J. and Woodcock, A. The impact of an integrated transport system: An ethnographic case study of the experiences of Metrocable users in Medellin, Colombia in Stanton, N.A. (ed) (2013): *Advances in Human Aspects of Road and Rail Transportation,* CRC Press: Taylor and Francis, USA, pp. 311–322

Susilo, Y.O. and Cats, O. 2014. Exploring Key Determinants of Travel Satisfaction for Multi-Modal Trips by Different Travellers' Groups, *Transportation Research A,* 67, 366–380

Thomson, D.M. 1994. The woman in the street: reclaiming the public space from sexual harassment, *Yale Journal of Law and Feminism,* 6, 313–348.

Turkheimer, D. 1997. Street harassment as sexual subordination: The phenomenology of gender-specific harm, *Wisconsin Women's Law Journal,* 12, 167–206

West Midlands Police. (2014). Public transport police team urges passengers to report sexual harassment. Available online: http://www.west-midlands.police.uk/your-local-police/birmingham-east/local-news/localnews.aspx?Id=609. Accessed 14 November 2014

Woodcock, A. 2012a. New insights, new challenges: Person centred transport design, *Triennial Congress of International Ergonomics Association 2012,* February, Recife, Brazil

Woodcock, A. 2012b, User centred transport design and user needs. In M.Tovey (ed). *A User Centred Approach to Vehicle Design and Travel,* Gower Press, pp. 21–70

Woodcock, A., Lenard, J. and Welsh, R. 2003. Women Drivers, Passengers, Cars and the Road, In L. Dorn (ed.). *Driver Training and Behaviour,* Ashgate Press, pp. 135–150

VISUAL SAMPLING IN A ROAD TRAFFIC MANAGEMENT CONTROL ROOM TASK

Sandra Starke[1], Neil Cooke[1], Andrew Howes[2],
Natan Morar[1] & Chris Baber[1]

[1] *School of Electronic, Electrical and Systems Engineering,*
University of Birmingham
[2] *School of Computer Science, University of Birmingham*

Control room tasks like road traffic management require continuous visual assessment paired with active interventions. This includes monitoring a range of information sources and resolution of live scenarios in real time using Standard Operating Procedures. We investigated how three operators resolved a simulated 'object in the road' task, requiring information gathering and integration from multiple displays. Visual sampling was quantified from eyetracking and related to a Hierarchical Task Analysis. Operators followed heterogeneous strategies to accomplish the same main goal, differing in the attended ROI sequence and task component weighting. Differing preferences for information sampling were accommodated by prescribed data entry fields in the incident log.

Introduction

In road traffic management control rooms, operators perform monitoring, supervisory and executive roles: after receiving an alert from an external source about incidents such as a congestion or accident, their role encompasses monitoring a multitude of information sources, assessing the severity of an incident, interacting with the road network *via* lane closures / speed limits to manage the situation, and recording an incident log. Little is known how different observers navigate multiple information sources to complete these activities.

Control room operators rely on visual processing of information, which is facilitated through 'visual sampling', the sequence of overt visual attention to regions of interest (ROIs) executed by head and eye movement: due to the physiological properties of the eye, humans have to perform continuous eye movements ('saccades') to build up a mental representation of the world around them using information from 'fixations' (gaze directed at stationary object) or 'smooth pursuit' (tracking a moving object). The use of eye tracking devices to understand the underlying cognitive processes has a long tradition especially under the consideration of 'active vision' (Findlay and Gilchrist, 2003), the active direction of gaze towards regions that hold task-relevant information. Visual search refers to individuals sampling a scene with the goal of finding information. Search patterns can be systematically changed in response to the goal (Yarbus, 1967) or training (Chapman et al., 2002) of the searcher. 'Bottom-up' search is typically feature driven, i.e., based on image salience (prominent visual features) or distractions; 'top-down' search is typically based on the user task (Doshi and Trivedi, 2012). For the task of this study, we assume that visual sampling is not exclusively guided by image saliency but by the goals of the observer, reflecting an 'internal agenda' (Yarbus, 1967, Hayhoe and Ballard, 2005). This assumption has been argued in context of many real-life tasks, where visual search patterns are closely linked to an observer's 'schema', or mental task representation, being controlled top-down (Land, 2009). This raises the challenge of developing a combination of eye-tracking metrics which can be related to the goals that the operator is seeking to achieve. This is the first aim of the study reported in this paper.

The redirection of gaze can be accomplished by a combination of eye-, head- and body movement depending on the desired saccade amplitude. While gaze shifts >40° commonly require a contribution from head movements *per se*, literature values on head contributions to smaller gaze shifts vary (Fuller, 1992). In general, eye-head coupling is highly subject specific (Thumser and Stahl, 2009) and even small gaze shifts <15° can have a head contribution (Goossens and Opstal, 1997). While knowledge about the cognitive reasons for eye-head coupling remains sparse, head movements have been linked to processing demands (Stein, 1992) and the temporal lag between eye- and head movement appears linked to top-down/bottom-up attention shifts (Doshi and Trivedi, 2012). Thus, the second aim of this study is to consider ways in which eye and head movements are combined in visual search.

It is proposed that the notion of an 'internal agenda' is beneficial to Ergonomics studies of operator performance because there ought to be correspondence between eye movement, head movement and the goal structure of a task. This pilot study provides a descriptive framework of operator behaviour in a road traffic management control centre performing a standard task. We investigated whether operators follow comparable workflows with respect to visual information sampling when resolving incidents, given the multitude of sources they can use, or whether there are individual differences in strategy between operators. Specifically, we were interested in the relationship between head movements, gaze shifts and their relation to goal structure. This work is part of a larger study into interactive human decision making where decisions are facilitated by continuous ongoing visual attention.

Methodology

Location and scenario

Data were collected from three expert operators (one female, two male) at the road traffic management facility at DIR Centre Est, Grenoble, France. The operators had a minimum of three years post-training experience. The protocol for data collection had been approved by the University of Birmingham Ethics Committee. Figure 1

**Figure 1.   Road traffic control room with participant wearing mobile eye-
tracker (Tobii glasses v. 1).**

shows the control room, which has, on the desk in front of the operator, five computer monitors displaying a user interface for incident logs (LG), a schematic map
of the traffic network (MP), a live CCTV feed linked to the camera that the operator
is currently controlling (TV) and access to the internet or other information (D1
and D5). In the background is a large display panel (BS) with a 4 × 4 colour array
of CCTV feeds from different traffic cameras and a smaller display panel (SS)
presenting colour CCTV feeds which can be interacted with from the desk. Also
located on the desk are standard PC peripherals and phones / radios for communication with stakeholders from outside the facility such as traffic patrol staff and
emergency services personnel.

The facility is a fully operational traffic control centre, hence forbidding the inter-
ruption of incoming data with a recorded scenario. For this reason, operators were
asked to engage in a pretend task, initiated by a call from a member of staff which
simulated a standard alert of 'object in the road'. The rationale behind this approach
was that operators were assumed to navigate this pretend task in a manner representative of their 'average' behaviour, hence reducing the confounding influence of
otherwise highly specific live events.

Collection and analysis of eyetracking data

Eyetracking data were recorded using Tobii glasses with a sampling frequency
of 30 Hz. Prior to each recording session, the eyetracker was calibrated to each
individual participant. To later relate the gaze data in the local glasses reference
frame to the global reference frame of the control room, infrared markers were
placed in strategic positions around the display units. All participants received an
explanation of the task and had the opportunity not to participate.

Following data collection (task duration: 3.8 to 4.5 min.), gaze data were automatically mapped onto ten regions of interests (ROIs, Figure 1) defined by the IR markers using Python. Head orientation was automatically inferred from the video data. Subsequent analysis was performed in Matlab (MathWorks). Data were pre-processed using conditional rules, such as the removal of data points in one data stream (e.g., eyes) if the other stream (e.g., head) had a dropout. Times at which observers attended to external stimuli (such as unrelated phone calls) were cropped out. The resulting dataset had a trackability ranging from 81% to 87%. For gaze data, cumulative percentage viewing time per ROI, frequency of switches between ROIs and viewing networks were calculated. Agreement between the ROI attended to by eye and head was calculated for each sample (excluding empty samples) and expressed in % total tracked viewing time.

Defining operator goals

Hierarchical task analysis (HTA) was performed using observations and interviews with subject matter experts, corroborated with eyetracking recordings across participants performing both real and pretend tasks to define goals, sub-goals and plans. Goals and subgoals of the operators were mapped to ROIs, i.e., the display which operators were most likely to use for a given subgoal.

Relating visual sampling to task goals

To relate the observer's navigation through the task to the HTA, video recordings of each observer where mapped to the identified steps in the task analysis. For the eyetracking data, 100 equally spaced bins were created for the duration of the task. For each bin, the ROI with the highest viewing time within that bin was extracted, arriving at a smoothed signal holding the main stimulus participants attended to, 'filtering out' high frequency switches between ROIs. Results from this mapped gaze data and HTA were then contrasted qualitatively.

Results

Attended ROIs

The cumulative % viewing time (Figure 2) for different displays varied largely: participant 1 spent most time viewing LG / SS, participant 2 favoured LG / MP and participant 3 favoured LG / TV. None of the participants allocated noteworthy time (viewing time for ROI $\leq 1.6\%$) to D1, D5, PH, BS and XX.

Sequential information sampling

Participants followed different search patterns when redirecting gaze between the different display options (Figure 3). The switch frequency between ROIs averaged

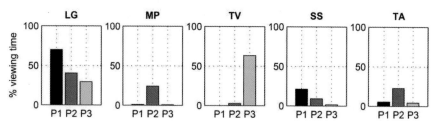

Figure 2. Cumulative percentage viewing time allocated to those five regions of interest (ROIs) within the display configuration of the control room that received noteworthy attention. Black – participant 1, dark grey – participant 2, light grey – participant 3.

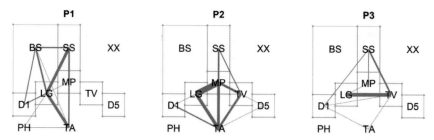

Figure 3. Viewing networks within the schematic control room layout of participant 1 (left), 2 (middle) and 3 (right), illustrating switches between ROIs. Line thickness proportional to switch count.

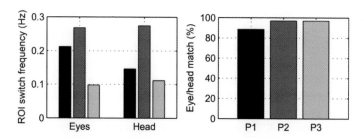

Figure 4. Left – average switch frequency between ROIs. Right – percentage of time for which head and eyes pointed at matching ROIs. For colour coding please refer to Figure 2.

1 switch every 5 to 10 seconds (0.10 to 0.27 Hz) for the eyes and 0.11 to 0.27 for the head (Figure 4). However, switches were not evenly distributed, often showing bursts. The match between ROIs attended to by eyes and head ranged from 88% to 97% viewing time (Figure 4).

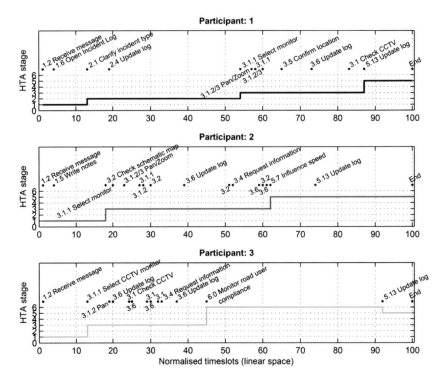

Figure 5. Stages of the hierarchical task analysis (HTA) normalized to 0-100% task time. For colour coding please refer to Figure 2.

Task navigation

Operators navigated through the task systematically, viewing information sources relevant to the task (Figure 5). Rather than frequently switching information sources during individual subtasks, operators tended to adhere to a single source. For each operator, task navigation was intersected by entries into the digital log (LG). After navigating the first three subgoals of the HTA, namely 1. Receive notification, 2. Determine incident type, 3. Determine incident location, participants differed in their simulation of 4. Determine incident impact, 5. Initiate response and 6. Monitor road user compliance. There were differences at further sub-task levels (Figure 5) in the style of task execution between operators (e.g. participant 2 took written notes on paper while participant 3 monitored the CCTV feed on TV).

Discussion and conclusions

In this study, operators favoured different information sources to complete the same goal. The difference in CCTV monitoring between participant 1 and 3 can be explained, as the selected CCTV feed on SS can also be viewed on TV – the two participants had developed different preferences regarding the viewing modality.

However, while participant 2 emphasised information gathering from the schematic road map on MP, the other two participants largely ignored this information source. The lack of attention to the large bank of CCTV feeds which we found matches a recent review in the CCTV crime monitoring domain: those authors highlighted that with increasing number of feeds (16 on BS in our study) both observer accuracy and confidence rapidly diminish (Stainer et al., 2013). Instead, our three observers relied on the interactive bank of smaller CCTV screens (SS), selecting a relevant camera and adjusting the view in order to extract desired information. We observed that operators may attend to one source of information for the majority of time, similar to operators searching a bank of CCTV screens attending primarily to a selected feed on a separate monitor (Stainer et al., 2013) or air traffic controllers attending to a preferred information source (Stein, 1992).

We found that operators aligned head and eye direction with respect to the attended ROIs for the majority of time. On the one hand, this finding may be due to chance (N = 3): the contribution of head movements to gaze shifts is generally highly subject specific; while some subjects move their head a lot, others do not (Boyer, 1995, Goossens and Opstal, 1997). For example, in an airtraffic control task the distribution of operators moving their head when shifting gaze from a monitor to an input device was approximately 50:50 (Boyer, 1995). On the other hand, we propose that our findings result from the top-down gaze control based on the internal task schema of the operator: humans executing domestic tasks typically move their body in the direction of objects of interests prior to moving the eyes (Land, 2009). In our study, operators were familiar with the control room layout and we conclude that head movement indicated their higher level goal structure similar to Land's (2009) task scenarios. Alignment of head and gaze direction can be affected by physiological factors, i.e., increasing gaze eccentricity causes stress on the eye muscles, which compensatory head movement helps relieve (Sanders, 1963). Hence, operators might seek to move their heads to minimise gaze eccentricity. However, we believe that the key factor in aligning eye and head in the control room is the arrangement of ROIs and their relation to operator goals. We noticed that occasionally the eyes briefly flick to a new ROI while the head remained directed at the old target. We hypothesise that this relates to the operator's goal structure, and that analysis of disjunctions between ROIs attended to by eyes and head serve to quantify changes between primary and secondary goals.

In static displays, re-sampling has been linked to refreshing memory content (Peterson and Beck, 2013) and has received attention in context of 'embedded' (Spivey and Dale, 2013) or distributed cognition, where the location of information is stored in memory to access detailed information when needed. This raises an interesting question of what needs to be encoded during visual search: do operators encode the content of the displays, perhaps in some form of mental model, or do they encode the location of pieces of information, in an approach more closely aligned with distributed cognition? In control rooms where information content continuously evolves, re-sampling is essential to maintain awareness of dynamic situations. In our study, operators clearly performed task-directed sequential sampling. The different information sampling strategies we found may indicate that in an over-determined system, operator decision making may lead to differences in task navigation and that there may be several rationally optimal solutions to a task. Since the task was a simulated event, operator behavior may differ to that observed in a real event (although initial comparison with data collected during performance of real tasks suggests that such differences may not be as clear cut as one might imagine). In the present scenario, we believe that the eye- and head-movement data can be used to infer observers' perceived importance of individual regions of interest relative to the goal structure that they were employing. Following the structured incident log ensured that despite heterogeneous sampling approaches, operators extracted information relevant to resolving the incident systematically. Our results suggest that operators, in a control room, have developed idiosyncratic strategies for information search, that these strategies reflect the emphasis that operators place on the ways in which they navigate the goal structure of the tasks they are performing, and that eye and head movement give insight into these strategies in ways that either recording operator activity through HTA or interviewing operators might miss.

References

Boyer, D. J. 1995, *The Relationship Among Eye Movements, Head Movements, and Manual Responses in a Simulated Air Traffic Control Task*, U.S. Department of Transportation Federal Aviation Administration.

Chapman, P., G. Underwood and K. Roberts. 2002, Visual search patterns in trained and untrained novice drivers, *Transportation Research Part F: Traffic Psychology and Behaviour* 5(2): 157–167.

Doshi, A. and M. M. Trivedi. 2012, Head and eye gaze dynamics during visual attention shifts in complex environments. *Journal of Vision* 12(2).

Findlay, J. M. and I. D. Gilchrist. 2003, *Active vision: The psychology of looking and seeing*, Oxford: Oxford University Press.

Fuller, J. 1992, Head movement propensity. *Experimental Brain Research* 92(1): 152–164.

Goossens, H. H. L. M. and A. J. V. Opstal. 1997, Human eye-head coordination in two dimensions under different sensorimotor conditions. *Experimental Brain Research* 114(3): 542–560.

Hayhoe, M. and D. Ballard. 2005, Eye movements in natural behavior. *Trends in Cognitive Sciences* 9(4): 188–194.

Land, M. F. 2009, Vision, eye movements, and natural behavior. *Visual Neuroscience* 26(01): 51–62.

Peterson, M. S. and M. R. Beck (2013) Eye movements and memory, In S. P. Liversedge and I. D. Gilchrist (eds.) *The Oxford Handbook of Eye Movements*, Oxford: Oxford University Press.

Sanders, A. F. 1963, *The selective process in the functional visual field*, Institute for perception RVO-TNO.

Spivey, M. J. and R. Dale (2013) Eye movements both reveal and influence problem solving, In S. P. Liversedge and I. D. Gilchrist (eds.) *The Oxford Handbook of Eye Movements*, Oxford: Oxford University Press.

Stainer, M. J., K. C. Scott-Brown and B. Tatler. 2013, Looking for trouble: A description of oculomotor search strategies during live CCTV operation. *Frontiers in Human Neuroscience* 7: article 615.

Stein, E. S. 1992, *Air Traffic Controller Visual Scanning*, U.S. Department of Transportation Federal Aviation Administration.

Thumser, Z. and J. Stahl. 2009, Eye–head coupling tendencies in stationary and moving subjects. *Experimental Brain Research* 195(3): 393–401.

Yarbus, A. L. 1967, *Eye movements and vision*, New York: Plenum Press.

YOUNG DRIVERS, RISKY DRIVING AND PEER INFLUENCE

L. Weston & E. Hellier

School of Psychology, University of Plymouth, Devon

Young drivers tend to have more peer passengers than older drivers; and passenger presence is a key risk factor in young drivers' collision rates (Aldridge et al, 1999). Peer pressure can be either indirect (drivers' perceptions of expected driving) or direct (active encouragement to behave in a certain way). We explored the relationship between peer influence and young drivers' (N = 163) self-reported risky driving behaviour. High susceptibility to peer influence (attaining social prestige and peers intervening in decisions) was related to more self-reported risky driving violations. The results provide a base from which to direct future initiatives to address young drivers' susceptibility to peer influence and the effect it has on driving behaviour.

Introduction

For teenage drivers, driving is often perceived as a means of socializing, and young drivers are more likely than older drivers to a) have passengers and b) have a greater number of passengers per trip (Shope & Bingham, 2008). The presence of passengers is a key factor implicated in the crash rate of drivers under 21 years old (Bedard & Meyers, 2004), and it appears to be the presence of same-age passengers that is most risky. Collision data shows that young drivers are most at risk when accompanied by teenage passengers (particularly male passengers); and their crash risk is much reduced in the presence of adult passengers (Ouimet et al, 2010). Driving with peer passengers is a factor consistently implicated in young drivers' over-representation in collisions (Doherty, Andrey & MacGregor, 1998), and recently researchers have focused on understanding why peer passengers increase the crash risk for young drivers.

Peer pressure

Allen and Brown (2008) suggested that factors such as trying to please one's peers and divided attention between driving and entertaining friends, may influence young drivers' propensity to engage in risky driving when accompanied by peer passengers. Horvath et al. (2012) suggested that these social risk factors can exert their influence on young drivers' behaviour directly, through verbal encouragement, or indirectly, through the drivers' perceptions of how others think they should drive. Allen and

Brown (2008) suggest that this perceived pressure is likely to stem from group norms that specify appropriate behaviour for members, developed through relationships within a group, and on which identity as a member of that group is based (Tajfel, 1982).

Social Identity Theory (SIT) suggests that individuals base their identity on group membership. To strengthen their feeling of in-group membership they are motivated to behave in accordance with the group's norms even when not explicitly instructed to (Tajfel, 1982). Thus it could be that young drivers are affected by passive peer pressure as the mere presence of the passengers enforces the group norms and implicitly encourages them to behave in accordance with them, without active persuasion from the passenger (Allen & Brown, 2008).

There is some evidence to support the idea that passive forms of influence correlate with young drivers' risky driving. Sela-Shayovitz (2008) found that only passive forms of influence: driver's apprehension about friends' evaluations and thoughts about attaining social prestige, were highly correlated with driving violations and crash involvement (Sela-Shayovitz, 2008). By contrast Horvath et al (2012) found similar levels of speeding intention for both active and passive influence, illustrating that there is uncertainty about the relative impact of passive vs. active influence on risky behaviours in young drivers. Passive peer influence is a critical factor in other teenage health behaviours, such as starting smoking – which appears to be a result of teenagers wanting to conform to the norms of the peer group, rather than through active pressure to smoke (Stewart-Knox et al, 2005). In accordance with SIT, engagement in smoking appears to be a channel through which young people can define their social group, identify with in-group members, and emphasize their commitment to the group's norms (Stewart-Knox et al, 2005).

This research demonstrates that peer influence is a complex and multi-dimensional concept that can exert effects in different ways. The present study aims to develop our understanding of how peers influence young drivers' risky driving, using a relatively new measure of driving behaviour. We predict that participants who report more risky driving behaviours will also report being influenced by their peers, and that indirect influence will be a bigger predictor of risky driving than direct influence. We also predict that participants who report high resistance to peer influence will report fewer risky driving behaviours.

Methodology

Participants

163 drivers aged 18–25 years old (Mean = 19.74 years) completed the study. There were 137 females, 26 males. Paid participants were recruited via the University of Plymouth participant pool and from advertisements around campus; undergraduate psychology students also participated for course credit.

The length of time participants had held their driving license varied from just 3 months to 8 years 8 months (Mean $= 2$ years 2 months). Participants made on average 2–5 trips per week.

Materials

Susceptibility to Peer Influence (SPI), (Sela-Shayovitz, et al, 2008)

Active and Passive Peer Influence were measured with items from Sela-Shayovitz's (2008) self report questionnaire for young drivers. We used four subscales, each of five items: attaining social prestige (ASP) e.g. 'driving allows me to impress others'; apprehension about friends' evaluations (AFE) e.g. 'what my friends think about my driving is important to me'; peer intervention in decisions (PID) e.g. 'when I'm driving my friends sometimes encourage me to speed to have fun'; and pressure to make traffic violations (PTV) e.g. 'my friends pressure me to drive after I've had an alcoholic drink'.

Resistance to Peer Influence (RPI), (Steinberg & Monahan, 2007)

RPI consists of 10 pairs of opposing statements about inter-individual interactions, e.g. "some people go along with their friends just to keep their friends happy, BUT, other people refuse to go along with what their friends want to do, even though they know it will make their friends unhappy". Respondents answered whether each statement sounded 'really true for them' or just 'sort of true for them'. It is designed for specifically adolescents, higher scores indicate greater resistance to peer influence.

Behaviour of Young Novice Drivers Scale (BYNDS), (Scott-Parker et al, 2012). BYNDS measures the risky driving behaviour of specifically young drivers and comprises five subscales: transient violations, fixed violations, misjudgement, risky driving exposure, and driving in response to mood. The survey uses a Likert scale ranging from 1 (never) to 5 (almost always). Higher scores indicate more risky driving.

Procedure

Participants completed the questionnaires online in the order listed above.

Results

Reliability of measures

Reliability analyses were conducted on each of the measures. The alpha scores indicate good/very good reliability for all measures: SPI (Sela-Shayovitz, 2008) $\alpha = .83$, RPI (Steinberg & Monahan, 2007) $\alpha = .74$ and BYNDS (Scott-Parker et al, 2012) $\alpha = .92$.

Table 1. Self-reported susceptibility to passive peer influence, active peer influence, resistance to peer influence and prior road traffic violations.

	All Participants Mean (SD)
Passive Peer Influence (sum)	28.7 (6.6)
Passive: Attaining Social Prestige	15.1 (4.6)
Passive: Apprehension of Friend's Evaluations	13.6 (3.0)
Active Peer Influence (sum)	23.6 (6.1)
Active: Peer Intervention in Decisions	12.6 (3.4)
Active: Pressure to make Traffic Violations	13.7 (3.5)
Resistance to Peer Influence	30.8 (4.1)
Road Traffic Violations (sum)	94.1 (18.7)

Analysis

Table 1 provides the descriptive information for participants' responses on all measures. Total road traffic violation scores were calculated by summing the responses from each item. Higher scores equated to more risky driving behavior. One-way ANOVAs revealed no statistically significant differences between males and females on measures of passive peer influence $(F(1, 152) = .97, p > .05)$, active peer influence $(F(1, 152) = .1.6, p > .05)$; RPI $(F(1, 152) = 2.6, p > .05)$ or road traffic violations $(F(1, 152) = .18, p > .05)$. Similarly, no significant differences were found between males and females on any of the susceptibility to peer influence subscales: ASP $(F(1, 152) = .67, p > .05)$, AFE $(F(1, 152) = .84, p > .05)$, PID $(F(1, 152) = .15, p > .05)$ or PTV $(F(1, 152) = 3.5, p > .05)$. As males' and females' scores did not differ the data was collapsed across gender.

Correlations

Correlations between the study variables are presented in Table 2. The results suggest that when young drivers report feeling influenced by their peers they also report higher rates of various road traffic violations. The results also indicate that susceptibility to both active peer influence and passive peer influence is associated with self-reported driving violations. Young drivers reporting high resistance to peer influence did not report significantly fewer road traffic violations.

As the peer influence measures were correlated with each other, variance inflation factors (VIF) were used to assess the degree of collinearity between variables. There are various recommendations in the literature regarding acceptable levels of VIF. Most commonly, a value of 10 has been regarded as the maximum level of VIF to accept (e.g. Hair et al, 1995; Kennedy, 1992; Marquardt, 1970; Neter et al, 1989); and this corresponds to the tolerance recommendation of .10. The VIFs for each of the variables in the present study (*ASP, AFE, PID, and PTV*) were 2.5, 1.7, 1.9, and 2.8 respectively. These VIF levels indicate that the degree of collinearity between the variables is low and they are suitable for use within a multiple regression analysis.

Table 2.   Correlations among Variables. Note *p < 0.5. **p $< .01$.

Variable	1	2	3	4	5	6	7	8
1. Passive Peer Influence	–							
2. Passive: Attaining Social Prestige	.93**	–						
3. Passive: Apprehension of Friend's Evaluations	.81**	.53**	–					
4. Active Peer Influence	.80**	.78**	.59**	–				
5. Active: Peer Intervention in Decisions	.64**	.66**	.41**	.89**	–			
6. Active: Pressure to make Traffic Violations	.79**	.73**	.64**	.90**	.60**	–		
7. Resistance to Peer Influence	−.35**	−.37**	−.20**	−.35**	−.28**	−.34**	–	
8. Road Traffic Violations	.53**	.56**	.32**	.57**	.59**	.43**	−.12	–

Regression analyses

Hierarchical regression analysis were conducted to assess relations between the four forms of peer influence (passive: ASP & AFE; active: PID & PTV) and reported road traffic violations. The first model incorporated all four variables and found that reported road traffic violations was predicted by ASP $\beta = .32$, $t(149) = 3.1$, $p < .01$) and by PID $\beta = .40$, $t(149) = 4.6$, $p < .01$, but not by AFE $\beta = .03$, ($t(149) = .35$, $p > .05$, or PTV $\beta = −.15$, $t(149) = −.05$, $p > .05$). ASP and PID also explained a significant proportion of the variance in reported road traffic violations, Adjusted $R^2 = .38$, $F(4, 149) = 24.5$.

The two non-significant measures of peer influence were removed from the model (AFE & PTV) to see whether this affected the amount of variance accounted for by the two remaining variables in the model. Adjusted R Square increased from .38 to .39, with an R Square change of −.001. This change was non-significant $F(2, 149) = .14$, $p > .05$; and therefore it can be assumed that AFE and PTV do not explain any additional variance. Figure 1 shows the adjusted model, whereby attaining social prestige and peer intervention in decisions explained a significant proportion of the variance in reported road traffic violations, Adjusted $R^2 = .39$, $F(2, 149) = 24.5$.

Conclusions

Young drivers' self-reported risky driving violations were examined in relation to measures of susceptibility to active and passive peer influence, and resistance to

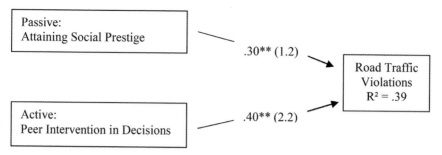

Figure 1. Attaining social prestige, peer intervention in decisions and road traffic violations. Values represent standardized estimates with unstandardized estimates in parentheses. **p < .01.

peer influence. It was found that high susceptibility to peer influence was related to more self-reported risky driving behaviours and that attaining social prestige (Passive) and peers intervening in decisions (Active) were significant predictors of more violations; rather than apprehension of friends evaluations or through drivers feeling pressure to make traffic violations. Resistance to peer influence was not associated with road traffic violations.

The findings are consistent with previous research (Sela-Shayovitz, 2008; Allen & Brown, 2008) and partially support the notion that young people may perform risky behaviours to be in accordance with the social norms of their peer group (Stewart-Knox et al, 2005). SIT (Tajfel, 1982) posits that members of social groups seek to strengthen their in-group membership by acting in ways to enhance their in-group similarity. For young drivers, driving is a part of their identity and a mode with which to transmit their norms and beliefs. If performing a risky driving behaviour is considered an appropriate way to sustain a good position within their social group, those susceptible to peer influence are likely to do it (Stewart-Knox et al, 2005). In terms of the active forms of peer influence, 'peers intervening in decisions' was associated with more risky driving, but 'pressure to make violations' was not. This suggests that young drivers perceive the input of their peer passengers to be collaborative, rather than coercive. This further supports SIT, with risky driving likely to be considered a shared interest between some young drivers' and their peers. Initiatives to confront the issue should focus on changing the perception that driving in a risky way is a means to attain social prestige.

The study was not able to explore potential gender differences between susceptibility to peer influence and risky driving as very few males took part. Previous research has found that the presence of young male passengers has a much greater negative impact on the risky driving behaviour of young drivers than female passengers (Simons-Morton et al, 2005); and so it would be helpful to understand how males and females are differentially influenced by their peers. Future research could therefore focus on identifying the mechanisms by which peer influence differentially affects males and females.

The findings from the present study are in line with research into peer influence and teenage health behaviours (Stewart-Knox et al, 2005); and provides further evidence that when young adults report being highly susceptible to peer influence, they also tend to report engaging in more risky driving behaviours. The results are important in terms of identifying risk factors for young drivers, and designing appropriate interventions to tackle this issue. Peer influence has been shown to be a critical factor in the risky driving behaviour of young drivers; and thus future initiatives should aim to address this issue and devise ways in which to counteract young people's desire to use driving as a means to attain social prestige within their peer group.

References

Allen, J.P., & Brown, B.B. 2008. Adolescents, peers, and motor vehicles: the perfect storm? *American Journal of Preventative Medicine,* 35, 289–293.

Bédard, M., & Meyers, J.R. 2004. The influence of passengers on older drivers involved in fatal crashes. *Experimental Aging Research,* 30, 205–215.

Doherty, S.T., Andrey, J.C., & MacGregor, C. 1998. The situational risks of young drivers: The influence of passengers, time of day, and day of week on accident rates. *Accident Analysis and Prevention,* 30, 45–52.

Hair, J.F. Jr., Anderson, R.E., Tatham, R.L. & Black, W.C. 1995. *Multivariate Data Analysis (3rd ed).* New York: Macmillan.

Horvath, C., Lewis, I., & Watson, B. 2012. Peer passenger identity and passenger pressure on young drivers' speeding intentions. *Transportation Research Part F: Traffic Psychology and Behaviour*, 15, 52–64.

Kennedy, P. 1992. *A Guide to Econometrics.* Oxford: Blackwell.

Marquardt, D.W. 1970. Generalized inverses, ridge regression, biased linear estimation, and nonlinear estimation. *Technometrics*, 12, 591–256.

Neter, J., Wasserman, W. & Kutner, M.H. 1989. *Applied Linear Regression Models.* Homewood, IL: Irwin.

Ouimet, M.C., Simons-Morton, B.G., Zador, P.L., Lerner, N.D., Freedman, M., & Duncan, G.D. 2010. Using the U.S. National Household Travel Survey to estimate the impact of passenger characteristics on young drivers' relative risk of fatal crash involvement. *Accident Analysis & Prevention*, 42, 689–694.

Scott-Parker, B., Watson, B., King, M.J., & Hyde, M.K. 2012. The influence of sensitivity to reward and punishment, propensity for sensation seeking, depression, and anxiety on the risky behaviour of novice drivers: a path model. *British Journal of Psychology*, 103, 248–267.

Sela-Shayovitz, R. 2008. Young drivers' perceptions of peer pressure, driving under the influence of alcohol and drugs, and involvement in road accidents. *Criminal Justice Studies*, 21, 3–14.

Shope, J.T., & Bingham, C.R. 2008. Teen driving: Motor vehicle crashes and factors that contribute. *American Journal of Preventative Medicine*, 35, 261–271.

Simons-Morton, B., Lerner, N., & Singer, J. 2005. The observed effects of teenage passengers on the risky driving behaviour of teenage drivers. *Accident Analysis and Prevention*, 37, 973–982.

Steinberg, L., & Monahan, K.C. 2007. Age differences in resistance to peer influence. *Developmental Psychology*, 43, 1531–1543.

Stewart-Knox, B.J., Sittlington, J., Rugkasa, J., Harrisson, S., Treacy, M. & Santos Abaunza, P. 2005. Smoking and peer groups: results from a longitudinal study of young people in Northern Ireland. *British Journal of Social Psychology*, 44, 397–414.

Tajfel, H. 1982. Social psychology of intergroup relations. In M.R. Rosensweig & L.W. Porter (Eds.), *Annual review of psychology* (Vol. 33, pp. 1–39). Palo Alto, CA: Annual Reviews.

TRANSPORT USERS; KNOWLEDGE GAPS AND THE POTENTIAL OF REAL TIME TRANSPORT INFORMATION

Andree Woodcock

Coventry School of Art and Design, Coventry University, UK

Interviews were held with 25, UK transport stakeholders in 2014 to assess the extent to which they collected and used information about transport service users, where existing gaps in knowledge were and the potential of real time passenger data to improve knowledge, effectiveness and efficiency of service provision. The interviews revealed substantial gaps in knowledge, especially in relation to origin – destination, trip chains, whole journey experiences and from traditional 'hard to reach' groups. All participants saw the value of real time data while acknowledging that there were barriers to its adoption. The paper examines the findings from the interviews and considers organisational and other barriers which need to be addressed.

Introduction

Transporting people, goods and services within a city is costly. Transport has a direct effect on the efficiency of the city, its environmental footprint and the health and well-being of the citizens. To reduce current problems relating to pollution, congestion and dissatisfaction with services, cities have to become smarter and develop inclusive sustainable transport (Hull, 2007). This includes getting people to work and supporting trip chaining, developing transport to support flexible working, making transport an enabler of other smart initiatives and managing the interdependency of various modes of transport. Making data available across system silos, for intelligent analysis and integration, is seen as a key enabler in meeting the cross-cutting challenges identified at city level (Department for Business, Innovation and Skills, 2013). A 'smart city' (Boehm, Flechl and Froetscher, 2013) needs to be able to apply data realized through diverse platforms for the benefit of its citizens.

The objective of this research was to understand the type of information collected by transport stakeholders about their passengers, their technological readiness to deal with the large data sets delivered by real time passenger data collection, and the barriers to using such information. It is timely because of the large investment in transport telematics (e.g. integrated, smart ticketing and journey planners) and the plethora of apps which can to automatically record user information and the need for improved mobility (Rehrl et al, 2007). The aims were to understand:

1. How real time data from citizens could provide local authorities, transport planners and operators with a rich dataset to better understand journeys and travel behaviour and inform more effective transport services.

2. How information on passenger movements and characteristics is currently collected and identify gaps in provision.
3. How such information might be used, its impact, ownership and barriers to use.

Methodology

The results are based on semi structured interviews with transport stakeholders. Stakeholders included representatives from transport (5) and local authorities (5), passenger action group (1), the academic community (1), government agencies (1), transport consultants (6) operators (3) and IT providers (2). Where appropriate, material was included from interviews conducted with similar stakeholders as part of the FP7 METPEX project (www.metpex.eu) in line with its stated objective to harmonize the collection of transport data across the EU. The small sample had a wealth of experience at local, regional and international levels, dealing with day to day running of transport services (bus, park and ride, trams and trains), organisation of transport provision across counties (e.g. Hampshire), cities (e.g. Brighton and Hove, Coventry, Bristol) and wider areas covered by the transport authorities (such as CENTRO and Merseytravel) including problems associated with international travel (e.g. Birmingham Airport). The geographic spread of the sample was across the UK. Together the sample had experience of commissioning, designing and using a wide range of transport related data to understand, model and improve transport services, transport infrastructure and the overall passenger experience and were therefore considered as sufficiently experienced to provide valuable insights into the use of traveller information. The transport consultants had special interests, knowledge and experience of UK transport provision, the use of IT and new forms of data gathering in particular. Other stakeholders selected had special concerns in relation to national passenger surveys, disability, old age and mobility.

Most modes of transport were discussed (air, rail and road) including private, public and active transport forms. The size of the operators varied from those offering a limited number of services to the larger operators and transport authorities (for example, Merseyside 2012/13 had a bus patronage of 136 million journeys and a rail patronage of 38.9 million journeys).

Results

There was great diversity in the extent to which the different stakeholders were informed about different aspects of the passenger journey (e.g. route, mode, quality, time, frequency, how passengers reach the service, problems and issues). Generally Local Authorities (LAs) were the least well informed on all aspects of passenger journeys, unless they were commissioning specific research (e.g. for redevelopments). Transport Authorities (TAs) had variable levels of knowledge about all issues, whereas Transport Operators (TOs) had detailed knowledge about a limited set of issues (namely route, quality and time of journey). Consultants were sceptical about the use of journey information by the smaller TOs and LAs, and between the capability of larger and smaller, urban and rural operators and authorities to collect and use the information, especially to increase the quality of passenger experience.

Scope of surveys

Journey information was collected by different methods, at different times and for different purposes by all groups. The TAs conduct a range of surveys e.g. cordon counts, counts on stations and stops, and annual audits of facilities. Smart ticketing is enabling continuous data collection and is used by both TOs and TAs. The larger TOs conduct annual surveys, and may be able to draw on large pools of passengers. LAs are more dependent on information being passed to them and have limited resources to pay for and conduct their own surveys. Both TAs and LAs mentioned budgetary constraints. TA budgets are around £100k per annum, whereas LA budgets are between £25–£50k, rising to over £100k if linked to a bid to the Department of Transport. TOs typically spend around £10k. As both LAs and TAs are seeking to reduce their costs and increase their knowledge of Origin-Destination (O-D), real time data, mobile apps and smart ticketing have the potential to enable them to achieve this.

The geographic area of travel survey depends on the stakeholder conducting the survey e.g. SYPTE (South Yorkshire Transport Authority) will cover Sheffield, Rotherham, Doncaster, Barnsley etc.; Network Rail is similarly divided into regions, each of which may have different approaches to data collection and storage. LAs will run individual campaigns based on their immediate interests (in line with strategic plans and transport schemes). The frequency and extent of rail surveys are stipulated in franchise agreements. Passenger Focus (the UK Passenger Watchdog) conducts annual surveys of trams, trains and buses on a regional basis. The number of people surveyed also varies according to the stakeholder and the purpose of the survey.

Surveys are conducted by stakeholders themselves or by consultants and academics. All use Passenger Focus and the National Travel Survey (https://www.gov.uk/government/collections/national-travel-survey-statistics) is the primary source of data on personal travel patterns in Great Britain. It is an established household survey which has been running continuously since 1988. It is designed to monitor long-term trends in personal travel and to inform the development of policy with approximately 16,000 individuals, in 7,000 households in England, participate in the NTS each year. Some TOs use travel surveys on a continual basis. However, the information collected will be specific to their operation.

A wide variety of data collection methods was used including smart ticketing, manual and automatic vehicle counts, use of cctv (but without any automatic analysis), loop circuits, interviews, twitter, follow up interviews and surveys (online retrospective, en-route, paper based and automatic). All stakeholders are looking at automation and in particular mobile apps.

Usefulness of information

Usefulness of was rated on a 5 point scale, from being very useful, to of variable use, depending on its purpose of the survey. Some surveys might have fulfilled their requirements but do not provide a detailed picture of overall transport. For example, one LA concluded that the data is relatively useful, but it is patchy and non-representative '*Difficulty is not in looking at it and understanding it, but in what you can do with it'*; and another saying

> '*But there are a lot of gaps. When they do do them, the quality is high and the response rate is good. Normally traffic counts or road side interviews – v useful because one of the few opportunities to ask people where they start and end their journeys. Very expensive and quite unpopular. Stop the majority of the cars, and they don't necessarily want to take part in the survey'*

Information has been used by TOs to change routes and bus design. For example

> '*Useful for punctuality; Useful to tell each depot about specific passenger requirements for a given route and creating depot action plans. An example of this was that customers were complaining about interior bus comfort on certain routes in Shrewsbury, so we re-trimmed the seats with better cushions and new upholstery to meet that specific routes' requirements. Since 2013, the last survey after the re-trim was done; there have been no more comfort problems'*

TAs have used information to develop customer strategies and for medium – long term planning:

Criticisms levelled at existing methods of data collection were as follows:

- too expensive
- poor coverage which only provides a snapshot of transport,
- skewed data, with omission of traditional hard to reach groups (e.g. travellers with disabilities or dependents)
- underrepresentation of certain modes of transport,
- omission of trip chains (a sequence of trips between given anchor points e.g., such as dropping children off on the journey to work)
- lack of essential information such as expectations, and reasons for the journey, and origin and destination information.

Additional problems related to speed of analysis, provision of the right information at the right time, and the ability to act on the data. Both the transport and local authorities admitted that the information that was collected was disconnected. This may be because of the use of different methods, times and places of data collection, who the data is collected by and for whom, and its final resting place within the organisation. This is an immediate concern issue for smaller authorities and operators who may be failing to capitalize on their existing knowledge and know-how, and will lead to legacy problems going forward as Smart Cities.

The extent to which organisations had a strategy for using technology was not clear. However, the benefits of better data integration and use of mobile technology was clear from the responses of all stakeholder groups. For example, TOs commented that future growth will depend on capturing data on travel info; TAs were 'looking *to expand use of real time data to provide passenger information, smart ticketing etc.'* LAs commenting 'we *want to use new tech where possible. Very interested in mobile technologies and crowdsource data and new types of sensor technology, keen to allow them to be tried'* and '*We are committed to making use of all possible areas where technology can support, widen and improve delivery'.*

However, one consultant commented unfavourably on the lack of national leadership and standards in the creation of smart transport technology. This has led to individual organisations creating their own apps (such as myPTP). These may not join at a national level, and may not deliver routes suitable for people's needs. A lack of publicity means that the National Transport Direct Journey Planner may not be known or used. This has meant that many organisations have had trouble in bringing products to market.

The pattern which emerged when more detailed questions were asked about journey characteristics was in accordance with the findings of Woodcock et al, 2013. The TOs valued information specifically about their service highest (e.g. number of journeys per week, travel time and waiting time), but were not so interested in related areas (such as modal transfer and purpose of journey). Their level of knowledge was low for all elements except those relating to their service, e.g. number of journeys per week, purpose of journey time of day.

The TAs rated their knowledge of all factors concerning journeys more highly than respondents from other groups, but had variable levels of knowledge about specific parts of the journey. For example one TA commented that, *'We especially focus on the peak times in order to get people to and from work smoothly i.e. for TIME OF DAY question'.* For TAs knowledge of

- personal passenger characteristics was important for the design of systems
- alternative transport mode availability was important for planning
- total distance and total journey time for bench-marking and planning.
- journey total travel time helps with variability

LAs rated journey characteristics in a similar way, but their actual level of knowledge was less. One commented, *'All [aspects] have a high level of importance, but actually level of knowledge is quite low.'*

Detailed journey information was felt to be highly relevant by all the TOs, but was rated as 'somewhat' and highly relevant' by the TAs and LAs. One LA explained that such information was important to public transport operators in order to help them manage their operations. Key omissions include information about whole journey experiences, journey desire, origin and destination information, information about transport interchanges. Information about disabled travelers, cycling and walking is still not being collected systematically, leading to worries about inclusivity of provision and the effects of poor transport behaviour on health and well-being.

In conclusion, all stakeholders believed that having more detailed transport information, such as that provided automatically from smart phones would lead to positive impact. The Transport and Local Authorities admit that their current data is skewed, with many transport models based on estimates. For example, the National Household Survey only collects information about travel on business, so omits information about leisure, chains and shorter trips.

The use of real time data

All stakeholders believed that real time data would make a valuable contribution to existing knowledge. They expressed few reservations about citizens' willingness to contribute such information as many users already divulge personal information on mobile devices. They were nevertheless able to point to a number of incentives which could be developed to encourage citizens to provide mobility data. Ownership of data, and how and who could use the data and for what purposes, was a concern amongst some stakeholders.

Many organisations are already embarking on collecting and using real time information. However, from the users' perspective there may appear to be a plethora of travel related apps, each of which needs to be accessed to provide the required information, and which will have different functionality/user interfaces. For example, operators and authorities are using apps to inform the travellers of travel delays and disruptions, which is very welcomed by groups such as Transport for All. It was not clear how operators and authorities would use real time data to change service provisions (especially in real time). Smart ticketing (and the next generation of machines) will provide increasingly sophisticated real time information, which will assist operators in increasing service efficiency, in conjunction with travel cards (such as Oyster card) which support multimodal journeys. However, these will not reduce gaps in understanding the whole journey or provide information about non users or specific disadvantaged groups.

Organizational issues

There were mixed opinions about the level of skills within organisations to process the amount of data, to develop a big picture from this and use it to make strategic decisions (especially for transport operators). Most transport authorities and larger local authorities were beginning to integrate data from different sources, including mobile applications. However, there is a worry that the provision of detailed information may 'swamp' authorities and operators. Systems and departments need to be in place to handle the data quickly and efficiently to maximize its potential to effect real time changes. Concerns were expressed about transport operators' capacity to use and understand this information. Tools are needed to help synthesize, visualise and interpret this data along with guidance to show how data can be used to inform strategic decisions.

The interviewees identified a number of internal and external barriers, but many (such as data security, leadership, silo mentality) issues were being addressed. However, in some organisations there is no champion for passenger information within the organisation, and information is not shared easily or effectively across the organisation. Consultants in particular, were very sceptical about the ability and willingness of transport operators to use real time information, to share resources and evolve from their 'cottage industry mentality'. Clearly those companies which are embracing new technology are already reaping rewards in terms of service efficiency, passenger satisfaction and new contracts.

Citizen engagement and willingness to use, or buy into, schemes were not seen as insurmountable problems, given the rising levels of IT literacy and wider adoption of smart phones. Although cybercrime continues to rise, there is less fear over data security and people willingly share location and other types of sensitive information. However, although mobile apps may fill in gaps about whole journey experiences, there are several areas of caution around gaps in data; firstly if data is only provided by those with smart phones, data will be skewed; secondly, the current 'hard to reach groups' (such as the elderly and disabled) will still be excluded. Attention will still be needed to design appropriate incentive schemes which match the citizens' needs, and which detail the direct benefits they and their cohort will receive if they share their data. Issues around how information is used (e.g. to push marketing material), and ownership of data will have to be addressed, with one consultant pointing out that people are beginning to see their mobility behaviour as an exploitable commodity. Those who have been involved in pilot projects also expressed the need to build in sustainability (i.e. after the pilot scheme had finished).

Conclusions and recommendations

The study has revealed and confirmed several issues such as:

- Institutional readiness to effectively process and handle large amounts of data as authorities are left with legacy systems and are still working in organisational silos. On the one hand this represents lost opportunities, but on the other, it presents large technology providers with opportunities to provide integrated systems. Cost implications and austerity measures may mean that budgets in authorities and small operators are squeezed.
- Ethical issues around the automatic collection and use of personal and locational data and social media need to be addressed, not only at individual but at city wide levels to ensure anonymity is protected and awareness of how and by whom the information will be used.

- Inclusivity and skewed data gathering are major issues which have still not been addressed, and which mean that transport services are still not modelled or optimised to take into account the needs of all users
- Technology providers have developed a plethora of competing services, e.g. satnav systems, journey planners. This means that users are often bewildered by choice, may not find the most useful app for their trip, or may be provided with out of date or redundant information. Authorities and operators may be wasting valuable resources commissioning systems that already exist. Standardisation is needed to ensure that users receive systems which are usable and helpful in planning their journeys.

Acknowledgments

The research was undertaken on behalf of TravelAI Ltd, funded by SBRI Future Cities – Project #97218 – Citizens at the City's Heart (Catch).

References

Boehm, M., Flechl, B. & Froetscher, A. (2013) *ICT concepts for optimization of mobility in Smart Cities*, Luxembourg: Publications Office.

Department for Business, Innovation and Skills (2013) *Smart cities, background paper*, Downloadable for www.gov.uk/bis

Hull, A. (2008) Policy integration: What will it take to achieve more sustainable transport, *Transport Policy*, **15**(2), 94–10

Rehrl, K., Göll, N., Leitinger, S., Bruntsch, S. & Mentz, H.-J. (2007) Smartphone-based information and navigation aids for public transport travelers, *Location Based Services and TeleCartography*, pp. 525–544

Woodcock, A., Berkeley, N., Cats, O., Susilo, Y., Hrin, G. R., O'Reilly, O., Markucevičiute, I. & Pimentel, T. (2014) Measuring quality across the whole journey, *Contemporary Ergonomics and Human Factors* 2014, 316–323

Author index